POCKET SPECS FOR INJECTION MOLDING

Compiled by
MICHAEL L. KMETZ
SUZY WITZLER

2nd Edition

A PUBLICATION FROM IDES INC. AND
INJECTION MOLDING MAGAZINE

THIS BOOK BELONGS TO:

NAME

CO.

ADDRESS

WORK PHONE HOME PHONE

INJECTION MOLDING MAGAZINE
ABBY COMMUNICATIONS INC.
3400 E. Bayaud Ave., Suite 230
Denver, CO 80209
(303) 321-2322
(303) 321-3552 (fax)

IDES, INC.
Box 2131
Laramie, WY 82070
(800) 788-4668
(307) 745-9339 (fax)

ISBN 0-9642570-1-7

Preface

Injection Molding Magazine and IDES, Inc. have made every effort possible to provide accurate and consistent information in this book. All data has been compiled from plastic material suppliers' information. The testing for the reported values has been done by the material suppliers or by their contractors. Therefore, the authors do not make any representations or warranties of any kind as to the accuracy of the information. Neither Injection Molding Magazine nor IDES shall be held responsible for any inaccuracies in *Pocket Specs for Injection Molding* that may lead to damages of any kind.

We would welcome any comments regarding the content or layout of this publication. Future editions of *Pocket Specs for Injection Molding* will undoubtedly incorporate additional information that our readers suggest.

Table of Contents

Preface .... **i**

Table of Contents .... **ii**

Introduction .... **iv**

Property Descriptions .... **v**

Generic Material Names .... **x**

Fillers and Additives .... **xvi**

Melt Flow Conditions .... **xviii**

*Property Tables** **1**

Supplier Directory **296**

Tradename Directory **303**

Troubleshooting Guide **313**

Index to Materials **326**

Ordering Information **332**

*Keys to the thumb tabs are found on the facing page, where the alphabetic range of materials is shown for each thumb tab.

ABS *to* ABS+PC Alloy

ABS+PTFE Alloy to Acetal Copolymer

Acrylic *to* DAP

EMA *to* Ionomer

LCP *to* Melamine Phenolic

Nylon *to* PAEK

PBT *to* PBT+PET Alloy

PC (polycarbonate) *to* PCTG

PE (polyethylene) *to* PEKEKK

PES *to* Polyarylsulfone

Polybutylene *to* POM

PP *to* PP Homopolymer

PPA *to* PPS

PPSS *to* PUREL-TP

PVC *to* SAN

SB *to* Urea Formaldehyde

Introduction

Pocket Specs for Injection Molding's purpose is to provide you with a compact guide for the injection molding of thermoplastic and thermoset materials. Data is provided for thousands of individual grades of molding materials from nearly 100 manufacturers. The data provided in the tables is intended to give you the basic information for determining material drying temperatures and times, and initial machine settings for barrel heat and mold temperature. Additional physical property data includes specific gravity, shrink data, processing temperature ranges, and melt flow. Following the property listings are other listings that may help:

- A directory of suppliers for all resins in the tables (p. 296);
- A directory of tradenames, resin names, and suppliers (p. 303);
- A troubleshooting guide with definitions and a chart of recommended actions (p. 313);
- An index to all resin family listings (p. 326).

In addidtion, number of keys are provided to help you with the abbreviations found in the tables:

- Abbreviations/generic names for the materials are found on p. x;
- Abbreviations/filler and additive names are found on p. xvi.
- Melt flow condition keys are found on p. xviii.

Property Descriptions

The following properties comprise the data for nearly 6000 grades of injection molding resins covered in this book. Here's what each property represents.
Note: Barrel temperatures are recommended machine settings.

Generic Name

Indicates the family that a material belongs to. Every material in the database will have one generic name.

Trade Name

Often called commercial name, this is the name given to a material by a manufacturer.

Manufacturer/Supplier

Indicates the organization that manufactures or supplies the material.

Grade

Uniquely identifies a material and is associated with a particular trade name.

Filler

The type of substance added to a plastic material; often used to make a resin less costly. Fillers are used to enhance material strength, stiffness, or other characteristics. Numbers following the abbreviation in the tables represent the percent of reinforcement or filler.

Examples: aluminum, calcium carbonate, carbon fiber, cellulose, chopped glass. A list of filler abbreviations used in the tables follows on p. xvi.

Specific Gravity

The ratio of the density of the material to the density of water where density is defined as the mass divided by the volume of the material. The specific gravity is exactly equal to the density in metric units (1 gram/ cubic centimeter = a specific gravity of 1).

Shrink

The ratio of the difference between the molded plastic part dimension and the mold dimension. Shrink is tested by obtaining a measurement of the length of the cavity of a standard bar mold, or the diameter of the cavity of a standard disk, to the nearest 0.001 inch. Measurements of the molded test specimens are then taken and the shrink value is the difference between the two.

For injection molding materials the test specimens consist of two types:

- Bars, ½ by ⅛ by 5 inches gated at the end to provide flow throughout the entire length. These are usually used to determine shrinkage in the direction of flow.
- Disks, ⅛ inch thick and 4 inches in diameter gated radially at a single point in the edge. These are usually used for measurements of shrinkage of diameters parallel and perpendicular to the flow.

Mold shrink can vary due to wall thickness, flow direction and molding conditions. Generally rigid amorphous and thermoset materials have lower mold shrink than semi-crystalline thermoplastics. Reinforced or filled materials tend to have lower shrink than unfilled materials.

Melt Flow

Rate of extrusion of molten resin through a die of a specified length and diameter under prescribed conditions of temperature and pressure.

An extrusion plastometer is used for this test. The test material is placed in the cylinder and the piston is loaded as prescribed by the test conditions. The extrudate issuing from the orifice is cut off flush and discarded at 5 minutes and again one minute later. Cuts for the test are taken at 1, 2, 3, or 6 minutes, depending on the material or its flow rate. The results are strongly dependent on temperature. Therefore, values of melt flow are reported at prescribed test conditions. Keys to melt flow conditions used in the data follows on p. xviii.

Any material form that can be introduced into the cylinder bore may be used. For example, powder, granules, strips of film, or molded slugs are common material forms. Reported values of melt flow help distinguish between the flow characteristics of different grades of material. Generally, high molecular weight materials are more resistant to flow than low molecular weight materials. In addition, melt flow results are useful to the manufacturer as a method of controlling material uniformity.

Melt Temperature

The temperature that a material turns from a solid to a liquid. Two basic methods are used for melting point determination. ASTM D789 is used for amorphous and crystalline polymers, and ASTM D2117 is used for semi-crystalline polymers.

Back Pressure

Resistance of a material, because of its viscosity, to continued flow when the mold is closing.

Drying Temperature

The recommended temperature for drying the material before molding.

Drying Time

The recommended time for drying the material before molding.

Injection Time

The recommended time for injecting the material into the mold cavity.

Front Barrel Temperature

The recommended temperature setting for the front

portion of the barrel, or cylinder, of the injection molding machine.

Middle Barrel Temperature

The recommended temperature setting for the middle portion of the barrel of the injection molding machine.

Rear Barrel Temperature

The recommended temperature setting for the rear portion of the barrel of the injection molding machine.

Nozzle Temperature

The recommended temperature setting for the nozzle of the injection molding machine.

Mold Temperature

The recommended temperature of the mold during molding.

Processing Temperature

The recommended overall processing temperature for a material, usually expressed as a range.

GENERIC MATERIALS

SYMBOL	LONG NAME
ABS	Acrylonitrile Butadiene Styrene
ABS+PA Alloy	Acrylonitrile Butadiene Styrene + Polyamide Alloy
ABS+PBT Alloy	Acrylonitrile Butadiene Styrene + PBT Alloy
ABS+PC Alloy	Acrylonitrile Butadiene Styrene + PC Alloy
ABS+PTFE Alloy	Acrylonitrile Butadiene Styrene + PTFE Alloy
ABS+PVC Alloy	Acrylonitrile Butadiene Styrene + PVC Alloy
ABS+TPU Alloy	Acrylonitrile Butadiene Styrene + TPU Alloy
Acetal	Acetal Homopolymer
Acetal Copoly	Acetal Copolymer
Acrylic	Acrylic
Acrylic + ABS Alloy	Acrylic + ABS Alloy
AS Copolymer	Acrylonitrile Styrene Copolymer
ASA	Acrylonitrile Styrene Acrylate
ASA+PC Alloy	Acrylonitrile Styrene Acrylate + PC Alloy
ASA+PVC Alloy	Acrylonitrile Styrene Acrylate + PVC Alloy
ASA/AES	Acrylonitrile Styrene Acrylate + AES Alloy
Bio Syn Poly	Biodegradable Synthetic Polymers

SYMBOL	LONG NAME
CA	Cellulose Acetate
DAP	Diallyl Phthalate
EMA	Ethylene Methacrylic Acid
EnBA	Ethylene n-Butyl Acrylate
Epoxy	Epoxy; Epoxide
ETFE	Ethylene Tetrafluoroethylene Copolymer
EVA	Ethylene Vinyl Acetate
FEP	Perfluoroethylene Propylene Copolymer
Fluorelast	Fluoroelastomer
HDPE	Polyethylene, High Density
HDPE Copolymer	Polyethylene, High Density Copolymer
HIPS	Polystyrene, High Impact
Ionomer	Ionomer
LCP	Liquid Crystal Polymer
LDPE	Polyethylene, Low Density
LLDPE	Polyethylene, Linear Low Density
MBS Terpolymer	Methyl Methacrylate Butadiene Styrene Terpolymer
Mel Formaldehyd	Melamine Formaldehyde
Mel Phenolic	Melamine Phenolic
Nylon	Polyamide (nylon)
Nylon 11	Polyamide (nylon 11)
Nylon 12	Polyamide (nylon 12)
Nylon 12 Elast.	Polyamide (nylon 12) Elastomer
Nylon 12/12	Polyamide (nylon 12/12)

SYMBOL	LONG NAME
Nylon 4/6	Polyamide (nylon 4/6)
Nylon 6	Polyamide (nylon 6)
Nylon 6 Alloy	Polyamide (nylon 6) Alloy
Nylon 6 Copoly.	Polyamide (nylon 6) Copolymer
Nylon 6 Elast.	Polyamide (nylon 6) Elastomer
Nylon 6/10	Polyamide (nylon 6/10)
Nylon 6/12	Polyamide (nylon 6/12)
Nylon 6/6	Polyamide (nylon 6/6)
Nylon 6/6 Alloy	Polyamide (nylon 6/6) Alloy
Nylon 6/66 Cop.	Polyamide (nylon 6/66) Copolymer
Nylon 6/6T	Polyamide (nylon 6/6T)
Nylon 6/9	Polyamide (nylon 6/9)
Nylon MXD6	Polyamide (nylon MXD6)
PAEK	Polyaryletherketone
PBT	Polybutylene Terephthalate
PBT Alloy	Polybutylene Terephthalate + Unspecified Alloy
PBT+PET Alloy	Polybutylene Terephthalate + PET Alloy
PC	Polycarbonate
PC+PBT Alloy	Polycarbonate + Polybutylene Terephthalate Alloy
PC+PET Alloy	Polycarbonate + Polyethylene Terephthalate Alloy
PC+Polyester	Polycarbonate + Polyester Alloy
PC+PUR Alloy	Polycarbonate + Polyurethane Alloy
PCT	Polycyclohexylene Terephthalate
PCTG	Polycyclohexylene Dimethylene Terephthalate

Symbol	**Long Name**
PE	Polyethylene
PEBA	Polyether Block Amide
PECTFE	Polyethylene, Chlorotrifluoroethylene
PEEK	Polyetheretherketone
PEI	Polyether Imide
PEI+PC Alloy	Polyether Imide + PC Alloy
PEKEKK	Polyetherketone Etherketone Ketone
PES	Polyether Sulfone
PET	Polyethylene Terephthalate
PETG	Polyethylene Terephthalate Glycol Comonomer
PFA	Perfluoralkoxy
Phenolic	Phenolic
Plastomer	Plastomer, ethylene-based
PMMA	Polymethylmethalcrylate
PMP	Polymethylpentene
Polyamide-Imide	Polyamide-Imide
Polyaryl Amide	Polyaryl Amide
Polyarylate	Polyarylate
Polyarylsulfone	Polyarylsulfone
Polybutylene	Polybutlyene
Polyester	Polyester
Polyester Alloy	Polyester + Unspecified Alloy
Polyester, ElCo	Polyester, Elastomer Copolymer
Polyester, Ela	Polyester, Elastomer
Polyester, TP	Polyester, Thermoplastic
Polyester, TS	Polyester, Thermoset
Polyether	Polyether
Polyimide (TP)	Polyimide, Thermoplastic

Symbol	Long Name
Polyolefin	Unspecified Polyolefin
Polyolefin Ela	Polyolefin elastomer
PP	Polypropylene
PP Copolymer	Polypropylene Copolymer
PP Homopolymer	Polypropylene Homopolymer
PPA	Polyphthalamide
PPC	Polyphthalate Carbonate
PPE	Polyphenylene Ether
PPO	Polyphenylene Oxide
PPO Alloy	Polyphenylene Oxide Alloy
PPS	Polyphenylene Sulfide
PPSS	Polyphenylene Sulfide Sulfone
PPSU	Polyphenylsulfone
PS	Polystyrene
PSU	Polysulfone
PSU+ABS	Polysulfone + ABS Alloy
PSU+Unspecified	Polysulfone + Unspecified Alloy
PUR	Polyurethane
PUR-capro	Polyurethane (Polycaprolactone based)
PUR-ether/MDI	Polyurethane (Polyether, MDI based)
PUR-Polyester	Polyurethane (Polyester based)
PUR-Polyether	Polyurethane (Polyether based)
PUREL-TP	Polyurethane Elastomer, Thermoplastic
PVC	Polyvinyl Chloride
PVC Alloy	Polyvinyl Chloride Alloy
PVC Elastomer	Polyvinyl Chloride Elastomer

Symbol	Long Name
PVC+NBR Alloy	Polyvinyl Chloride + Nitrile Butadiene Rubber Alloy
PVC+NR Alloy	Polyvinyl Chloride + Nitrile Rubber Alloy
PVC+PUR Alloy	Polyvinyl Chloride + Polyurethane Alloy
PVDF	Polyvinylidene Fluoride
SAN	Styrene Acrylonitrile
SB	Styrene Butadiene
SBS Copolymer	Styrene Butadiene Styrene Block Copolymer
SEBS Copolymer	Styrene Ethylene Butylene Styrene Block Copolymer
SMA	Styrene Maleic Anhydride
SMA+PBT Alloy	Styrene Maleic Anhydride + PBT Alloy
SMMA Copolymer	Styrene Methyl Methacrylate Copolymer
SVA	Styrenic + Vinyl + Acrylonitrile Alloy
Styrene Elast.	Styrene Elastomer
TPE	Thermoplastic Elastomer
TPO	Thermoplastic Olefin Elastomer
UHMWPE	Polyethylene, Ultra High Molecular Weight
Urea Compound	Urea Compound
Urea Formal.	Urea Formaldehyde

Fillers and Additives

Symbol	Long Name
AC	Alpha cellulose
BR	Bronze
CAC	Calcium carbonate ($CaCO_3$)
CBI	Carbonyl iron powder
CBL	Carbon black
CE	Cellulose fiber
CF	Carbon fiber
CFN	Carbon fiber, nickel-coated
CGM	Carbon\glass\mineral
CH	Chalk
CP	Carbon powder
CRF	Ceramic fiber
CSO	Wollastonite ($CaSiO_3$)
DAC	Dacron
FLK	Flock
GL	Glass
GLB	Glass bead (sphere)
GLC	Glass fiber, chopped
GLF	Glass fiber
GLM	Glass fabric
GMI	Glass\mica
GMN	Glass\mineral
GRF	Graphite fiber
GRK	Graphite flake
GRN	Graphite fiber, nickel-coated

Symbol	Long Name
GRP	Graphite powder
GSA	Glass\silica reinforcement
KEV	Kevlar fiber
MI	Mica
MN	Mineral
MNF	Mineral fiber
MOS	Molybdenum disulfide
MS	Molysulfide
NY6	Nylon 6
ORL	Orlon
PE	Polyethylene
PTF	Polytetrafluoroethylene (PTFE)
PU	Polyurethane
SGL	S-glass fiber
SI	Silicone
SIF	Silicone fluid
STS	Stainless steel fiber
TAL	Talc
TIN	Tin stabilizing agent
TIO	Titanium dioxide (TiO_2)
UNS	Unspecified filler\reinforcement
WDF	Wood flour
ZIS	Zinc stearate

Melt Flow Conditions

Note: These melt flow conditions are footnotes to the property tables that follow.

Condition	°C	Load, kg	Load, kPa	Load, psi
A	125	0.325	44.8	6.50
B	125	2.16	298.2	43.25
C	150	2.16	298.2	43.25
D	190	0.325	44.8	6.50
E	190	2.16	298.2	43.25
F	190	21.6	2982.2	432.5
G	200	5.0	689.5	100.0
H	230	1.2	165.4	24.0
I	230	3.8	524.0	76.0
J	265	12.5	1723.7	250.0
K	275	0.325	44.8	6.5
L	230	2.16	298.2	43.25
M	190	1.05	144.7	21.0
N	190	10.0	1379.0	200.0
O	300	1.2	165.4	24.0
P	190	5.0	689.5	100.0
Q	235	1.0	138.2	20.05
R	235	2.16	298.2	43.25
S	235	5.0	689.5	100.0
T	250	2.16	298.2	43.25
U	310	12.5	1723.7	250.0
V	210	2.16	298.2	43.25
W	285	2.16	298.2	43.25
X	315	5.0	698.5	100.0

PROPERTY TABLES

The organization of materials in the tables that follow is in this sequence:

• Materials are arranged alphabetically by generic name

• Within generic name, they are alphabetical by trade name.

• Within trade name, they are sorted numerically first, then alphabetically.

• When known, the number following the filler type is the percent of loading.

• The letters beside the melt flow numbers refer to the conditions giving on p. xviii (opposite).

While every effort has been made to gather data from all material suppliers, data on some materials has not been made available to us. We hope to have more grades included in future editions.

Please, let us hear from you how we can improve this publication.

Grade	Filler	Sp Grav	Shrink, mils/in	Melt flow, g/10 min	Melt temp, °F	Back pres, psi	Drying temp, °F
ABS		**AVP**			**Polymerland**		
FCCS0		1.22	5.0- 9.0	3.50 I		50- 100	190
GCC01		1.05	5.0- 9.0	3.00 I		50- 100	190
GCC20		1.05	5.0- 9.0	3.00 I		50- 100	190
GCC30		1.05	5.0- 9.0	3.00 I		50- 100	190
GCCHG		1.05	5.0- 9.0	4.00 I		50- 100	190
GCMHG		1.05	5.0- 9.0	4.00 I		50- 100	190
RCC10	GL 10	1.10	2.0- 3.0	2.50 I		50- 200	190
RCC20	GL 20	1.19	1.0- 2.0	2.00 I		50- 200	190
RCC30	GL 30	1.29	1.0- 2.0	1.00 I		250- 500	190
ABS		**Celstran S**			**Hoechst**		
ABSS10-01-4	STS 10	1.14	4.0- 6.0				200
ABSS10-02-4	STS 10	1.32	4.0- 6.0				200
ABSS6-01-4	STS 6	1.10	4.0- 6.0				200
ABSS6-02-4	STS 6	1.28	4.0- 6.0				200
ABS		**Cevian**			**Hoechst**		
SAG20	GLF	1.35	2.0- 5.0				176-185
SAG30	GLF	1.41	1.0- 4.0				176-185
SEHG20	GLF	1.34	2.0- 5.0				176-185
SEHG30	GLF	1.40	1.0- 4.0				176-185
SER20		1.18	4.0- 6.0	13.00 I		50- 100	176-185
SER90		1.17	4.0- 6.0	5.00 I		50- 100	176-194
SER91		1.19	4.0- 6.0	7.00 I		50- 100	176-194
SER91X		1.25	4.0- 6.0	12.00 I		50- 100	176-194
SERG1	GLF	1.30	2.0- 5.0				176-185
SERG2	GLF	1.36	2.0- 5.0				176-185
SFG20	GLF	1.33	2.0- 5.0				176-185
SFG30	GLF	1.41	1.0- 4.0				176-185
SUG20	GLF	1.34	2.0- 5.0				176-185
SUG30	GLF	1.42	1.0- 4.0				176-185
VF191		1.15	4.0- 6.0	8.00 I		50- 100	176-194
VF512		1.10	4.0- 6.0	11.00 I		50- 100	176-185
VF790		1.09	4.0- 6.0	12.00 I		50- 100	176-194
ABS		**Cevian-V**			**Hoechst**		
320		1.05	4.0- 6.0	5.00 I		100- 300	176-185
400N		1.05	5.0- 8.0	0.60 I		100- 300	176-185
400T		1.05	4.0- 6.0	1.00 I		100- 300	176-185
420		1.05	4.0- 6.0	0.60 I		100- 300	176-185
500		1.05	4.0- 6.0	5.00 I		100- 300	176-185
510		1.06	4.0- 6.0	6.00 I		100- 300	176-185
660SF		1.05	5.0- 7.0	16.00 I		100- 300	176-185
ABS		**CTI ABS**			**CTI**		
AS-10GF	GLF 10	1.10	3.0				190-200
AS-15CF/000	CF 15	1.10	2.0				190-200
AS-20GF	GLF 20	1.20	2.0				190-200
AS-20GF/HI UV	GLF 20	1.19	2.0				190-200
AS-20NCF	CFN 20	1.25	0.9				190-200
AS-30GF	GLF 30	1.29	1.0				190-200
AS-40GF	GLF 40	1.38	1.0				190-200
AS-7SS	STS 7	1.12	5.0				190-200

Drying time, hr	Inj time, sec	Front temp, °F	Mid temp, °F	Rear temp, °F	Nozzle temp, °F	Mold temp, °F	Proc temp, °F
2.0						130-160	380-450
2.0						120-150	450-500
3.0						120-150	425-500
2.0						120-150	425-500
2.0						120-150	425-500
2.0						120-150	390-480
4.0						150-200	475-500
4.0						150-200	475-500
4.0						150-200	475-500
4.0		400	390	380	390	160	
4.0		400	390	380	390	160	
4.0		400	390	380	390	160	
4.0		400	390	380	390	160	
2.0- 4.0						140-176	446-482
2.0- 4.0						140-176	446-482
2.0- 4.0						140-176	428-464
2.0- 4.0						140-176	428-464
2.0- 4.0		410	392	356		104-140	356-446
2.0- 4.0		464	428	392		104-140	392-482
2.0- 4.0		428	392	356		104-140	356-446
2.0- 4.0		428	392	356		104-140	356-446
2.0- 4.0						140-176	428-464
2.0- 4.0						140-176	428-464
2.0- 4.0						140-176	482-518
2.0- 4.0						140-176	482-518
2.0- 4.0						140-176	482-518
2.0- 4.0						140-176	482-518
2.0- 4.0		428	392	356		104-140	356-446
2.0- 4.0		410	392	356		104-140	356-446
2.0- 4.0		410	374	338		104-140	356-446
2.0- 4.0		430-450	410-430	370-390		120-140	450-470
2.0- 4.0		450-470	430-450	390-410		120-140	470-490
2.0- 4.0		450-470	430-450	390-410		120-140	470-490
2.0- 4.0		450-470	430-450	390-410		120-140	470-490
2.0- 4.0		430-450	410-430	370-390		120-140	450-470
2.0- 4.0		430-450	410-430	370-390		120-140	450-470
2.0- 4.0		430-450	410-430	370-390		120-140	450-470
2.0		490-520	450-490	420-450	450-500	140-200	
2.0		490-520	450-490	420-450	450-500	140-200	
2.0		490-520	450-490	420-450	450-500	140-200	
2.0		490-520	450-490	420-450	450-500	140-200	
2.0		490-520	450-490	420-450	450-500	140-200	
2.0		490-520	450-490	420-450	450-500	140-200	
2.0		490-520	450-490	420-450	450-500	140-200	
2.0		490-520	450-490	420-450	450-500	140-200	

ABS Cycolac GE

Grade	Filler	Sp Grav	Shrink, mils/in	Melt flow, g/10 min	Melt temp, °F	Back pres, psi	Drying temp, °F
AM		1.04	6.0- 8.0	4.00 G		50- 100	180-200
AR		1.05	5.0- 8.0			50- 100	180-190
CKM1		1.21				100	190-200
CKM2		1.21				100	190-200
DFAR		1.04	5.0- 8.0	2.10 G		50- 100	190-200
DFB		1.05	5.0- 8.0	4.00 G		50- 60	190-200
DFN1200		1.04	5.0- 8.0	1.60 G		0- 10	190-200
DFS		1.04	5.0- 8.0			0- 100	190-200
DH		1.05	5.0- 8.0			50- 100	180-190
DSK		1.05	6.0- 8.0	17.20		0- 100	190-200
EP		1.04	5.0- 8.0				190-200
EPB		1.05	5.0- 8.0				190-200
ETC		1.04	5.0- 8.0				190-200
ETS		1.05	5.0- 8.0				180-200
GDM6500		1.06	5.0- 8.0	1.90 G			180-190
GDT2510		1.05	5.0- 8.0				200-220
GDT6400		1.05	5.0- 8.0	2.00 G		50- 100	180-190
GHM3510		1.05	5.0- 8.0				200
GHT3510		1.05	5.0- 8.0				200-220
GHT4320		1.05	5.0- 8.0				200-220
GHT4400		1.02	5.0- 8.0				190-200
GLM3840		1.03					180-190
GPM4000		1.04	5.0- 8.0			0- 100	190-200
GPM4700		1.04	5.0- 8.0			100	190-200
GPM5500		1.05	5.0- 8.0	5.60 I		100	190-200
GPM5500F		1.05	5.0- 8.0			0- 100	190-200
GPM550M		1.05	6.0- 8.0	14.00 I		50- 100	190-200
GPM5600		1.05	5.0- 8.0			100	190-200
GPM5600F		1.05	5.0- 8.0			100	190-200
GPM5601		1.05	5.0- 8.0			0- 100	190-200
GPM6300		1.05	5.0- 8.0			0- 100	190-200
GPM6300F		1.05	5.0- 8.0			0- 100	190-200
GPT1900		1.02	5.0- 8.0			50- 100	190-200
GPT3800		1.02	5.0- 8.0				180-190
GPT4600		1.02	5.0- 8.0				190-200
GPT4800		1.03	5.0- 8.0	0.60 G			180-190
GPT5500		1.05	5.0- 8.0	1.70 G		0- 100	190-200
GSM		1.04	5.0- 8.0	2.00 G		0- 100	180-200
GT		1.05	5.0- 8.0			50- 100	190-200
GTM5300		1.07	5.0- 7.0			100	160-170
HMM		1.05	5.0- 8.0			0- 100	190-200
HTF		1.20	5.0- 8.0			0- 100	190-200
KF5		1.22				100	190-200
KJB		1.22	5.0- 8.0			0- 100	190-200
KJL		1.22	5.0- 8.0			100	190-200
KJM		1.20	5.0- 8.0			100	190-200
KJT		1.20	5.0- 8.0			100	190-200
KJU		1.22	5.0- 8.0			100	190-200
KJW		1.23	5.0- 8.0			100	190-200
L		1.02	5.0- 8.0	0.50 G			190-200
LDI		1.03	5.0- 8.0	3.30 I		0- 100	180-200
PLN4000		1.05	5.0- 8.0			0- 100	190-200
PLN6000		1.05	5.0- 8.0			0- 100	190-200
PLN7000		1.02	5.0- 8.0				180-190

Drying time, hr	Inj time, sec	Front temp, °F	Mid temp, °F	Rear temp, °F	Nozzle temp, °F	Mold temp, °F	Proc temp, °F
2.0- 4.0		430	410	380		120-150	425-500
2.0- 4.0		450	420	390		120-150	450-500
2.0	2.0- 15.					120-140	
2.0	2.0- 15.					120-140	
2.0- 4.0	2.0- 15.	430	410	380	430	120-150	425-500
2.0- 4.0		430	410	380	430	120-150	425-500
2.0- 8.0		430-450	410-430	370-390	450-500	120-150	450-500
2.0		430	410	380		120-150	425-500
2.0	2.0- 15.	460	420	390		140-180	450-500
2.0	2.0- 10.	430	410	380	500	120-150	425-500
2.0- 4.0		480	450	400		100-170	450-500
4.0		500	450	400		125-180	475-530
2.0- 4.0		480	450	400		100-170	450-500
2.0- 4.0		480	450	400		100-170	470-510
2.0- 4.0		460	430	400		80-140	450-500
2.0- 4.0		475	460	440		120-180	475-525
2.0- 4.0		450-470	430-450	410-430		80-140	475-525
2.0- 4.0		460	450	440		120-180	475-525
2.0- 4.0		460	450	440		120-180	475-525
2.0- 4.0		460	450	440		120-180	475-525
2.0- 4.0		450	425	400		140-180	450-500
2.0		450	420	390		120-150	450-500
2.0- 4.0	2.0- 10.	440	420	380		120-160	425-500
2.0- 4.0	2.0- 15.	440	420	380	395-490	120-160	425-500
2.0- 4.0	2.0- 10.	430	410	380	395-490	120-160	425-500
2.0- 4.0	2.0- 10.	430	410	380		120-160	425-500
2.0		430	410	380		120-150	425-500
2.0- 4.0	2.0- 10.	440	420	380	395-490	120-160	425-500
2.0- 4.0		440	420	380		120-160	425-500
2.0- 4.0	2.0- 10.	440	420	380		120-160	425-500
2.0- 4.0	2.0- 10.	430	410	380	395-490	120-160	425-500
2.0- 4.0	2.0- 10.	430	410	380		120-160	425-500
2.0- 4.0		440-460	410-430	370-390		120-150	475-500
2.0- 8.0		450	420	390		120-150	450-500
2.0- 4.0		440	420	380		120-160	425-475
2.0- 4.0		450	420	390		120-150	450-500
2.0- 4.0		440	420	380		120-160	450-475
2.0- 4.0	2.0- 15.	440	420	380		100-180	450-500
2.0- 4.0		430	410	380	430	120-150	425-500
2.0		440	420	380	370-450	120-160	400-460
2.0- 4.0	2.0- 10.	440	420	380		120-160	425-500
2.0		420	400	350		130-160	380-450
2.0	2.0- 15.					120-140	
2.0	2.0- 15.	420	400	350		130-160	380-450
2.0	2.0- 15.	420	400	350		130-160	380-450
2.0	2.0- 15.					120-140	
2.0		420	400	350		130-160	380-450
2.0	2.0- 15.	420	400	350		130-160	380-450
2.0	2.0- 15.	420	400	350		130-160	380-450
2.0- 4.0	2.0- 15.	450	420	380		120-150	475-500
2.0- 4.0		380-400	350-400	350-380	400-430	100-120	400-430
2.0- 4.0	2.0- 10.	430	410	380		120-160	425-500
2.0- 4.0	2.0- 10.	440	420	380		120-160	425-500
2.0- 4.0		450	420	390		120-150	450-500

Grade	Filler	Sp Grav	Shrink, mils/in	Melt flow, g/10 min	Melt temp, °F	Back pres, psi	Drying temp, °F
R32		1.04	5.0- 8.0	2.20 G		100	190-200
R33		1.05	5.0- 8.0			100	190-200
REC150		1.05	5.0- 9.0	3.00 I		50- 100	190
SDB		1.07				100	160-170
T		1.04	5.0- 8.0	2.20 G		100	190-200
TA		1.04	5.0- 8.0			100	180-190
TE		1.04	5.0- 8.0	2.00 G		50- 100	190-200
TN		1.04	5.0- 8.0	1.60 G		50- 100	190-200
V100		1.21	5.0- 7.0			50- 100	180
V200		1.19	5.0- 7.0			50- 100	180
VW300		1.20	5.0- 7.0			50- 100	180
X11		1.04	5.0- 8.0			50- 100	190-200
X15		1.05	5.0- 8.0				210-220
X17		1.05	5.0- 8.0			50- 100	200-210
X37		1.06	5.0- 8.0	1.00 I		50- 100	200-210
Z48		1.06	5.0- 8.0			100	200-210
Z86		1.20	4.0- 6.0			100	190-200
ZA2		1.04	5.0- 8.0			100	180-190
ABS		**Diamond ABS**			**Diamond**		
1501		1.06		2.20 G			176-185
1501HF		1.06		4.00 G			176-185
1510		1.06		6.00 G			176-185
2510		1.06		4.00 G			176-185
3001-MC		1.05		2.00 G			176-185
3500		1.05	5.0	2.00 G			176-185
3500 Black LG1200		1.05	5.0	1.50 G			176-185
3501		1.05	5.0	1.00 G			176-185
3501 GF-10	GLF 10			2.50 I			176-185
3501E		1.05		1.00 G			176-185
3510 HF		1.05		3.20 G			176-185
4000 HH		1.07		0.45 G			185-194
4500R		1.05	5.0	2.00 G			176-185
7500E		1.04		0.60 G			176-185
7501		1.04		1.40 G			176-185
9501		1.04		1.00 G			176-185
VPII		1.21		2.00 G			185-194
ABS		**DSM ABS**			**DSM**		
G-1200/40	GLF 40	1.36	1.0				170-190
J-1200/10	GLF 10	1.11	3.0				170-190
J-1200/20	GLF 20	1.23	3.0				170-190
J-1200/30	GLF 30	1.29	2.0				170-190
ABS		**Electrafil**			**DSM**		
ABS-1200/SD		1.08	6.0			50- 100	150-170
ABS-1250/SD		1.08	6.0			50- 100	150-170
G-1204/SS/3	STS 3	1.08	5.0				170-190
G-1204/SS/5/FR	STS 5	1.29	4.0				170-190
G-1204/SS/7	STS 7	1.12	4.0				170-190
G-1204/SS/7/FR	STS 7	1.30					170-190
J-1200/CF/10	CF 10	1.10	1.0				170-190
J-1200/CF/10/FR	CF 10	1.25	1.5				170-190
J-1200/CF/20	CF 20	1.14	0.5				170-190
J-1200/CF/40	CF 40	1.24	0.5				170-190

Drying time, hr	Inj time, sec	Front temp, °F	Mid temp, °F	Rear temp, °F	Nozzle temp, °F	Mold temp, °F	Proc temp, °F
2.0- 4.0		420-480	400-450	380-420	425-500	130-160	425-500
2.0- 4.0		420-440	400-420	370-390	425-500	120-150	425-500
2.0						120-150	425-500
2.0	2.0- 15.					120-150	380-460
2.0- 4.0	2.0- 15.	420-480	400-450	380-420	425-500	130-160	425-500
2.0- 4.0		450	420	390		120-150	450-500
2.0- 8.0		430-450	410-430	370-390	450-500	120-150	450-500
2.0- 4.0		440-460	410-432	370-390	475-500	120-150	475-500
2.0- 4.0		420	400	350	410	120-140	400-450
2.0- 4.0		420	400	350	410	120-140	400-450
2.0- 4.0		420	400	350	410	120-140	400-450
4.0	2.0- 15.	460	420	390		140-180	450-500
4.0		480	440	400		140-180	475-525
4.0	2.0- 15.	480	440	400		140-180	475-525
4.0	2.0- 15.	480	440	400		140-200	475-500
4.0		480	440	400		140-180	475-515
2.0		420	400	350		130-160	380-460
2.0- 4.0		460	440	400		120-180	450-525
2.0- 4.0		374-482	374-482	374-482		104-176	
2.0- 4.0		356-482	356-482	356-482		104-176	
2.0- 4.0		356-482	356-482	356-482		104-176	
2.0- 4.0		356-482	356-482	356-482		104-176	
2.0- 4.0		392-482	392-482	392-482		104-176	
2.0- 4.0		374-482	374-482	374-482		104-176	
2.0- 4.0		374-482	374-482	374-482		104-176	
2.0- 4.0		374-482	374-482	374-482		104-176	
2.0- 4.0		374-482	374-482	374-482		104-176	
2.0- 4.0		374-482	374-482	374-482		104-176	
2.0- 4.0		356-482	356-482	356-482		104-176	
2.0- 4.0		392-482	392-482	392-482		104-176	
2.0- 4.0		374-482	374-482	374-482		104-176	
2.0- 4.0		374-482	374-482	374-482		104-176	
2.0- 4.0		374-482	374-482	374-482		104-176	
2.0- 4.0		374-482	374-482	374-482		104-176	
2.0- 4.0		392-482	392-482	392-482		104-176	
2.0-16.0		410-430	430-460	420-450	390-430	160-190	450-500
2.0-16.0		410-430	430-460	420-450	390-430	160-190	450-500
2.0-16.0		410-430	430-460	420-450	390-430	160-190	450-500
2.0-16.0		410-430	430-460	420-450	390-430	160-190	450-500
4.0-16.0		390	400	380	390	90	385
4.0-16.0		390	400	380	390	90	385
2.0-16.0		410-430	430-460	420-450	390-430	160-190	450-500
2.0-16.0		410-430	430-460	420-450	390-430	160-190	450-500
2.0-16.0		410-430	430-460	420-450	390-430	160-190	450-500
2.0-16.0		410-430	430-460	420-450	390-430	160-190	450-500
2.0-16.0		410-430	430-460	420-450	390-430	160-190	450-500
2.0-16.0		410-430	430-460	420-450	390-430	160-190	450-500
2.0-16.0		410-430	430-460	420-450	390-430	160-190	450-500
2.0-16.0		410-430	430-460	420-450	390-430	160-190	450-500

Grade	Filler	Sp Grav	Shrink, mils/in	Melt flow, g/10 min	Melt temp, °F	Back pres, psi	Drying temp, °F
ABS		**Fiberfil**			**DSM**		
J-1200/10	GLF 10	1.11	3.0				170-190
J-1200/20	GLF 20	1.23	3.0				170-190
J-1200/30	GLF 30	1.29	2.0- 3.0				170-190
ABS		**Fiberfil VO**			**DSM**		
J-1200/20/VO	GLF 20	1.37	0.8- 1.8				160-180
J-1200/25/VO	GLF 25	1.39					160-180
J-1200/30/VO	GLF 30	1.41	0.6- 1.3				160-180
ABS		**Fiberstran**			**DSM**		
G-1200/20	GLF 20	1.23	1.0- 2.0			10- 50	165-180
G-1200/40	GLF 40	1.36	1.0			10- 50	165-180
G-1204/SS/3	STS 3	1.08	5.0				170-190
G-1204/SS/7	STS 7	1.12	4.0				170-190
G-1204/SS/7/FR	STS 7	1.30					170-190
ABS		**Lupos**			**Lucky**		
GP-2100	GLF 10	1.10	2.0- 3.0			0- 568	176-194
GP-2101F	GLF 10	1.24	1.0- 3.0			0- 568	176-194
GP-2200	GLF 20	1.21	2.0- 3.0			0- 568	176-194
GP-2201F	GLF 20	1.35	1.0- 3.0			0- 568	176-194
GP-2206F	GLF 20	1.35	1.0- 3.0			0- 568	176-194
GP-2300	GLF 30	1.28	1.0- 2.0			0- 568	176-194
GP-2301F	GLF 30	1.42	1.0- 2.0			0- 568	176-194
HR-2207	GLF 20	1.21	2.0- 3.0			0- 568	176-194
ABS		**Lustran ABS**			**Monsanto**		
1146		1.05	6.0- 8.0			0- 200	170-190
1148		1.05	6.0- 8.0			0- 200	170-190
137		1.06	4.0- 6.0			0- 200	170-190
246		1.07	4.0- 6.0			0- 200	170-190
248		1.06	4.0- 6.0			0- 200	170-190
446		1.06	4.0- 6.0			0- 200	170-190
448		1.05	4.0- 6.0			0- 200	170-190
633		1.05	4.0- 6.0	4.50 I			180-190
648		1.04	4.0- 6.0			0- 200	170-190
743		1.04	4.0- 6.0			0- 200	170-190
746		1.05	5.0- 8.0			0- 200	170-190
750-10802		1.05				0- 200	170-190
911		1.21	4.0- 6.0			0- 200	170-190
914 HM		1.20	4.0- 6.0			0- 200	170-190
921 UV		1.20	4.0- 6.0			0- 200	160-190
G6220	GL 20	1.23	2.0	1.30 I		0- 200	170-190
PG-298 Plated						0- 200	170-190
PG-298 Unplated		1.06	4.0- 6.0			0- 200	170-190
PG-299 Plated						0- 200	170-190
PG-299 Unplated		1.06	4.0- 6.0			0- 200	170-190
PG-300 Plated						0- 200	170-190
PG-300 Unplated		1.06	4.0- 6.0			0- 200	170-190
ABS		**Lustran Elite**			**Monsanto**		
HH 1627		1.04	5.0	2.30 I			160-190
HH 1677		1.04	5.0	2.60 I			160-190
HH 1827		1.05	4.0- 7.0	3.00 I			160-190

Drying time, hr	Inj time, sec	Front temp, °F	Mid temp, °F	Rear temp, °F	Nozzle temp, °F	Mold temp, °F	Proc temp, °F
2.0-16.0		410-430	430-460	420-450	390-430	160-190	450-500
2.0-16.0		410-430	430-460	420-450	390-430	160-190	450-500
2.0-16.0		410-430	430-460	420-450	390-430	160-190	450-500
2.0-16.0		400-420	420-450	410-440	380-420	150-180	440-490
2.0-16.0		400-420	420-450	410-440	380-420	150-180	440-490
2.0-16.0		400-420	420-450	410-440	380-420	150-180	440-490
2.0-16.0		430-460	440-470	410-430	430-460	120-180	440-470
2.0-16.0		430-460	440-470	410-430	430-460	120-180	440-470
2.0-16.0		410-430	430-460	420-450	390-430	160-190	450-500
2.0-16.0		410-430	430-460	420-450	390-430	160-190	450-500
2.0-16.0		410-430	430-460	420-450	390-430	160-190	450-500
2.0- 3.0		410-446	392-428	374-410	410-446	140-194	428-464
2.0- 3.0		410-446	392-428	374-410	410-446	140-194	428-464
2.0- 3.0		410-446	392-428	374-410	410-446	140-194	428-464
2.0- 3.0		410-446	392-428	374-410	410-446	140-194	428-464
2.0- 3.0		410-446	392-428	374-410	410-446	140-194	428-464
2.0- 3.0		410-446	392-428	374-410	410-446	140-194	428-464
2.0- 3.0		410-446	392-428	374-410	410-446	140-194	428-464
2.0- 3.0		446-482	428-464	410-446	446-482	140-194	464-500
2.0- 4.0		400-475	400-475	400-475	425-500	120-150	425-500
2.0- 4.0		400-475	400-475	400-475	425-500	120-150	425-500
2.0- 4.0		400-475	400-475	400-475	425-500	120-150	425-500
2.0- 4.0		400-475	400-475	400-475	425-500	120-150	425-500
2.0- 4.0		400-475	400-475	400-475	425-500	120-150	425-500
2.0- 4.0		400-475	400-475	400-475	425-500	120-150	425-500
2.0- 4.0		400-475	400-475	400-475	425-500	120-150	425-500
2.0						110-150	475-525
2.0- 4.0		400-475	400-475	400-475	425-500	120-150	425-500
2.0- 4.0		400-475	400-475	400-475	425-500	120-150	425-500
2.0- 4.0		400-475	400-475	400-475	425-500	120-150	425-500
2.0- 4.0		400-475	400-475	400-475	425-500	120-150	425-500
2.0- 4.0		400-475	400-475	400-475	425-500	120-150	425-500
2.0- 4.0		400-475	400-475	400-475	425-500	120-150	425-500
2.0- 4.0		400-475	400-475	400-475	425-500	110-150	380-450
2.0- 4.0		400-475	400-475	400-475	425-500	120-150	425-500
2.0- 4.0		475	475	475	500	120-150	500
2.0- 4.0		475	475	475	500	120-150	500
2.0- 4.0		475	475	475	500	120-150	500
2.0- 4.0		475	475	475	500	120-150	500
2.0- 4.0		475	475	475	500	120-150	500
2.0- 4.0		475	475	475	500	120-150	500
2.0- 4.0						60-160	460-540
2.0- 4.0						100-160	460-540
2.0- 4.0						100-160	460-540

Grade	Filler	Sp Grav	Shrink, mils/in	Melt flow, g/10 min	Melt temp, °F	Back pres, psi	Drying temp, °F
HH 1829				6.00 I			160-190
HH 1891		1.05	4.0- 7.0	1.50 I			160-190
LGA		1.05	4.0- 7.0				160-190
LGA-SF		1.05	4.0- 7.0				160-180
LGM		1.05	4.0- 7.0				180-190

ABS Lustran Ultra Monsanto

Grade	Filler	Sp Grav	Shrink, mils/in	Melt flow, g/10 min	Melt temp, °F	Back pres, psi	Drying temp, °F
HDX		1.04	4.0- 6.0			0- 200	170-190
HX		1.04	4.0- 6.0			0- 200	170-190
MCX		1.05	4.0- 6.0			0- 200	170-190
MX		1.05	4.0- 6.0			0- 200	170-190
MXA		1.05	4.0- 6.0			0- 200	170-190

ABS Magnum Dow

Grade	Filler	Sp Grav	Shrink, mils/in	Melt flow, g/10 min	Melt temp, °F	Back pres, psi	Drying temp, °F
1040		1.04	4.0- 7.0	2.00		150- 500	180-190
213		1.05	5.0	5.50 I		150- 500	180-190
240		1.05	5.0	5.50 I		150- 500	180-190
2610		1.04	5.5	7.00 I		150- 500	180-190
2620		1.04	5.5	5.50 I		150- 500	180-190
2630		1.04	5.5	4.00 I		150- 500	180-190
2642		1.05	5.0	6.00 I		150- 500	180-190
2645		1.04	5.0	5.50 I		150- 500	180-190
270		1.05		4.20 I		150- 500	180-190
340		1.05	5.0	2.50 I		150- 500	180-190
341		1.05	5.0	5.00 I		150- 500	180-190
342EZ		1.05	4.0- 7.0	6.50			180-185
343		1.05	5.0	8.00 I		150- 500	180-190
344HP		1.05	4.0- 7.0	3.00			180-185
3490		1.04	5.5	8.00 I		150- 500	180-190
357HP		1.05	4.0- 7.0	2.00			180-185
3650		1.04		5.00 I		150- 500	180-190
3661		1.24	6.0	6.10 I		150- 500	180-190
3690		1.04	5.5	5.50 I		150- 500	180-190
3890		1.04	5.5	5.00 I		150- 500	180-190
4410		1.23	4.0- 7.0	11.00 I		50- 200	160-180
4420		1.23	4.0- 7.0	10.50 I		50- 200	160-180
4430		1.23	4.0- 7.0	10.00 I		50- 200	160-180
445 HQ		1.05	4.0- 6.0	5.00		150- 500	180-190
541		1.05	4.0- 7.0	5.50			180-185
545		1.04	5.0	5.50 I		150- 500	180-190
750		1.05	5.0	2.50 I		150- 500	180-190
780HP		1.05	5.0	3.00 I		150- 500	180-190
788HP		1.05	5.0	1.60 I		150- 500	180-190
9010		1.04	5.5	7.00 I		150- 500	180-190
9020		1.04	5.5	5.50 I		150- 500	180-190
9030		1.04	5.5	4.00 I		150- 500	180-190
9085		1.04	5.0	3.00 I		150- 500	180-190
9095		1.04	5.0	6.00 I		150- 500	180-190
941		1.05	5.0	2.00 I		150- 500	180-190
9450P		1.04	5.5	7.00 I		150- 500	180-190
9555		1.04		5.00 I		150- 500	180-190
9650		1.04		5.00 I		150- 500	180-190
9651		1.04		4.00 I		150- 500	180-190
9700 HF		1.04	5.0	6.00 I		150- 500	180-190
AG 700		1.05		4.20 I		150- 500	180-190
FG960		1.04	5.0	2.60 I		150- 500	180-190

Drying time, hr	Inj time, sec	Front temp, °F	Mid temp, °F	Rear temp, °F	Nozzle temp, °F	Mold temp, °F	Proc temp, °F
2.0- 4.0						80-160	460-540
2.0- 4.0						100-160	460-540
2.0- 4.0						90-120	475-525
2.0- 4.0						80-120	440-520
2.0						90-120	475-550
2.0- 4.0		400	400	400	425	120-150	425
2.0- 4.0		400	400	400	425	120-150	425
2.0- 4.0		400	400	400	425	120-150	425
2.0- 4.0		400	400	400	425	120-150	425
2.0- 4.0		400	400	400	425	120-150	425
2.0						80-140	425-525
2.0						80-140	425-525
2.0						80-140	425-525
2.0						80-140	425-525
2.0						80-140	425-525
2.0						80-140	425-525
2.0						80-140	425-525
2.0						80-140	425-525
2.0						80-140	425-525
2.0						80-140	425-525
2.0						80-140	425-525
2.0						80-120	450-475
2.0						80-140	425-525
2.0						100-140	470-520
2.0						80-140	425-525
2.0						100-180	500-540
2.0						80-140	425-525
2.0						80-140	425-525
2.0						80-140	425-525
2.0						80-140	425-525
2.0- 3.0		410	400	390	410	100-150	390-450
2.0- 3.0		410	400	390	410	100-150	390-450
2.0- 3.0		410	400	390	410	100-150	390-450
2.0						80-140	425-525
2.0						80-120	450-475
2.0						80-140	425-525
2.0						80-140	425-525
2.0						80-140	425-525
2.0						80-140	425-525
2.0						80-140	425-525
2.0						80-140	425-525
2.0						80-140	425-525
2.0						80-140	425-525
2.0						80-140	425-525
2.0						80-140	425-525
2.0						80-140	425-525
2.0						80-140	425-525
2.0						80-140	425-525
2.0						80-140	425-525
2.0						80-140	425-525
2.0						80-140	425-525
2.0						80-140	425-525

Grade	Filler	Sp Grav	Shrink, mils/in	Melt flow, g/10 min	Melt temp, °F	Back pres, psi	Drying temp, °F
HPC 650		1.04		1.00 I		150- 500	180-190
HPC 950		1.05		0.80 I		150- 500	180-190
PG912		1.05		4.20 I		150- 500	180-190
PG914		1.05		2.50 I		150- 500	180-190
ABS		**PermaStat**			**RTP**		
600		1.06	6.5- 8.5				
600 FR		1.28	6.5- 7.5				
603	GLF 20						
ABS		**Porene**			**Thai Petrochem**		
GA201			4.0- 6.0	37.00			176-185
GA300			4.0- 6.0	30.00			176-185
GA400			4.0- 6.0	45.00			176-185
GA703			4.0- 6.0	50.00			176-185
GA800			4.0- 6.0	20.00			176-185
IH			4.0- 6.0	12.00			194-212
IM-11			4.0- 6.0	15.00			194-212
KU600			4.0- 6.0	2.80			194-212
KU650			4.0- 6.0	1.80			194-212
MH			4.0- 6.0	25.00			194-212
MH-1			4.0- 6.0	18.00			176-185
MHB			4.0- 6.0	14.00			176-185
MHR			4.0- 6.0	10.00			194-212
MTH			4.0- 6.0	4.00			194-212
MVF			4.0- 6.0	18.00			176-185
SHF			4.0- 6.0	37.00			176-185
SP-100			4.0- 6.0	18.00			176-185
ABS		**Ronfalin**			**DSM**		
VE-10		1.22		14.00			176-194
VE-17		1.30		77.00			176-194
VE-18		1.22		55.00			176-194
VE-19		1.30		77.00			176-194
VE-30		1.22		17.00			176-194
VE-31		1.22		35.00			176-194
VX-10		1.24		35.00			176-194
VX-50		1.24		19.00			176-194
ABS		**RTP Polymers**			**RTP**		
600		1.04	6.0			50	180
600 SI2	SIF	1.05	7.0			50	180
600 TFE 10 FR	PTF	1.28	7.0			50	180
601	GLF 10	1.11	4.0			50	180
601 FR	GLF	1.29	3.0			50	180
601Z	GLF 10	1.11	4.0			50	180
601Z	GLF 10	1.11	3.0			50	180
603	GLF 20	1.18	2.0			50	180
603 FR	GLF	1.37	2.0			50	180
603Z	GLF 20	1.20	2.0			50	180
603Z	GLF 20	1.20	1.0			50	180
605	GLF 30	1.28	2.0			50	180
605 FR	GLF	1.45	1.0			50	180
605Z	GLF 30	1.28	2.0			50	180
605Z	GLF 30	1.28	1.0			50	180
607	GLF 40	1.38	1.0			50	180

Drying time, hr	Inj time, sec	Front temp, °F	Mid temp, °F	Rear temp, °F	Nozzle temp, °F	Mold temp, °F	Proc temp, °F
2.0						80-140	425-525
2.0						80-140	425-525
2.0						80-140	425-525
2.0						80-140	425-525
		390-480	390-480	390-480		150-200	
		390-480	390-480	390-480		150-200	
		390-480	390-480	390-480		150-200	
2.0- 3.0						104-140	356-482
2.0- 3.0						104-140	356-482
2.0- 3.0						104-140	356-482
2.0- 3.0						104-140	356-482
2.0- 3.0						104-140	356-482
2.0- 3.0						104-176	428-500
2.0- 3.0						104-176	428-500
2.0- 3.0						104-176	428-500
2.0- 3.0						104-176	428-500
2.0- 3.0						104-176	428-500
2.0- 3.0						104-140	356-482
2.0- 3.0						104-140	356-482
2.0- 3.0						104-176	428-500
2.0- 3.0						104-176	428-500
2.0- 3.0						104-140	356-482
2.0- 3.0						104-140	356-482
2.0- 3.0						104-140	356-482
2.0- 4.0						104-140	
2.0- 4.0						104-140	
2.0- 4.0						104-140	
2.0- 4.0						104-140	
2.0- 4.0						104-140	
2.0- 4.0						104-140	
2.0- 4.0						104-140	
2.0- 4.0						104-140	
2.0		450-550	430-530	420-520		150-200	
2.0		420-500	400-480	380-460		150-200	
2.0		420-500	400-480	380-460		150-200	
2.0		450-550	430-530	420-520		150-200	
2.0		420-500	400-480	380-460		150-200	
2.0		450-550	430-530	420-520		150-200	
2.0		450-550	430-530	420-520		150-200	
2.0		450-550	430-530	420-520		150-200	
2.0		420-500	400-480	380-460		150-200	
2.0		450-550	430-530	420-520		150-200	
2.0		450-550	430-530	420-520		150-200	
2.0		450-550	430-530	420-520		150-200	
2.0		420-500	400-480	380-460		150-200	
2.0		450-550	430-530	420-520		150-200	
2.0		450-550	430-530	420-520		150-200	
2.0		450-550	430-530	420-520		150-200	

Grade	Filler	Sp Grav	Shrink, mils/in	Melt flow, g/10 min	Melt temp, °F	Back pres, psi	Drying temp, °F
681 HB	CF	1.08	1.0			50	180
682 FR	CF	1.28	2.0			50	180
683 FR	CF	1.31	1.0			50	180
683 HB	CF	1.13	1.0			50	180
685 FR	CF	1.37	1.0			50	180
685 HB	CF	1.18	1.0			50	180
687 FR	CF	1.41	1.0			50	180
687 HB	CF	1.24	1.0			50	180
EMI-661	STS	1.16	5.0				
EMI-681	GRN	1.14	2.0				
ESD-A-600	CF	1.12	5.0				
ESD-A-680	CF	1.07	2.0				
ESD-C-600	CF	1.14	5.0				
ESD-C-660	STS	1.10	5.0				
ESD-C-680	CF	1.09	2.0				
ABS		**Shuman ABS**			**Shuman Plastics**		
710		1.04		1.20			170-200
720		1.04		1.20			170-200
730		1.04		1.20			170-200
ABS		**Stat-Kon**			**LNP**		
PDX-A-91512		1.03	4.0- 7.0			25- 50	160-180
ABS		**Terluran**			**BASF**		
877M/ME		1.06	4.0- 7.0				176
877T/TE		1.06	4.0- 7.0				176
967K		1.05	4.0- 7.0				176
KR2889		1.05	4.0- 7.0				176
ABS		**Toyolac**			**Toray**		
100		1.04		13.00			176-194
100G10	GLF 10	1.10					176-194
100G20	GLF 20	1.23					176-194
100G30	GLF 30	1.30					176-194
125		1.03		16.50			176-194
180-X18		1.04		2.00			176-194
300		1.03		11.00			176-194
360-X11		1.03		10.00			176-194
360-X39		1.03		20.00			176-194
440-345		1.05		6.00			176-194
450-X21		1.05		3.50			176-194
470-X60		1.05		2.00			176-194
500		1.05		17.00			176-194
700		1.05		20.00			176-194
900		1.07		14.50			176-194
920		1.08		21.00			176-194
930		1.07		14.50			176-194
ABS		**Tufrex**			**Monsanto**		
SI		1.22	4.0- 6.0				180-190
STG-2		1.20	4.0- 6.0				180-190
ABS+PA Alloy		**Triax**			**Monsanto**		
1120		1.06	10.0-12.0			50- 100	190
1120 Q010		1.06	10.0-12.0			50- 100	190

Drying time, hr	Inj time, sec	Front temp, °F	Mid temp, °F	Rear temp, °F	Nozzle temp, °F	Mold temp, °F	Proc temp, °F
2.0		450-550	430-520	400-500		120-200	
2.0		420-500	400-480	380-460		150-200	
2.0		420-500	400-480	380-460		150-200	
2.0		450-550	430-520	400-500		120-200	
2.0		420-500	400-480	380-460		150-200	
2.0		450-550	430-520	400-500		120-200	
2.0		420-500	400-480	380-460		150-200	
2.0		450-550	430-520	400-500		120-200	
		450-550	450-550	450-550		100-200	
		400-525	400-525	400-525		175-225	
		450-550	450-550	450-550		150-200	
		400-525	400-525	400-525		175-225	
		450-550	450-550	450-550		150-200	
		450-550	450-550	450-550		100-200	
		400-525	400-525	400-525		175-225	
2.0-24.0							400-440
2.0-24.0							400-440
2.0-24.0							400-440
4.0- 6.0		400-450	400-450	400-450		50-120	440
2.0- 4.0						122-176	446-536
2.0- 4.0						122-176	464-536
2.0- 4.0						122-176	410-518
2.0- 4.0						122-176	410-518
2.0- 4.0		464	428	392		104-140	
2.0- 4.0		464	428	392		104-140	
2.0- 4.0		464	428	392		104-140	
2.0- 4.0		464	428	392		104-140	
2.0- 4.0		464	428	392		104-140	
2.0- 4.0		464	428	392		104-140	
2.0- 4.0		464	428	392		104-140	
2.0- 4.0		464	428	392		104-140	
2.0- 4.0		464	428	392		104-140	
2.0- 4.0		464	428	392		104-140	
2.0- 4.0		464	428	392		104-140	
2.0- 4.0		464	428	392		104-140	
2.0- 4.0		464	428	392		104-140	
2.0- 4.0		464	428	392		104-140	
2.0- 4.0		464	428	392		104-140	
2.0- 4.0		464	428	392		104-140	
2.0- 4.0		464	428	392		104-140	
2.0		350-420	350-420	350-420	380-450	110-150	380-450
2.0		350-420	350-420	350-420	380-450	110-150	380-450
3.0		450-510	450-510	450-510	480-500	70-200	485-525
3.0		450-510	450-510	450-510	480-500	70-200	485-525

Grade	Filler	Sp Grav	Shrink, mils/in	Melt flow, g/10 min	Melt temp, °F	Back pres, psi	Drying temp, °F
1125		1.06	8.0-10.0	1.30 S		50- 100	190
1180		1.07	9.0	7.20 S		50- 100	190
1315GF	GLF 15	1.17	3.0	0.10 S		100- 200	190

ABS+PBT Alloy Cevian Hoechst

Grade	Filler	Sp Grav	Shrink, mils/in	Melt flow, g/10 min	Melt temp, °F	Back pres, psi	Drying temp, °F
B1500		1.17	7.0- 9.0			70- 210	180-250
B2504	GL 20	1.31	3.0- 5.0			70- 210	180-250
B2506	GL 30	1.39	2.0- 4.0			70- 210	180-250
B2703	GL 15	1.33	4.0- 7.0			70- 210	180-250
B2706	GL 30	1.44	3.0- 6.0			70- 210	180-250
B45M0		1.32	6.0- 8.0			70- 210	180-250
B5503	GL 15	1.38	3.0- 5.0			70- 210	180-250
B5504	GL 20	1.42	2.5- 4.5			70- 210	180-250
B5506	GL 30	1.50	2.0- 4.0			70- 210	180-250
B5526	GL 30	1.50	2.0- 4.0			70- 210	180-250
B5536	GL 30	1.56	2.0- 4.0			70- 210	180-250
B5706	GL 30	1.55	3.0- 5.0			70- 210	180-250
B5723	GL 15	1.43	4.0- 6.0			70- 210	180-250
B5726	GL 30	1.55	3.0- 5.0			70- 210	180-250
B5736	GL 30	1.64	3.0- 5.0			70- 210	180-250
B6506	GMN 30	1.51	2.0- 4.0			70- 210	180-250
B6508	GMN 40	1.61	2.0- 3.0			70- 210	180-250

ABS+PBT Alloy CTI General CTI

Grade	Filler	Sp Grav	Shrink, mils/in	Melt flow, g/10 min	Melt temp, °F	Back pres, psi	Drying temp, °F
CTX-308	GLF 20	1.27	3.5				190-200

ABS+PBT Alloy Cycolac GE

Grade	Filler	Sp Grav	Shrink, mils/in	Melt flow, g/10 min	Melt temp, °F	Back pres, psi	Drying temp, °F
GCM1900		1.12				50- 100	160-170
GCM2900		1.13				50- 100	210-230

ABS+PBT Alloy Cycolin GE

Grade	Filler	Sp Grav	Shrink, mils/in	Melt flow, g/10 min	Melt temp, °F	Back pres, psi	Drying temp, °F
GCM1900		1.12	6.0- 8.0			50- 100	210-230
GCM2900		1.13	8.0- 9.0			50- 100	210-230
GCT1900		1.12	6.0- 8.0			50- 100	210-230
GCT2900		1.13	8.0- 9.0			50- 100	210-230

ABS+PBT Alloy Lupoy Lucky

Grade	Filler	Sp Grav	Shrink, mils/in	Melt flow, g/10 min	Melt temp, °F	Back pres, psi	Drying temp, °F
GP-5001AF		1.22	5.0- 7.0			0- 568	212-230
GP-5001BF		1.22	5.0- 7.0			0- 568	212-230
GP-5006AF		1.22	5.0- 7.0			0- 568	212-230
GP-5006BF		1.22	5.0- 7.0			0- 568	212-230
GP-5008AF		1.19	5.0- 7.0			0- 568	212-230
GP-5008BF		1.20	5.0- 7.0			0- 568	212-230
GP-5150	UNS 15	1.25	2.5- 3.5			0- 568	212-230
GP-5300	UNS 30	1.36	2.0- 3.5			0- 568	212-230
GP-5300F	UNS 30	1.44	2.0- 3.5			0- 568	212-230
GP-5306F	UNS 30	1.44	2.0- 3.5			0- 568	212-230
HI-5002A		1.12	5.0- 7.0			0- 568	212-230
HI-5006A		1.12	5.0- 7.0			0- 568	212-230
HR-5007A		1.13	5.0- 7.0			0- 568	212-230
HR-5009A		1.10	5.0- 7.0			0- 568	212-230
LT-1A		1.12	5.0- 7.0			0- 568	212-230
MP-5000A		1.12	5.0- 7.0			0- 568	212-230
MP-5001AF		1.22	5.0- 7.0			0- 568	212-230

Drying time, hr	Inj time, sec	Front temp, °F	Mid temp, °F	Rear temp, °F	Nozzle temp, °F	Mold temp, °F	Proc temp, °F
3.0		475-510	475-510	475-510	490-510	70-200	500-525
2.0- 4.0						100-150	460-520
2.0- 4.0						100-200	500-525
3.0- 5.0						140-176	465-500
3.0- 5.0						140-176	465-500
3.0- 5.0						140-176	465-500
3.0- 5.0						140-176	465-500
3.0- 5.0						140-176	465-500
3.0- 5.0						140-176	465-500
3.0- 5.0						140-176	465-500
3.0- 5.0						140-176	465-500
3.0- 5.0						140-176	465-500
3.0- 5.0						140-176	465-500
3.0- 5.0						140-176	465-500
3.0- 5.0						140-176	465-500
3.0- 5.0						140-176	465-500
3.0- 5.0						140-176	465-500
3.0- 5.0						140-176	465-500
3.0- 5.0						140-176	465-500
3.0- 5.0						140-176	465-500
2.0		490-520	450-490	420-450	450-500	140-200	
1.0- 2.0	2.0- 15.	450-475	440-465	430-455	475	120-150	475-520
4.0	2.0- 15.	450-475	440-465	430-455	475	120-150	475-520
4.0		450-475	440-465	430-455	475	120-150	475-520
4.0		450-475	440-465	430-455	475	120-150	475-520
4.0		450-475	440-465	430-455	475	120-150	475-520
4.0		450-475	440-465	430-455	475	120-150	475-520
3.0- 5.0		455-473	446-464	428-455	455-482	140-212	464-518
3.0- 5.0		455-473	446-464	428-455	455-482	140-212	464-518
3.0- 5.0		455-473	446-464	428-455	455-482	140-212	464-518
3.0- 5.0		455-473	446-464	428-455	455-482	140-212	464-518
3.0- 5.0		455-473	446-464	428-455	455-482	140-212	464-518
3.0- 5.0		455-473	446-464	428-455	455-482	140-212	464-518
3.0- 5.0		455-473	446-464	428-455	455-482	140-212	464-518
3.0- 5.0		455-473	446-464	428-455	455-482	140-212	464-518
3.0- 5.0		455-473	446-464	428-455	455-482	140-212	464-518
3.0- 5.0		455-473	446-464	428-455	455-482	140-212	464-518
3.0- 5.0		455-473	446-464	428-455	455-482	140-212	464-518
3.0- 5.0		455-473	446-464	428-455	455-482	140-212	464-518
3.0- 5.0		455-473	446-464	428-455	455-482	140-212	464-518
3.0- 5.0		455-473	446-464	428-455	455-482	140-212	464-518
3.0- 5.0		455-473	446-464	428-455	455-482	140-212	464-518
3.0- 5.0		455-473	446-464	428-455	455-482	140-212	464-518
3.0- 5.0		455-473	446-464	428-455	455-482	140-212	464-518

Grade	Filler	Sp Grav	Shrink, mils/in	Melt flow, g/10 min	Melt temp, °F	Back pres, psi	Drying temp, °F
ABS+PC Alloy		**Bayblend**			**Miles**		
DP2-1443		0.91				50- 100	175-195
DP2-1448		1.19				50- 100	175-195
FR 110		1.19	4.0- 6.0	35.00		50- 100	175-210
FR 1439		1.17	4.3- 5.5	24.00		50- 100	175-195
FR 1440		1.18	4.5- 5.6	19.00		50- 100	175-210
FR 1441		1.18	4.8- 5.8	16.00		50- 100	175-230
FR 90		1.17	3.0- 5.0	35.00		50- 100	175-195
T 44		1.10	5.0- 7.0	13.00		50- 100	230
T 45 MN		1.10	5.0- 7.0	8.00		50- 100	230
T 64		1.13	5.0- 7.0	13.00		50- 100	230
T 65 MN		1.13	5.0- 7.0	8.00 S		50- 100	230
T 84		1.15	5.0- 7.0	13.00		50- 100	230
T 85 MN		1.15	5.0- 7.0	8.00 S		50- 100	230
ABS+PC Alloy		**Celstran**			**Hoechst**		
PCG25-02-4	GLF 25	1.36					200
PCG40-02-4	GLF 40	1.50					200
ABS+PC Alloy		**Cycoloy**			**GE**		
C1110		1.14	5.0- 7.0	2.50 I		25- 50	230
C1110HF		1.14	5.0- 7.0	14.00		50- 100	230
C1200		1.15	5.0- 7.0	7.00		50- 100	230
C1950		1.12	5.0- 7.0	7.00 I		50- 60	230
C2800		1.17	4.0- 6.0	16.00		50- 100	170-180
C2950		1.18	4.0- 6.0	10.00		50- 100	200-210
C2950HF		1.18	4.0- 6.0	22.00		50- 100	195-205
MC8002		1.14	5.0- 7.0	8.00		50- 60	230
MC9000		1.14	5.0- 7.0				230
ABS+PC Alloy		**Iupilon**			**Mitsubishi Gas**		
GP-1		1.14	3.0- 7.0				230-248
GP-1L		1.10	3.0- 7.0				230-248
GP-2		1.14	3.0- 7.0				230-248
GP-2L		1.21	3.0- 7.0				230-248
GP-3L		1.25	3.0- 7.0				230-248
ABS+PC Alloy		**Koblend**			**EniChem**		
PCA 447		1.10	5.0- 7.5	10.00			
PCA 538		1.11	5.0- 7.5	9.00			
PCA 638		1.13	5.0- 7.5	9.00			
PCA 839		1.15	5.0- 7.5	9.00			
ABS+PC Alloy		**PermaStat**			**RTP**		
2500		1.24	6.0- 9.0				
2500FR							
2503	GLF 20	1.40	1.0- 3.0				
ABS+PC Alloy		**Pulse**			**Dow**		
1310		1.12	5.0- 7.0	5.00 I		50	200
1350		1.14	5.0- 7.0	3.00 I		50	200
1370		1.14	5.0- 7.0	3.00 I		50	200
1460		1.13	5.0- 7.0	2.00 I		50	200
1550	GLF	1.19	3.0- 5.0	5.00 I			200
1670	MN	1.18	5.0- 7.0	2.00 I			210

Drying time, hr	Inj time, sec	Front temp, °F	Mid temp, °F	Rear temp, °F	Nozzle temp, °F	Mold temp, °F	Proc temp, °F
4.0		445-465	435-455	430-445	485-505	120-175	430-520
4.0		445-465	435-455	430-445	485-505	120-175	430-520
3.0- 4.0		445-465	435-455	430-445	485-505	120-175	430-520
4.0		445-465	435-455	430-445	485-505	120-175	430-520
4.0		445-465	435-455	430-445	485-505	120-175	430-520
4.0		445-465	435-455	430-445	485-505	120-175	430-520
3.0- 4.0		445-465	435-455	430-445	485-505	120-175	430-520
4.0		500-530	480-520	460-500	480-510	155-212	465-540
4.0		500-530	480-520	460-500	480-510	155-212	465-540
4.0		500-530	480-520	460-500	480-510	155-212	465-540
4.0		500-530	480-520	460-500	480-510	155-212	465-540
4.0		500-530	480-520	460-500	480-510	155-212	465-540
4.0		500-530	480-520	460-500	480-510	155-212	465-540
4.0		520-540	510-530	500-520	520-540	170-190	520-540
4.0		530-550	520-540	510-530	530-550	170-190	530-550
3.0- 4.0		500-560	490-550	480-540	525-575	180-200	525-575
4.0- 6.0		490-560	470-550	460-540	500-575	170-210	500-575
4.0- 6.0		490-560	470-550	460-540	500-575	175-210	500-575
3.0- 4.0		465-525	455-505	445-495	475-525	175-200	475-525
2.0- 4.0		440-525	420-500	410-490	450-525	120-180	450-525
2.0- 4.0		470-545	450-525	440-510	480-545	140-180	480-545
2.0- 4.0		450-525	420-500	420-490	460-525	140-180	460-525
3.0- 4.0		500-560	490-550	480-540	525-575	180-200	525-575
4.0- 6.0		490-560	470-550	460-540	500-575	175-210	500-575
5.0						122-176	446-500
5.0						122-176	446-500
5.0						122-176	446-500
5.0						122-176	446-500
5.0						122-176	446-500
						176-203	464-509
						176-203	464-509
						185-212	473-518
						185-212	482-527
		390-450	390-450	390-450		150-200	
		390-450	390-450	390-450		150-200	
		390-450	390-450	390-450		150-200	
3.0- 4.0		480-520	470-510	450-490	480-520	175-200	490-540
3.0- 4.0		500-540	490-530	470-510	500-540	180-205	525-575
3.0- 4.0		500-540	490-530	470-510	500-540	180-205	525-575
3.0- 4.0		525-550	475-500	400-450	525-550	108-205	480-500
3.0- 4.0						180-205	525-575
3.0- 4.0						180-200	530-540

Grade	Filler	Sp Grav	Shrink, mils/in	Melt flow, g/10 min	Melt temp, °F	Back pres, psi	Drying temp, °F
1725		1.18	5.0- 7.0	25.00 O		50- 150	210
ABS+PC Alloy		**Stapron C**			**DSM**		
CM 204		1.12	5.0- 7.0	23.00			212
CM 205		1.12	5.0- 7.0	19.00			212
CM 404		1.13	5.0- 7.0	20.00			212-230
CM 405		1.13	5.0- 7.0	16.00			212-230
CM 504		1.14	5.0- 7.0	16.00			230
CM 505		1.14	5.0- 7.0	16.00			230
ABS+PC Alloy		**Triax**			**Monsanto**		
2122		1.10	7.0	3.40 S		50- 200	190-200
2153		1.13	7.0	5.60 S		50- 200	190-200
2173		1.14	7.0	5.20 S		25- 50	210
2722		1.10	7.0	3.40 S		25- 50	190
2753		1.13	7.0	5.60 S		25- 50	200
2773		1.14	8.0	4.20 S		25- 50	220
2783		1.15	7.0	3.80 S		25- 50	220
2976		1.13	8.0	2.00 S		50- 200	190-200
ABS+PTFE Alloy		**RTP Polymers**			**RTP**		
600 TFE 10 FR	PTF	1.28	7.0			50	180
ABS+PVC Alloy		**Shuman ABS**			**Shuman Plastics**		
780		1.17		1.50			160-170
790		1.20		1.20			170-200
791		1.20		1.20			170-200
ABS+PVC Alloy		**Triax**			**Monsanto**		
CBE\1		1.20	4.0- 6.0			0- 50	150-170
ABS+TPU Alloy		**Prevail**			**Dow**		
3050		1.12	6.0	58.00 I			170
3100		1.10	6.0	40.00 I			170
3150		1.09		25.00 I			170
Acetal		**Acetron**			**DSM**		
Molding Resin		1.44	25.0		347	50- 100	
Acetal		**Celcon**			**Hoechst**		
AS270		1.41	22.0				180
AS450		1.41	22.0				180
EC90PLUS		1.40	22.0				180
EF25	CF 25	1.43	22.0				180
EP90		1.43	22.0			0- 50	180
GB25	GLB 25	1.62	16.0		331		180
GC25A	GLF 25	1.62	4.0		331		180
LW90		1.43	22.0	9.00			180
LW90F2	PTF	1.40	22.0				180
LW90S2	SI 2	1.40	22.0	9.00			180
LWGCS2	GLC 25	1.62	4.0				180
M140		1.41	22.0	14.00			180
M140L1				14.00			180
M25		1.41	22.0	2.50	329		180
M270		1.41	22.0	27.00	329		180
M450		1.41	22.0	27.00	329		180

Drying time, hr	Inj time, sec	Front temp, °F	Mid temp, °F	Rear temp, °F	Nozzle temp, °F	Mold temp, °F	Proc temp, °F
3.0- 4.0		475	465	450	475	140-180	525-575
4.0							464-518
4.0							464-518
4.0							482-536
4.0							482-536
4.0							500-554
4.0							500-554
2.0- 3.0		430-520	430-520	430-500	470-490	70-185	485-510
2.0- 3.0		430-520	430-520	430-500	470-490	70-185	485-510
4.0						160-200	520-540
4.0						160-180	510-530
4.0						160-180	510-530
4.0						160-200	520-560
4.0						160-200	510-530
2.0- 3.0		430-520	430-520	430-500	470-490	70-185	485-510
2.0		420-500	400-480	380-460		150-200	
2.0- 4.0							360-400
2.0-24.0							400-440
2.0-24.0							400-440
2.0- 4.0		390	370	350	380	120-160	400-420
4.0						80	410
4.0						80	410
4.0						80	410
		360	370	350	350	180	380
3.0		370-390	370-390	370-390	400-420	170-200	370-390
3.0		370-390	370-390	370-390	400-420	170-200	370-390
3.0		370-390	370-390	370-390	400-420	170-200	370-390
3.0		370-390	370-390	370-390	400-420	170-200	370-390
3.0	2.0- 5.	370	360	340	380	170-250	360-390
3.0		370-390	370-390	370-390	400-420	190-230	380-410
3.0		370-390	370-390	370-390	400-420	190-230	380-410
3.0		370-390	370-390	370-390	400-420	170-200	370-390
3.0		370-390	370-390	370-390	400-420	170-200	370-390
3.0		370-390	370-390	370-390	400-420	170-200	370-390
3.0		370-390	370-390	370-390	400-420	170-200	370-390
3.0		370-390	370-390	370-390	400-420	170-200	370-390
3.0		370-390	370-390	370-390	400-420	170-200	370-390
3.0		370-390	370-390	370-390	400-420	170-200	370-390
3.0		370-390	370-390	370-390	400-420	170-200	370-390
3.0		370-390	370-390	370-390	400-420	170-200	370-390

Grade	Filler	Sp Grav	Shrink, mils/in	Melt flow, g/10 min	Melt temp, °F	Back pres, psi	Drying temp, °F
M50		1.41		27.00			180
M520		1.41	18.0-22.0	52.00 F	329	0- 50	180
M90		1.41	22.0	9.00	329		180
M900		1.41	20.0	90.00	338		180
MC270	MN	1.48	19.0	6.50			180
MC270HM	MN	1.60	15.0				180
MC750	MN			46.00			180
MC90	MN	1.48	19.0	8.50			180
MC90HM	MN	1.60	15.0	6.50			180
TX90		1.39	21.0	8.00 E			180
TX90PLUS		1.37	17.0				180
U10		1.41	20.0				180
UV25		1.41		2.50 E			180
UV25Z		1.41	22.0	2.50 E			180
UV270Z		1.41	22.0	9.00 E			180
UV90		1.41		9.00 E			180
UV90Z		1.41	22.0	9.00 E			180
WR25		1.41		2.50 E			180
WR25Z		1.41	22.0	2.50 E			180
WR90		1.41		9.00 E			180
WR90Z		1.41	22.0				180

Acetal Celstran Hoechst

Grade	Filler	Sp Grav	Shrink, mils/in	Melt flow, g/10 min	Melt temp, °F	Back pres, psi	Drying temp, °F
ACG40-01-4	GLF 40	1.72					180
ACK30-01-2	KEV 30	1.42					180

Acetal CTI Acetal CTI

Grade	Filler	Sp Grav	Shrink, mils/in	Melt flow, g/10 min	Melt temp, °F	Back pres, psi	Drying temp, °F
AT		1.41	20.0				180
AT-000/20T	PTF 20	1.51	20.0				180
AT-000/2S	PTF 20	1.42	20.0				180
AT-20GF	GLF 20	1.55	5.0				180
AT-30GF	GLF 30	1.63	4.5				180
AT-40GF	GLF 40	1.73	4.0				180

Acetal Delrin Du Pont

Grade	Filler	Sp Grav	Shrink, mils/in	Melt flow, g/10 min	Melt temp, °F	Back pres, psi	Drying temp, °F
100		1.42		1.00 M	347		
100 AF	PTF	1.54		1.00 M	347		
100 CL		1.42					
100 D		1.42					
100 ST		1.34		1.00 M	347		
100F		1.42					
107		1.42		1.00 M	347		
1700 HP		1.42					
200 PL		1.42					
500		1.42		6.00 M	347		
500 AF	PTF	1.54		6.00 M			
500 CL		1.42		6.00 M	347		
500 D		1.42					
500 HP		1.42					
500 T		1.39		6.00 M	347		
500 TL	PTF 2	1.42		6.00 M	347		
500F		1.42					
507		1.42		6.00 M	347		
570	GLF	1.56		6.00 M	347		
577	GLF	1.56		6.00 M	347		
900		1.42		11.00 M			

Drying time, hr	Inj time, sec	Front temp, °F	Mid temp, °F	Rear temp, °F	Nozzle temp, °F	Mold temp, °F	Proc temp, °F
3.0		370-390	370-390	370-390	400-420	170-200	370-390
3.0	2.0- 5.	370	360	340	380	170-250	360-390
3.0		370-390	370-390	370-390	400-420	170-200	370-390
3.0		370-390	370-390	370-390	400-420	170-200	370-390
3.0		370-390	370-390	370-390	400-420	170-200	370-390
3.0		370-390	370-390	370-390	400-420	170-200	370-390
3.0		370-390	370-390	370-390	400-420	170-200	370-390
3.0		370-390	370-390	370-390	400-420	170-200	370-390
3.0		370-390	370-390	370-390	400-420	170-200	370-390
3.0		370-390	370-390	370-390	400-420	170-200	370-390
3.0		370-390	370-390	370-390	400-420	170-200	370-390
3.0		370-390	370-390	370-390	400-420	170-200	370-390
3.0		370-390	370-390	370-390	400-420	170-200	370-390
3.0		370-390	370-390	370-390	400-420	170-200	370-390
3.0		370-390	370-390	370-390	400-420	170-200	370-390
3.0		370-390	370-390	370-390	400-420	170-200	370-390
3.0		370-390	370-390	370-390	400-420	170-200	370-390
3.0		370-390	370-390	370-390	400-420	170-200	370-390
3.0		370-390	370-390	370-390	400-420	170-200	370-390
3.0		370-390	370-390	370-390	400-420	170-200	370-390
3.0		370-390	370-390	370-390	400-420	170-200	370-390
3.0		390-410	380-400	370-390	390-410	170-190	390-410
3.0		400-420	390-410	380-400	400-420	140-160	400-420
2.0		390	380	350	380	190	
2.0		390	380	350	380	190	
2.0		390	380	350	380	190	
2.0		390-430	380-390	350-370	380-400	190-250	
2.0		390-430	380-390	350-370	380-400	190-250	
2.0		390-430	380-390	350-370	380-400	190-250	
		375	375	425	400	180-220	400-440
		400	400	400	400	180-220	400-440
		400	400	400	400	180-220	400-440
		400	400	400	400	180-220	400-440
		375	375	425	390	50-160	380-430
		400	400	400	400	180-220	400-440
		400	400	400	400	180-220	400-440
		400	400	400	400	180-220	400-440
		400	400	400	400	180-220	400-440
		400	400	400	400	180-220	400-440
		400	400	400	400	180-220	400-440
		400	100	400	400	180-220	400-440
		400	400	400	400	180-220	400-440
		400	400	400	400	180-220	400-440
		375	375	425	390	50-160	380-430
		400	400	400	400	180-220	400-440
		400	400	400	400	180-220	400-440
		400	400	400	400	180-220	400-440
		400	400	400	400	180-220	400-440
		400	400	400	400	180-220	400-440
		400	400	400	400	180-220	400-440

Grade	Filler	Sp Grav	Shrink, mils/in	Melt flow, g/10 min	Melt temp, °F	Back pres, psi	Drying temp, °F
900 D		1.42					
900 HP		1.42					
900F		1.42					
907F		1.42					
Acetal		**Delrin II**			**Du Pont**		
100 NC10		1.42		1.00 M	347		
500 NC10		1.42		6.00 M	347		
900 NC10		1.42		11.00 M	347		
Acetal		**Delrin P**			**Du Pont**		
100P		1.42		1.00 M	347		
1700P		1.42		16.00 M	347		
500P		1.42		6.00 M	347		
900P		1.42		11.00 M	347		
Acetal		**DSM Acetal**			**DSM**		
J-80/10	GLF 10	1.48	6.0				
J-80/20	GLF 20	1.55	5.0				
J-80/30	GLF 30	1.62	4.0				
Acetal		**Electrafil**			**DSM**		
J-80/CF/10/TF/10	CF 10	1.49	2.0				160
Acetal		**Fiberfil**			**DSM**		
J-80/10	GLF 10	1.48	6.0- 8.0				
J-80/20	GLF 20	1.55	5.0- 6.0				
J-80/30	GLF 30	1.62	4.0- 5.0				
Acetal		**Iupital**			**Mitsubishi Gas**		
ET-20	CF	1.41	16.0	11.00	329		
F10-01		1.41	22.0	2.50	329		
F10-02		1.41	22.0	2.50	329		
F20-01		1.41	20.0	9.00	329		
F20-02		1.41	20.0	9.00	329		
F20-03		1.41	20.0	9.00	329		
F20-51		1.41	20.0	9.00	329		
F20-52		1.41	20.0	9.00	329		
F20-61		1.41	20.0	9.00	329		
F25-02		1.41	20.0	16.00	329		
F30-01		1.41	20.0	27.00	329		
F30-02		1.41	20.0	27.00	329		
F30-03		1.41	20.0	27.00	329		
F40-03		1.41	20.0	45.00	329		
FA-2010		1.39	19.0	9.00	329		
FA-2020		1.37	18.0	9.00	329		
FB2025	GLB 25	1.59	16.0	5.00	329		
FC2020D	CF 20	1.46	4.0	3.50	329		
FC2020H	CF 20	1.46	4.0	6.00	329		
FE-21		1.34	20.0	5.50	329		
FG2025	GLF 25	1.59	6.0		329		
FG2025L	GLF 25	1.59	6.0		329		
FL2010	PTF 10	1.46	20.0	7.50	329		
FL2020	PTF 20	1.51	21.0	6.00	329		
FS2022	SIF	1.41	20.0	10.00	329		
FT2010	UNS 10	1.49	17.0	7.50	329		

Drying time, hr	Inj time, sec	Front temp, °F	Mid temp, °F	Rear temp, °F	Nozzle temp, °F	Mold temp, °F	Proc temp, °F
		400	400	400	400	180-220	400-440
		400	400	400	400	180-220	400-440
		400	400	400	400	180-220	400-440
		400	400	400	400	180-220	400-440
		390-410	380-410	370-430	410	180-220	395-450
		390-410	380-410	370-420	410	180-220	395-450
		390-410	380-410	370-420	410	180-220	395-450
		400	400	400	410	180-220	400-440
		390	390	390	400	180-220	400-440
		390	390	390	400	180-220	400-440
		390	390	390	400	180-220	400-440
		360-390	370-410	350-380	350-400	180-250	380-420
		360-390	370-410	350-380	350-400	180-250	380-420
		360-390	370-410	350-380	350-400	180-250	380-420
2.0- 4.0		360-390	370-410	350-380	350-400	180-250	380-420
		360-390	370-410	350-380	350-400	180-250	380-420
		360-390	370-410	350-380	350-400	180-250	380-420
		360-390	370-410	350-380	350-400	180-250	380-420
		374	356	338	356-410		392
		374	356	338	356-410		392
		374	356	338	356-410		392
		374	356	338	356-410		392
		374	356	338	356-410		392
		374	356	338	356-410		392
		374	356	338	356-410		392
		374	356	338	356-410		392
		374	356	338	356-410		392
		374	356	338	356-410		392
		374	356	338	356-410		392
		374	356	338	356-410		392
		374	356	338	356-410		392
		374	356	338	356-410		392
		356	347	338	356-410		392
		356	347	338	356-410		392
		392	374	338	356-410		392
		392	374	338	356-410		392
		392	374	338	356-410		392
		374	356	338	356-410		392
		392	374	338	356-410		392
		392	374	338	356-410		392
		374	356	338	356-410		392
		374	356	338	356-410		392
		374	356	338	356-410		392
		374	356	338	356-410		392

Grade	Filler	Sp Grav	Shrink, mils/in	Melt flow, g/10 min	Melt temp, °F	Back pres, psi	Drying temp, °F
FT2020	UNS 20	1.59	9.0	5.50	329		
FU2025		1.35	17.0	6.00	329		
FU2050		1.29	12.0	4.50	329		
FV-30		1.41	20.0	31.00	329		
FW-21	UNS	1.42	20.0	9.50	329		
FW-24	UNS	1.41	20.0	9.50	329		
MF3020	GL 20	1.55	17.0	20.00	329		
TC3015	TAL 15	1.52	19.0	10.50	329		
TC3030	TAL 30	1.63	15.0	9.50	329		
Acetal		**Lubricomp**			**LNP**		
Fulton 404	PTF 20	1.50	16.0-22.0		324	50- 100	200
Acetal		**PermaStat**			**RTP**		
800		1.35	20.0-30.0				
Acetal		**Plaslube**			**DSM**		
J-80/20/TF/15	GLF 20	1.63	3.0				
J-80/30/TF/15	GLF 30	1.71	2.0				
J-80/CF/10/TF/10	CF 10	1.49	2.0				
Acetal		**PMC Acetal**			**PMC**		
HP-GF-20	GLF 20	1.56			350	0- 700	230
HP-MF1		1.41		1.00	347	0-1500	230
HP-MF5		1.41		5.00	347	0-1500	230
HP-MF9		1.41		9.00	347	0-1500	230
Acetal		**RTP Polymers**			**RTP**		
800		1.41	30.0			25- 50	250
800 GB 10	GLB 10	1.48	30.0			25- 50	250
800 GB 20	GLB 20	1.55	30.0			25- 50	250
800 GB 30	GLB 30	1.62	30.0			25- 50	250
800 GB 40	GLB 40	1.71	27.0			25- 50	250
800 SI2	SIF	1.40	18.0			25- 50	250
800 SI2 HB	SIF	1.40	18.0			25- 50	250
800 TFE 10	PTF	1.49	20.0			25- 50	250
800 TFE 20	GLF	1.51	19.0			25- 50	250
801	GLF 10	1.48	10.0			25- 50	250
801CC	GLF 10	1.48	10.0			25- 50	250
803	GLF 20	1.55	7.0			25- 50	250
803CC	GLF 20	1.55	7.0			25- 50	250
804CC	GLF 25	1.58	6.0			25- 50	250
805	GLF 30	1.61	5.0			25- 50	250
805 TFE 15	GLF	1.74	2.0			25- 50	250
805CC	GLF 30	1.63	5.0			25- 50	250
807	GLF 40	1.70	4.0			25- 50	250
Acetal		**Texapol Acetal**			**Texapol**		
5102		1.42	20.0	2.50 E	354		230
5114		1.42	20.0	14.00 E	354		230
5130		1.42	20.0	27.00 E	354		230
Acetal Copoly		**Celcon**			**Hoechst**		
MM3.5S		1.41	20.0	75.00			180

Drying time, hr	Inj time, sec	Front temp, °F	Mid temp, °F	Rear temp, °F	Nozzle temp, °F	Mold temp, °F	Proc temp, °F
		374	356	338	356-410		392
		374	356	338	356-410		392
		374	356	338	356-410		392
		374	356	338	356-410		392
		374	356	338	356-410		392
		374	356	338	356-410		392
		392	374	338	356-410		392
		374	356	338	356-410		392
		374	356	338	356-410		392
3.0		350-425	350-425	350-425		190-225	390-415
		360-410	360-410	360-410		150-200	
		360-390	370-410	350-380	350-400	180-250	380-420
		360-390	370-410	350-380	350-400	180-250	380-420
		360-390	370-410	350-380	350-400	180-250	380-420
2.0	2.0- 10.	385-395	365-380	340-360	400-425	140-200	
2.0	1.0- 6.	385-395	365-380	340-360	400-425	140-195	
2.0	1.0- 6.	385-395	365-380	340-360	400-425	140-195	
2.0	1.0- 6.	385-395	365-380	340-360	400-425	140-195	
2.0		395-410	385-400	375-390		200-250	
2.0		370-390	360-380	350-370		150-200	
2.0		370-390	360-380	350-370		150-200	
2.0		370-390	360-380	350-370		150-200	
2.0		370-390	360-380	350-370		150-200	
2.0		370-410	360-400	350-390		150-250	
2.0		370-410	360-400	350-390		150-250	
2.0		370-410	360-400	350-390		150-250	
2.0		370-390	360-380	350-370		150-200	
2.0		370-390	360-380	350-370		150-200	
2.0		370-390	360-380	350-370		150-200	
2.0		370-390	360-380	350-370		150-200	
2.0		370-390	360-380	350-370		150-200	
2.0		370-390	360-380	350-370		150-200	
2.0		370-390	360-380	350-370		150-200	
2.0		370-390	360-380	350-370		150-200	
2.0		370-390	360-380	350-370		150-200	
2.0		370-390	360-380	350-370		150-200	
2.0		340	340	335	350	140-200	340-430
2.0		340	340	335	350	140-200	340-430
2.0		340	340	335	350	140-200	340-430
3.0		370-390	370-390	370-390	400-420	170-200	370-390

Acetal Copoly Lucel Lucky

Grade	Filler	Sp Grav	Shrink, mils/in	Melt flow, g/10 min	Melt temp, °F	Back pres, psi	Drying temp, °F
CF-620	CF 20	1.44	4.0- 9.5			0- 568	194-230
CR-620	CF 20	1.44	3.0- 9.0			0- 568	194-230
EC-600B	CBL	1.42	18.0-21.0			568-1137	194-230
FW-700M	MOS	1.44	20.0-22.0			568-1137	194-230
FW-700S	SIF	1.40	20.0-22.0			568-1137	194-230
FW-710F	PTF 10	1.46	14.0-17.0			568-1137	194-230
FW-720F	PTF 20	1.52	13.0-16.0			568-1137	194-230
GB-325	GLB 25	1.59	13.0-17.0			0- 568	194-230
GC-225	GLF 25	1.59	4.5- 8.5			0- 568	194-230
GR-220	GLF 20	1.54	4.5- 5.5			0- 568	194-230
HI-510		1.38	20.0-22.0			568-1137	194-230
HI-520		1.36	20.0-22.0			568-1137	194-230
MP-109		1.45	17.0-20.0	9.00 E		568-1137	194-230
MR-320	MN 20	1.53	13.0-17.0			0- 568	194-230
N103-03		1.41	20.0-23.0	3.00 E		568-1137	194-230
N109-01		1.41	18.0-21.0	9.00 E		568-1137	194-230
N109-02		1.41	18.0-21.0	9.00 E		568-1137	194-230
N109-03		1.41	18.0-21.0	9.00 E		568-1137	194-230
N109-AS		1.40	18.0-21.0	9.00 E		568-1137	194-230
N109-HR		1.41	18.0-21.0	9.00 E		568-1137	194-230
N109-LD		1.41	18.0-21.0	9.00 E		568-1137	194-230
N109-WR		1.41	18.0-21.0	9.00 E		568-1137	194-230
N127-02		1.41	18.0-21.0	27.00 E		568-1137	194-230
N127-03		1.41	18.0-21.0	27.00 E		568-1137	194-230
N127-AS		1.40	18.0-21.0	27.00 E		568-1137	194-230
N127-LD		1.41	18.0-21.0	27.00 E		568-1137	194-230
N127-WR		1.41	18.0-21.0	27.00 E		568-1137	194-230
N145		1.41	18.0-21.0	45.00 E		568-1137	194-230
N145-AS		1.40	18.0-21.0	45.00 E		568-1137	194-230
N145-LD		1.41	18.0-21.0	45.00 E		568-1137	194-230
ST-550		1.28	20.0-23.0			568-1137	194-230
VC-127		1.40	18.0-20.0	27.00 E		568-1137	194-230

Acetal Copoly Lucet Lucky

Grade	Filler	Sp Grav	Shrink, mils/in	Melt flow, g/10 min	Melt temp, °F	Back pres, psi	Drying temp, °F
CF-620	CF 20	1.48	4.0- 9.5			0- 568	176-212
CR-620	CF 20	1.48	3.0- 9.0			0- 568	176-212
EC-600B	CBL	1.42	18.0-21.0			568-1137	176-212
FW-700M	MOS	1.44	20.0-22.0			568-1137	176-212
FW-700S	SIF	1.40	20.0-22.0			568-1137	176-212
FW-710F	PTF 10	1.46	14.0-17.0			568-1137	176-212
FW-720F	PTF 20	1.52	13.0-16.0			568-1137	176-212
GB-325	GLB 25	1.59	13.0-17.0			0- 568	176-212
GC-225	GLF 25	1.59	4.5- 8.5			0- 568	176-212
GR-220	GLF 20	1.54	4.5- 5.5			0- 568	176-212
HI-510		1.38	20.0-22.0			568-1137	176-212
HI-520		1.36	20.0-22.0			568-1137	176-212
MP-109		1.45	17.0-20.0			568-1137	176-212
MR-320	MN 20	1.53	13.0-17.0			0- 568	176-212
N103-01		1.41	20.0-23.0	3.00 E		568-1137	176-212
N103-03		1.41	20.0-23.0	3.00 E		568-1137	176-212
N109-01		1.41	18.0-21.0	9.00 E		568-1137	176-212
N109-02		1.41	18.0-21.0	9.00 E		568-1137	176-212
N109-03		1.41	18.0-21.0	9.00 E		568-1137	176-212
N109-AS		1.40	18.0-21.0	9.00 E		568-1137	176-212

Drying time, hr	Inj time, sec	Front temp, °F	Mid temp, °F	Rear temp, °F	Nozzle temp, °F	Mold temp, °F	Proc temp, °F
3.0- 6.0		374-419	356-392	320-356	374-419	140-248	392-437
3.0- 6.0		374-419	356-392	320-356	374-419	140-248	392-437
3.0- 6.0		356-383	347-374	302-320	356-383	140-212	356-392
3.0- 6.0		356-383	347-374	302-320	356-383	140-212	356-392
3.0- 6.0		356-383	347-374	302-320	356-383	140-212	356-392
3.0- 6.0		356-383	347-374	302-320	356-383	140-212	356-392
3.0- 6.0		356-383	347-374	302-320	356-383	140-212	356-392
3.0- 6.0		374-419	356-392	320-356	374-419	140-248	392-437
3.0- 6.0		374-419	356-392	320-356	374-419	140-248	392-437
3.0- 6.0		374-419	356-392	320-356	374-419	140-248	392-437
3.0- 6.0		356-383	347-374	302-320	356-383	140-212	356-392
3.0- 6.0		356-383	347-374	302-320	356-383	140-212	356-392
3.0- 6.0		356-383	347-374	302-320	356-383	140-212	356-392
3.0- 6.0		374-419	356-392	320-356	374-419	140-248	392-437
3.0- 6.0		356-383	347-374	302-320	356-383	140-212	356-392
3.0- 6.0		356-383	347-374	302-320	356-383	140-212	356-392
3.0- 6.0		356-383	347-374	302-320	356-383	140-212	356-392
3.0- 6.0		356-383	347-374	302-320	356-383	140-212	356-392
3.0- 6.0		356-383	347-374	302-320	356-383	140-212	356-392
3.0- 6.0		356-383	347-374	302-320	356-383	140-212	356-392
3.0- 6.0		356-383	347-374	302-320	356-383	140-212	356-392
3.0- 6.0		356-383	347-374	302-320	356-383	140-212	356-392
3.0- 6.0		356-383	347-374	302-320	356-383	140-212	356-392
3.0- 6.0		356-383	347-374	302-320	356-383	140-212	356-392
3.0- 6.0		356-383	347-374	302-320	356-383	140-212	356-392
3.0- 6.0		356-383	347-374	302-320	356-383	140-212	356-392
3.0- 6.0		356-383	347-374	302-320	356-383	140-212	356-392
3.0- 6.0		356-383	347-374	302-320	356-383	140-212	356-392
3.0- 6.0		356-383	347-374	302-320	356-383	140-212	356-392
3.0- 6.0		356-383	347-374	302-320	356-383	140-212	356-392
3.0- 6.0		356-383	347-374	302-320	356-383	140-212	356-392
3.0- 6.0		356-383	347-374	302-320	356-383	140-212	356-392
3.0- 6.0		392-437	374-410	338-374	392-437	140-176	410-428
3.0- 6.0		392-437	374-410	338-374	392-437	140-176	410-428
3.0- 6.0		374-392	356-392	320-356	374-410	140-176	374-392
3.0- 6.0		374-392	356-392	320-356	374-410	140-176	374-392
3.0- 6.0		374-392	356-392	320-356	374-410	140-176	374-392
3.0- 6.0		374-392	356-392	320-356	374-410	140-176	374-392
3.0- 6.0		374-392	356-392	320-356	374-410	140-176	374-392
3.0- 6.0		392-437	374-410	338-374	392-437	140-176	410-428
3.0- 6.0		392-437	374-410	338-374	392-437	140-176	410-428
3.0- 6.0		392-437	374-410	338-374	392-437	140-176	410-428
3.0- 6.0		374-392	356-392	320-356	374-410	140-176	374-392
3.0- 6.0		374-392	356-392	320-356	374-410	140-176	374-392
3.0- 6.0		374-392	356-392	320-356	374-410	140-176	374-392
3.0- 6.0		392-437	374-410	338-374	392-437	140-176	410-428
3.0- 6.0		374-392	356-392	320-356	374-410	140-176	374-392
3.0- 6.0		374-392	356-392	320-356	374-410	140-176	374-392
3.0- 6.0		374-392	356-392	320-356	374-410	140-176	374-392
3.0- 6.0		374-392	356-392	320-356	374-410	140-176	374-392
3.0- 6.0		374-392	356-392	320-356	374-410	140-176	374-392
3.0- 6.0		374-392	356-392	320-356	374-410	140-176	374-392

Grade	Filler	Sp Grav	Shrink, mils/in	Melt flow, g/10 min	Melt temp, °F	Back pres, psi	Drying temp, °F
N109-HR		1.41	18.0-21.0	9.00 E		568-1137	176-212
N109-LD		1.41	18.0-21.0	9.00 E		568-1137	176-212
N109-WR		1.41	18.0-21.0	9.00 E		568-1137	176-212
N127-02		1.41	18.0-21.0	27.00 E		568-1137	176-212
N127-03		1.41	18.0-21.0	27.00 E		568-1137	176-212
N127-AS		1.40	18.0-21.0	27.00 E		568-1137	176-212
N127-LD		1.41	18.0-21.0	27.00 E		568-1137	176-212
N127-WR		1.41	18.0-21.0	27.00 E		568-1137	176-212
N145		1.41	18.0-21.0	45.00 E		568-1137	176-212
N145-AS		1.40	18.0-20.0	45.00 E		568-1137	176-212
N145-LD		1.41	18.0-21.0	45.00 E		568-1137	176-212
ST-550		1.28	20.0-23.0			568-1137	176-212
VC-127		1.40	18.0-21.0			568-1137	176-212
Acetal Copoly		**PMC Acetal**			**PMC**		
CP-G25	GLF 25	1.58			347	0- 700	230
CP-MF5		1.41		5.00	347	0-1500	230
CP-MF9		1.41		9.00	347	0-1500	230
Acetal Copoly		**Texapol Acetal**			**Texapol**		
5203 FPL-20	PTF 20	1.52	22.0	3.00 E	338		230
5203		1.41	27.0	3.00 E	329		230
5203-S2	SI	1.39	30.0	3.00 E	338		230
5209 FPL-15	PTF 15	1.48	25.0	9.00 E	338		230
5209 FPL-20	PTF 20	1.52	22.0	10.00 E	338		230
5209		1.41	27.0	9.00 E	329		230
5209 UV BK-30-A		1.40	22.0	5.00 E	329		230
5209 UV		1.40	27.0	9.00 E	329		230
5230		1.41	27.0	27.00 E	329		230
5506		1.35	13.0	6.50 E	329		230
GB 5209-20	GLB 20	1.54	19.0	6.00 E	338		230
GB 5230-20	GLB 20	1.54	19.0	25.00 E	338		230
GF 5209-25	GLF 25	1.58	3.0	6.00 E	338		230
MF 5209-25	MNF 25	1.59	16.0	7.00 E	338		230
Acrylic		**Acrylite**			**Cyro**		
H-12		1.19	4.0- 6.0	7.00 I			
H-15		1.19	4.0- 7.0	2.20 I			
L-40		1.19	3.0- 5.0	28.00 I			
M-30		1.19	3.0- 6.0	24.00 I			
Acrylic		**Acrylite Plus**			**Cyro**		
H-16		1.15		1.80 I		20- 100	
Acrylic		**Cyrolite**			**Cyro**		
G-20		1.11				0- 100	
G-20 HIFLO		1.11		12.00 S		20- 100	175-180
Acrylic		**Plexiglas**			**AtoHaas**		
DR		1.15	3.0- 8.0	1.00 I		10- 20	180
HFI-10		1.15	2.0- 8.0	3.30 I			175
HFI-7		1.17	2.0- 6.0	11.00 I			175
MI-7		1.17	3.0- 6.0	3.20 I		50- 100	185
V052		1.19	4.0- 7.0	2.80 I		100- 200	180
V825		1.19	4.0- 7.0	3.70 I		100- 200	190
V826		1.19	4.0- 7.0	1.60 I		100- 200	190

Drying time, hr	Inj time, sec	Front temp, °F	Mid temp, °F	Rear temp, °F	Nozzle temp, °F	Mold temp, °F	Proc temp, °F
3.0- 6.0		374-392	356-392	320-356	374-410	140-176	374-392
3.0- 6.0		374-392	356-392	320-356	374-410	140-176	374-392
3.0- 6.0		374-392	356-392	320-356	374-410	140-176	374-392
3.0- 6.0		374-392	356-392	320-356	374-410	140-176	374-392
3.0- 6.0		374-392	356-392	320-356	374-410	140-176	374-392
3.0- 6.0		374-392	356-392	320-356	374-410	140-176	374-392
3.0- 6.0		374-392	356-392	320-356	374-410	140-176	374-392
3.0- 6.0		374-392	356-392	320-356	374-410	140-176	374-392
3.0- 6.0		374-392	356-392	320-356	374-410	140-176	374-392
3.0- 6.0		374-392	356-392	320-356	374-410	140-176	374-392
3.0- 6.0		374-392	356-392	320-356	374-410	140-176	374-392
3.0- 6.0		374-392	356-392	320-356	374-410	140-176	374-392
3.0- 6.0		374-392	356-392	320-356	374-410	140-176	374-392
2.0	2.0- 10.	385-395	365-380	340-360	400-425	140-200	
2.0	1.0- 6.	385-395	365-380	340-360	400-425	140-195	
2.0	1.0- 6.	385-395	365-380	340-360	400-425	140-195	
2.0		340	340	335	350	140-200	340-430
2.0		340	340	335	350	140-200	340-430
2.0		340	340	335	350	140-200	340-430
2.0		340	340	335	350	140-200	340-430
2.0		340	340	335	350	140-200	340-430
2.0		340	340	335	350	140-200	340-430
2.0		340	340	335	350	140-200	340-430
2.0		340	340	335	350	140-200	340-430
2.0		340	340	335	350	140-200	340-430
2.0		340	340	335	350	140-200	340-430
2.0		360	350	350	380	140-200	340-430
2.0		360	350	350	380	140-200	340-430
2.0		370	360	360	390	140-200	340-430
2.0		360	350	350	380	140-200	340-430
		464-482	464-482	464-482			
		464-482	464-482	464-482			
		464-482	464-482	464-482			
		464-482	464-482	464-482			
		450-500	450-480	425-450	450-500	100-175	450-500
						90-150	400-475
3.0- 4.0		440-475	425-475	375-425	440-475	110-150	400-450
4.0	4.0- 25.	450-500	460-510	430-480	450-500	60-170	450-500
3.0- 4.0	1.0- 2.	450-490	460-500	440-480	450-490	90-170	470-510
2.0- 4.0	1.0- 2.	420-460	420-460	420-460	420-460	100-190	460
4.0	1.0- 2.	420-460	420-460	420-460	420-460	100-190	460
1.0- 4.0		440	430	420	430	150-190	
1.0- 4.0		440	430	420	430	150-200	
1.0- 4.0		450	435	425	435	150-200	

Grade	Filler	Sp Grav	Shrink, mils/in	Melt flow, g/10 min	Melt temp, °F	Back pres, psi	Drying temp, °F
V920		1.19	4.0- 7.0	8.00 I		100- 200	175
VH		1.18	4.0- 7.0	23.00 I		100- 200	165
VM		1.18	4.0- 7.0	14.50 I		100- 200	165
VS		1.18	4.0- 7.0	24.00 I		100- 200	150
Acrylic		**XT Polymer**			**Cyro**		
250		1.12	4.0- 8.0	3.50 S		0- 100	
255		1.11		4.00 S		0- 100	
375		1.11	4.0- 8.0	2.10 S		0- 100	
X800RG		1.11		14.00 S		0- 100	
X800RH		1.11		8.00 S		0- 100	
Acrylic		**Zylar**			**Novacor**		
93-541		1.06	4.0	8.40 N			
93-546		1.05	4.0	11.00 N			
Acrylic+ABS Aly		**Terlux**			**BASF**		
2802TR		1.08	4.0- 7.0			50- 100	160
2802TR 27768		1.08	4.0- 7.0			50- 100	160
KR2804		1.08	4.0- 7.0			50- 100	160
KR2804 Trans.		1.08	4.0- 7.0			50- 100	160
AS Copolymer		**Toyolac**			**Toray**		
ASG10	GLF 10	1.10					176-194
ASG20	GLF 20	1.22					176-194
ASG30	GLF 30	1.30					176-194
ASA		**Centrex**			**Monsanto**		
811 Black		1.05	4.0- 6.0	1.80 I			180-190
811		1.05	5.0- 7.0	1.80 I			180-190
821 Black		1.05	4.0- 6.0	1.40 I			180-190
821		1.05	5.0- 7.0	1.40 I			180-190
ASA		**Geloy**			**GE**		
XP1001		1.06	4.0- 6.0	1.30 I		30- 80	180
ASA		**Hitachi ASA**			**Hitachi**		
V6700A		1.06	4.0- 6.0	5.50 G			176-185
V6701A		1.06	4.0- 6.0	3.50 G			176-185
V6702A		1.07	4.0- 6.0	2.50 G			176-185
V6810		1.07	4.0- 6.0	1.20 G			185-212
V6815		1.08	4.0- 6.0	1.00 G			185-212
V6820		1.08	4.0- 6.0	0.30 G			203-221
ASA		**Luran S**			**BASF**		
757 R/RE		1.07	4.0- 7.0	2.24 I		50- 100	185-200
776 S/SE		1.07	4.0- 7.0	0.85 I		50- 100	185-200
778 T		1.07	4.0- 7.0	1.28 I		50- 100	185-200
786 R		1.07	4.0- 7.0	2.36 I		50- 100	185-200
797 S/SE		1.07	4.0- 7.0	0.85 I		50- 100	185-200
KR2853/1		1.07	4.0- 7.0	2.24 I		50- 100	185-200
KR2855		1.07	4.0- 7.0	6.84 I		50- 100	185-200
KR2856		1.07	4.0- 7.0	4.28 I		50- 100	185-200
ASA+PC Alloy		**Geloy**			**GE**		
XP4025		1.14	5.0- 7.0	1.60 I		50- 100	200

Drying time, hr	Inj time, sec	Front temp, °F	Mid temp, °F	Rear temp, °F	Nozzle temp, °F	Mold temp, °F	Proc temp, °F
1.0- 4.0		420	410	400	410	150-185	
1.0- 4.0		400	390	380	390	130-160	
1.0- 4.0		400	390	380	390	130-160	
1.0- 4.0		380	370	360	370	120-150	
						90-150	400-475
						90-150	400-475
						90-150	400-475
						90-150	400-475
						90-150	400-475
		400	390	380		130	420
		400	390	380		130	420
2.0- 4.0						120-165	445-500
2.0- 4.0						120-165	445-500
2.0- 4.0						120-165	445-500
2.0- 4.0						120-165	445-500
2.0- 4.0		464	428	392		104-140	
2.0- 4.0		464	428	392		104-140	
2.0- 4.0		464	428	392		104-140	
2.0		460-525	460-525	460-525	485-550	110-160	485-550
2.0		460-525	460-525	460-525	485-550	110-160	485-550
2.0		460-525	460-525	460-525	485-550	110-160	485-550
2.0		460-525	460-525	460-525	485-550	110-160	485-550
3.0- 6.0		430-460	420-450	400-420	440-470	120-150	440-490
2.0- 4.0		428-482	428-482	428-482		104-176	
2.0- 4.0		428-482	428-482	428-482		104-176	
2.0- 4.0		428-482	428-482	428-482		104-176	
2.0- 4.0		446-518	446-518	446-518		104-176	
2.0- 4.0		446-518	446-518	446-518		104-176	
2.0- 4.0		464-536	464-536	464-536		104-176	
2.0- 4.0						100-175	450-550
2.0- 4.0						100-175	450-550
2.0- 4.0						100-175	450-550
2.0- 4.0						100-175	450-550
2.0- 4.0						100-175	450-550
2.0- 4.0						105-175	454-535
2.0- 4.0						105-175	454-535
2.0- 4.0						105-175	454-535
3.0- 4.0		430-480	420-460	400-430	440-500	130-170	460-520

Grade	Filler	Sp Grav	Shrink, mils/in	Melt flow, g/10 min	Melt temp, °F	Back pres, psi	Drying temp, °F
XP4034		1.15	5.0- 7.0	1.80 I		50- 150	210-220
ASA+PVC Alloy		**Geloy**			**GE**		
XP2003		1.21	3.0- 5.0	4.50 I		25- 50	160
ASA/AES		**Centrex**			**Monsanto**		
813		1.06	4.0- 6.0	2.50 I			180-190
ASA/AES		**Kodar**			**Eastman**		
5445		1.23					
Bio Syn Poly		**Mater-Bi**			**Novamont**		
AI 05 H		1.28		2.50	302		
AI 35 H				4.50			
CA		**Rotuba Cellulose Acetate**				**Rotuba**	
H		1.28				0- 50	150-170
H2		1.28				0- 50	150-170
M		1.27				0- 50	150-170
MH		1.27				0- 50	150-170
MS		1.26				0- 50	150-170
DAP		**Cosmic DAP**			**Cosmic Plastics**		
224	DAC	1.60	1.0- 4.0				
224V	DAC	1.42	7.0- 9.0				
306	ORL	1.52	1.0- 4.0				
6120	GLC	1.81	1.0- 4.0				
6120F	GLC	1.84	1.0- 4.0				
6130	GLF	1.82	1.0- 4.0				
6130F	GLF	1.82	1.0- 4.0				
6160	MN	1.84	1.0- 4.0				
6220	GLC	1.79	1.0- 4.0				
6220F	GLC	1.79	1.0- 4.0				
6230	GLF	1.82	1.0- 4.0				
6230F	GLF	1.82	1.0- 4.0				
D33	GLC	1.80	1.0- 4.0				
D44	MN	1.74	1.0- 4.0				
D45	MN	1.70	3.0- 7.0				
D62	GLF	1.82	1.0- 4.0		500		
D62CM	GL	1.94					
D63	GLF	1.77	1.0- 4.0				
D69	GLF	1.72	1.0- 4.0				
D72	GLC	1.82	1.0- 4.0				
D73	GMN	1.95					
D73F	GMN	1.97					
ID-40	ORL	1.52	1.0- 4.0				
ID-50	DAC	1.39	1.0- 4.0				
K31	GLC	1.76	1.0- 4.0				
K43	MN	1.80	3.0- 7.0				
K43V	MN	1.67	3.0- 7.0				
K61	GLF	1.72	1.0- 4.0				
K66	GLF	1.82	1.0- 4.0				
K77	GLC	1.78	1.0- 4.0				
DAP		**Rogers DAP**			**Rogers**		
51-01CAFR	MN	1.75	5.0			25	

Drying time, hr	Inj time, sec	Front temp, °F	Mid temp, °F	Rear temp, °F	Nozzle temp, °F	Mold temp, °F	Proc temp, °F
3.0- 4.0		480-550	460-520	400-480	480-550	130-170	500-550
2.0- 3.0		360-385	340-370	330-350	370-400	80-140	380-410
2.0		460-525	460-525	460-525	485-550	110-160	485-550
							380-525
		329-347	311-329	284-302	338-356	122-158	
		329-347	311-329	284-302	338-356	59- 77	
2.0- 4.0	8.0- 12.	400-420	400-420	400-420	410-430	110-130	
2.0- 4.0	8.0- 12.	420-450	420-450	420-450	430-450	110-130	
2.0- 4.0	8.0- 10.	360-380	360-380	360-380	380-400	110-130	
2.0- 4.0	8.0- 10.	380-400	380-400	380-400	390-410	110-130	
2.0- 4.0	8.0- 10.	340-360	340-360	340-360	360-380	110-130	
							275-330
							275-350
							275-330
							275-330
							275-330
							275-330
							275-330
							275-330
							275-330
							275-330
							275-330
							275-330
							275-330
							275-300
							275-300
							275-330
							275-350
							275-350
							275-330
							275-330
							275-350
							275-350
							275-330
							275-330
							275-330
							275-300
							275-350
							275-330
							275-330
							275-330
	5.0		170	140	190	320-340	230-240

Grade	Filler	Sp Grav	Shrink, mils/in	Melt flow, g/10 min	Melt temp, °F	Back pres, psi	Drying temp, °F
52-01	GLC	1.93	2.0- 4.0			25	
52-70-70 VO	GLC	1.91	2.0- 4.0			25	
73-70-70 C	GLC	1.91	2.0- 4.0			25	
73-70-70 R	GLC	1.85	2.0- 4.0			25	
775 CAF	MN	1.67	6.5			25	
FS-10 VO	GLC	1.91	2.0- 4.0			25	
FS-5	GLC	1.91	2.0- 4.0			25	
FS-6 CAF	MN	1.86	3.5			25	
RX 1-501AN	MN	1.73	7.0-10.0			25	
RX 1-501N	MN	1.80	3.0- 6.0			25	
RX 1-510N	MN	1.66	3.0- 6.0			25	
RX 1-520	GLC	1.83	1.0- 3.0			25	
RX 1-540	GMN	1.85	1.0- 3.0			25	
RX 1310	GLC	1.83	1.0- 3.0			25	
RX 1366FR	GLC	1.87	1.0- 3.0			25	
RX 2-501N	MN	1.80	3.0- 6.0			25	
RX 2-520	GLC	1.75	1.0- 3.0			25	
RX 3-1-501N	MN	1.83	3.0- 6.0			25	
RX 3-1-525F	GLC	1.87	1.0- 3.0			25	
RX 3-2-520F	GLC	1.90	1.0- 3.0			25	

EMA Emac Chevron

Grade	Filler	Sp Grav	Shrink, mils/in	Melt flow, g/10 min	Melt temp, °F	Back pres, psi	Drying temp, °F
SP 2257		0.94		2.00 E	180		

EMA Optema Exxon

Grade	Filler	Sp Grav	Shrink, mils/in	Melt flow, g/10 min	Melt temp, °F	Back pres, psi	Drying temp, °F
TC-111		0.94		2.00	167		

EnBA Enathene Quantum

Grade	Filler	Sp Grav	Shrink, mils/in	Melt flow, g/10 min	Melt temp, °F	Back pres, psi	Drying temp, °F
EA 720-009				6.00			

Epoxy Cosmic Epoxy Cosmic Plastics

Grade	Filler	Sp Grav	Shrink, mils/in	Melt flow, g/10 min	Melt temp, °F	Back pres, psi	Drying temp, °F
CP7311		2.00	1.0- 4.0				
CP7312		2.00	1.0- 4.0				
CP7314		2.00	1.0- 4.0				
CP7318		2.00	1.0- 4.0				
E484	GL	1.95	1.0- 4.0				
E486	GL	1.85	1.0- 4.0				
E487	GL	1.85	1.0- 4.0				
E4905	GL	1.95	1.0- 4.0				
E4920	GMN	1.90	1.0- 4.0				
E4930	GMN	2.00	1.0- 4.0				
E4940	GMN	1.85	1.0- 4.0				

Epoxy Rogers Epoxy Rogers

Grade	Filler	Sp Grav	Shrink, mils/in	Melt flow, g/10 min	Melt temp, °F	Back pres, psi	Drying temp, °F
1904B	GMN	2.05	1.0- 3.0			100- 250	
1906	GMN	1.90	1.0- 3.0			100- 250	
1907	GLC	1.95	2.0- 4.0			100- 250	
1908	GLC	1.85	2.0- 4.0			100- 250	
1908B	GLC	1.85	2.0- 4.0			100- 250	
1914	GLC	1.94	1.0- 3.0			100- 250	
1960B	GMN	1.75	4.0- 6.0			100- 250	
1961B	GMN	1.89	2.0- 4.0			100- 250	

Drying time, hr	Inj time, sec	Front temp, °F	Mid temp, °F	Rear temp, °F	Nozzle temp, °F	Mold temp, °F	Proc temp, °F
	5.0		170	140	190	320-340	230-240
	5.0		170	140	190	320-340	230-240
	5.0		170	140	190	320-340	230-240
	5.0		170	140	190	320-340	230-240
	5.0		170	140	190	320-340	230-240
	5.0		170	140	190	320-340	230-240
	5.0		170	140	190	320-340	230-240
	5.0		170	140	190	320-340	230-240
	5.0		170	140	190	320-340	230-240
	5.0		170	140	190	320-340	230-240
	5.0		170	140	190	320-340	230-240
	5.0		170	140	190	320-340	230-240
	5.0		170	140	190	320-340	230-240
	5.0		170	140	190	320-340	230-240
	5.0		170	140	190	320-340	230-240
	5.0		170	140	190	320-340	230-240
	5.0		170	140	190	320-340	230-240
	5.0		170	140	190	320-340	230-240
	5.0		170	140	190	320-340	230-240
	5.0		170	140	190	320-340	230-240
							350-450
							360-625
							450-600
							275-330
							275-330
							275-330
							275-330
							300-350
							300-350
							300-350
							250-350
							250-350
							250-350
							250-350
	3.0- 12.		200	170	200	345-360	220-230
	3.0- 12.		200	170	200	345-360	220-230
	3.0- 12.		200	170	200	345-360	220-230
	3.0- 12.		200	170	200	345-360	220-230
	3.0- 12.		200	170	200	345-360	220-230
	3.0- 12.		200	170	200	345-360	220-230
	3.0- 12.		200	170	200	345-360	220-230
	3.0- 12.		200	170	200	345-360	220-230

Grade	Filler	Sp Grav	Shrink, mils/in	Melt flow, g/10 min	Melt temp, °F	Back pres, psi	Drying temp, °F
2004B	GMN	1.95	2.0- 4.0			100- 250	
2008	GLC	1.85	3.0- 5.0			100- 250	
2060B	GMN	1.80	4.0- 6.0			100- 250	
2061B	GMN	1.95	2.0- 4.0			100- 250	
ETFE		**Halon**			**Ausimont**		
101		1.70	15.0-20.0		500		
ETFE		**Tefzel**			**Du Pont**		
200		1.70			520		
210		1.70			512		
280		1.70			520		
HT-2004	GL 25	1.86			520		
EVA		**Elvax**			**Du Pont**		
150		0.96		43.00			120
250		0.95		25.00			140
260		0.96		6.00			140
265		0.95		3.00			140
350		0.95		19.00			140
360		0.95		2.00			140
450		0.94		8.00			140
460		0.94		2.50			140
470		0.94		0.70			140
550		0.94		8.00			140
560		0.94		2.50			140
565		0.94		1.50			140
650		0.93		8.00			140
660		0.93		2.50			140
670		0.93		0.30			140
750		0.93		7.00			140
760		0.93		2.00			140
EVA		**Petrothene**			**Quantum**		
XL 05421	CBL 26						
EVA		**Ultrathene**			**Quantum**		
UE 630-000		0.93		1.80			
UE 630-002				1.80			
UE 632-000		0.93		8.20			
UE 634-000				3.10			
UE 635-000		0.92		9.80			
UE 637-000		0.93		3.20			
UE 655-000		0.93		2.20			
FEP		**Teflon FEP**			**Du Pont**		
100		2.14			507		
Fluorelast		**FTPE**			**3M**		
L-12783C		1.57	15.0-20.0		410-420	50	150-176
HDPE		**Alathon**			**OxyChem**		
7030		0.96		2.80 E		50- 100	150-160
7040		0.96		6.00 E		50- 100	150-160
7046		0.96		8.00		50- 100	150-160
7050		0.96		17.50 E		50- 100	150-160

Drying time, hr	Inj time, sec	Front temp, °F	Mid temp, °F	Rear temp, °F	Nozzle temp, °F	Mold temp, °F	Proc temp, °F
	3.0- 12.		200	170	200	345-360	220-230
	3.0- 12.		200	170	200	345-360	220-230
	3.0- 12.		200	170	200	345-360	220-230
	3.0- 12.		200	170	200	345-360	220-230
						250-300	575-650
		575-625	575-625	525-575	650	73-375	575-625
		575-625	575-625	525-575	650	73-375	575-625
		575-625	575-625	525-575	650	73-375	575-625
		575-625	575-625	525-575	650	73-375	575-625
8.0		350	300	250	300-400	60-100	425
8.0		350	300	250	300-400	60-100	425
8.0		350	300	250	300-400	60-100	425
8.0		350	300	250	300-400	60-100	425
8.0		350	300	250	300-400	60-100	425
8.0		350	300	250	300-400	60-100	425
8.0		350	300	250	300-400	60-100	425
8.0		350	300	250	300-400	60-100	425
8.0		350	300	250	300-400	60-100	425
8.0		350	300	250	300-400	60-100	425
8.0		350	300	250	300-400	60-100	425
8.0		350	300	250	300-400	60-100	425
8.0		350	300	250	300-400	60-100	425
8.0		350	300	250	300-400	60-100	425
8.0		350	300	250	300-400	60-100	425
8.0		350	300	250	300-400	60-100	425
8.0		350	300	250	300-400	60-100	425
	15.0	340	340	320	340	70	
						50-100	430
							430
						50-100	430
							425
						50-100	450
						50-100	430
						50-100	430
		700	625-650	600-625	700	200	650-720
1.0- 4.0		455-465	455-465	445-455	455-465	100-180	
0.5- 1.0	15.0	350-400	350-400	325-350	350-425	80-100	400-550
0.5- 1.0	15.0	350-400	350-400	325-350	350-425	80-100	400-550
0.5- 1.0	15.0	350-400	350-400	325-350	350-425	80-100	400-550
0.5- 1.0	15.0	350-400	350-400	325-350	350-425	80-100	400-550

Grade	Filler	Sp Grav	Shrink, mils/in	Melt flow, g/10 min	Melt temp, °F	Back pres, psi	Drying temp, °F
7245		0.95		20.00 E		50- 100	150-160
7260		0.95		33.00 E		50- 100	150-160
7265		0.95		45.00 E		50- 100	150-160
7270-1		0.95		57.00 E		50- 100	150-160
7270-2		0.95		52.00 E		50- 100	150-160
7340		0.95		4.00 E		50- 100	150-160
H5232		0.95		32.00 E		50- 100	150-160
H5234		0.95		34.00 E		50- 100	150-160
H5244		0.95		44.00 E		50- 100	150-160
H5244-M		0.95		44.00 E		50- 100	150-160
H5512		0.96		12.00		50- 100	150-160
H6017		0.96		17.50 E		50- 100	150-160
M4845		0.95		4.50		50- 100	150-160
M5352		0.95		5.20 E		50- 100	150-160
M6028		0.96		2.80 E		50- 100	150-160
M6062		0.96		6.00 E		50- 100	150-160
M6080		0.96		8.00 E		50- 100	150-160
M6450		0.96		5.00 E		50- 100	150-160
M6580		0.97		8.00 E		50- 100	150-160
PE5510		0.95		1.20 E		50- 100	150-160
HDPE		**Celstran**			**Hoechst**		
PEG58-01-4	GLF 58	1.50					200
HDPE		**Dow HDPE**			**Dow**		
04352N		0.95		4.00 E			
08354N		0.95		7.00 E			
10062N		0.96		10.00 E			
10262N		0.96		10.00 E			
10362N		0.96		10.00 E			
12065		0.96		0.90 E			
12350N		0.96		12.00 E			
17350N		0.95		17.00 E			
25355N		0.95		25.00 E			
30360M		0.96		30.00 E			
38152M		0.95		38.00 E			
40360M		0.96		38.00 E			
42047N		0.94		42.00 E			
52053N		0.95		52.00 E			
65053N		0.95		65.00 E			
75053N		0.95		8.00 E			
IP-60							
HDPE		**DSM Polyethlylene**			**DSM**		
J-90/20	GLF 20	1.09	3.0				
J-90/30	GLF 30	1.18	2.0				
HDPE		**Electrafil**			**DSM**		
PE-90/EC	CBL	1.04	18.0				
HDPE		**Fiberfil**			**DSM**		
J-90/20	GLF 20	1.09	3.0- 4.0				
J-90/30	GLF 30	1.18	2.0- 3.0				
HDPE		**Fiberstran**			**DSM**		
G-90/20	GLF 20	1.09	3.0				

Drying time, hr	Inj time, sec	Front temp, °F	Mid temp, °F	Rear temp, °F	Nozzle temp, °F	Mold temp, °F	Proc temp, °F
0.5- 1.0	15.0	350-400	350-400	325-350	350-425	80-100	400-550
0.5- 1.0	15.0	350-400	350-400	325-350	350-425	80-100	400-550
0.5- 1.0	15.0	350-400	350-400	325-350	350-425	80-100	400-550
0.5- 1.0	15.0	350-400	350-400	325-350	350-425	80-100	400-550
0.5- 1.0	15.0	350-400	350-400	325-350	350-425	80-100	400-550
0.5- 1.0	15.0	350-400	350-400	325-350	350-425	80-100	400-550
0.5- 1.0	15.0	350-400	350-400	325-350	350-425	80-100	400-550
0.5- 1.0	15.0	350-400	350-400	325-350	350-425	80-100	400-550
0.5- 1.0	15.0	350-400	350-400	325-350	350-425	80-100	400-550
0.5- 1.0	15.0	350-400	350-400	325-350	350-425	80-100	400-550
0.5- 1.0	15.0	350-400	350-400	325-350	350-425	80-100	400-550
0.5- 1.0	15.0	350-400	350-400	325-350	350-425	80-100	400-550
0.5- 1.0	15.0	350-400	350-400	325-350	350-425	80-100	400-550
0.5- 1.0	15.0	350-400	350-400	325-350	350-425	80-100	400-550
0.5- 1.0	15.0	350-400	350-400	325-350	350-425	80-100	400-550
0.5- 1.0	15.0	350-400	350-400	325-350	350-425	80-100	400-550
0.5- 1.0	15.0	350-400	350-400	325-350	350-425	80-100	400-550
0.5- 1.0	15.0	350-400	350-400	325-350	350-425	80-100	400-550
0.5- 1.0	15.0	350-400	350-400	325-350	350-425	80-100	400-550
0.5- 1.0	15.0	350-400	350-400	325-350	350-425	80-100	400-550
2.0		400-420	395-415	390-410	390-410	140-160	400-420
		375-550	375-550	375-550			
		375-550	375-550	375-550			
		375-550	375-550	375-550			
		375-550	375-550	375-550			
		375-550	375-550	375-550			
		375-550	375-550	375-550			
		375-550	375-550	375-550			
		375-550	375-550	375-550			
		375-550	375-550	375-550			
		375-550	375-550	375-550			
		375-550	375-550	375-550			
		375-550	375-550	375-550			
		375-550	375-550	375-550			
		375-550	375-550	375-550			
		375-550	375-550	375-550			
		375-550	375-550	375-550			
		375-550	375-550	375-550			
		400-450	420-500	390-430	390-490	50-110	380-500
		400-450	420-500	390-430	390-490	50-110	380-500
		400-450	420-500	390-430	390-490	50-110	380-500
		400-450	420-500	390-430	390-490	50-110	380-500
		400-450	420-500	390-430	390-490	50-110	380-500
		400-450	420-500	390-430	390-490	50-110	380-500

Grade	Filler	Sp Grav	Shrink, mils/in	Melt flow, g/10 min	Melt temp, °F	Back pres, psi	Drying temp, °F
G-90/30	GLF 30	1.18	2.0				
HDPE		**Marlex**			**Phillips**		
HMN 4550		0.95		5.00			
HMN 5060		0.95		6.00			
HMN 54140		0.95		14.00			
HMN 55180		0.96		20.00			
HMN 5580		0.96		8.50			
HMN 6060		0.96		7.50			
HDPE		**Nortuff**			**Polymerland**		
RA 7020-KO	CAC 20	1.07		15.00			
RA 7040-KO	CAC 40	1.25		8.00			
HDPE		**Petrothene**			**Quantum**		
KR 21017		0.95		0.85			
KR 22780				0.92			
KR 50076		0.97		0.42			
KR 91028	CBL 3			0.42			
KR 92561	CBL 3			0.25			
KR 92828	CBL 3			0.15			
LB 5003-00				0.30			
LB 5003-10				0.30			
LB 5003-23				0.30			
LB 5003-42	ZIS			0.30			
LB 5602-00				0.30			
LB 5602-10				0.30			
LB 5602-12				0.30			
LB 5602-15				0.30			
LB 5602-43	ZIS			0.30			
LB 5604-00				0.35			
LB 5604-10				0.35			
LB 5604-54				0.35			
LB 5605-00				0.30			
LB 5605-10				0.30			
LB 5605-12				0.30			
LB 5703-00				11.50			
LB 6001-00		0.96		0.25			
LB 6002-00		0.96		0.14			
LB 6003-00		0.96		0.14			
LB 6004-00		0.96		0.32			
LB 7420-00		0.96		0.68			
LB 7480-00		0.96		0.65			
LB 8300-00		0.95		0.14			
LB 8320-00		0.95		0.26			
LB 8500-00				0.14			
LB 8510-00		0.96	15.0-45.0	1.30			
LB 8520-00				0.26			
LF 5712-00				12.50			
LF 6030-00			15.0-45.0	3.00			
LF 8610-00		0.96	15.0-45.0	2.00			
LM 6005-00				0.35			
LM 6007-00				0.70			
LM 6085-00		0.96		0.80			
LM 6186-00				0.80			
LM 6187-00				1.15			

Drying time, hr	Inj time, sec	Front temp, °F	Mid temp, °F	Rear temp, °F	Nozzle temp, °F	Mold temp, °F	Proc temp, °F
		400-450	420-500	390-430	390-490	50-110	380-500
							425-525
							425-525
							400-500
							400-500
							425-525
							400-500
						110-125	410-450
						110-125	410-450
	15.0	400	400	350	425	100	
	15.0	400	400	350	425	100	
	15.0	400	400	350	425	100	
	15.0	400	400	350	425	100	
	15.0	400	400	350	425	100	
	15.0	400	400	350	425	100	
	15.0	400	400	350	425	100	
	15.0	400	400	350	425	100	
	15.0	400	400	350	425	100	
	15.0	400	400	350	425	100	
	15.0	400	400	350	425	100	
	15.0	400	400	350	425	100	
	15.0	400	400	350	425	100	
	15.0	400	400	350	425	100	
	15.0	400	400	350	425	100	
	15.0	400	400	350	425	100	
	15.0	400	400	350	425	100	
	15.0	400	400	350	425	100	
	15.0	400	400	350	425	100	
	15.0	400	400	350	425	100	
	15.0	400	400	350	425	100	
	15.0	400	400	350	425	100	
	15.0	400	400	350	425	100	
	15.0	400	400	350	425	100	
	15.0	400	400	350	425	100	
	15.0	400	400	350	425	100	
	15.0	400	400	350	425	100	
	15.0	400	400	350	425	100	
	15.0	400	400	350	425	100	
	15.0	400	400	350	425	100	
	15.0	400	400	350	425	100	
	15.0	400	400	350	425	100	
	15.0	400	400	350	425	100	
	15.0	400	400	350	425	100	
	15.0	400	400	350	425	100	
	15.0	400	400	350	425	100	
	15.0	400	400	350	425	100	
	15.0	400	400	350	425	100	
	15.0	400	400	350	425	100	
	15.0	400	400	350	425	100	
	15.0	400	400	350	425	100	
	15.0	400	400	350	425	100	

Grade	Filler	Sp Grav	Shrink, mils/in	Melt flow, g/10 min	Melt temp, °F	Back pres, psi	Drying temp, °F
LN 5520-00		0.95		8.30			
LN 5530-00				8.30			
LN 5560-00		0.95		10.10			
LP 20820				0.17			
LP 5002-00				0.21			
LP 5002-10				0.21			
LP 5101-00				0.15			
LP 5102-00				0.21			
LP 5200-00		0.95		0.14			
LP 5711-00		0.95		11.00			
LP 5713-00		0.95		13.00			
LP 9230-00		0.94		0.45			
LR 5700-00				0.37			
LR 5800-00				0.43			
LR 5860-00				0.82			
LR 5900-00				0.70			
LR 7230-00		0.95		9.50			
LR 7320-00				0.30			
LR 7320-10				0.30			
LR 7320-12				0.30			
LR 7340-00				0.38			
LR 7340-10				0.38			
LR 7340-12				0.38			
LR 7340-43	ZIS			0.38			
LR 7350-00				0.33			
LR 7350-10				0.33			
LR 7350-12				0.33			
LR 7810-00				0.20			
LR 7810-10				0.20			
LR 9200-00				0.45			
LR 9200-10				0.45			
LR 9230-00				0.45			
LS 20812				1.40			
LS 20813				5.00			
LS 20842				65.00			
LS 20846				1.40			
LS 3150-00			15.0-45.0	17.00			
LS 3250-00			15.0-45.0	31.00			
LS 3420-00			15.0-45.0	44.00			
LS 3550-00			15.0-45.0	48.00			
LS 3601-00		0.96	15.0-45.0	5.00			
LS 3801-00		0.96	15.0-45.0	8.00			
LS 4040-00			15.0-45.0	4.00			
LS 4050-00			15.0-45.0	5.50			
LS 5060-00			15.0-45.0	8.00			
LS 5090-00		0.95	15.0-45.0	12.50			
LS 5110-00		0.95	15.0-45.0	16.00			
LS 5140-00			15.0-45.0	23.00			
LS 5230-00		0.96	15.0-45.0	23.00			
LS 5250-00		0.95	15.0-45.0	33.00			
LS 5280-00		0.95	15.0-45.0	31.00			
LS 5300-00		0.95	15.0-45.0	40.00			
LS 5450-00		0.96	15.0-45.0	45.00			
LS 5560-00			15.0-45.0	7.00			
LS 6040-00			15.0-45.0	5.50			
LS 6060-00			15.0-45.0	10.00			

Drying time, hr	Inj time, sec	Front temp, °F	Mid temp, °F	Rear temp, °F	Nozzle temp, °F	Mold temp, °F	Proc temp, °F
	15.0	400	400	350	425	100	
	15.0	400	400	350	425	100	
	15.0	400	400	350	425	100	
	15.0	400	400	350	425	100	
	15.0	400	400	350	425	100	
	15.0	400	400	350	425	100	
	15.0	400	400	350	425	100	
	15.0	400	400	350	425	100	
	15.0	400	400	350	425	100	
	15.0	400	400	350	425	100	
	15.0	400	400	350	425	100	
	15.0	400	400	350	425	100	
	15.0	400	400	350	425	100	
	15.0	400	400	350	425	100	
	15.0	400	400	350	425	100	
	15.0	400	400	350	425	100	
	15.0	400	400	350	425	100	
	15.0	400	400	350	425	100	
	15.0	400	400	350	425	100	
	15.0	400	400	350	425	100	
	15.0	400	400	350	425	100	
	15.0	400	400	350	425	100	
	15.0	400	400	350	425	100	
	15.0	400	400	350	425	100	
	15.0	400	400	350	425	100	
	15.0	400	400	350	425	100	
	15.0	400	400	350	425	100	
	15.0	400	400	350	425	100	
	15.0	400	400	350	425	100	
	15.0	400	400	350	425	100	
	15.0	400	400	350	425	100	
	15.0	400	400	350	425	100	
	15.0	400	400	350	425	100	
	15.0	400	400	350	425	100	
	15.0	400	400	350	425	100	
	15.0	400	400	350	425	100	
	15.0	400	400	350	425	100	
	15.0	400	400	350	425	100	
	15.0	400	400	350	425	100	
	15.0	400	400	350	425	100	
	15.0	400	400	350	425	100	
	15.0	400	400	350	425	100	
	15.0	400	400	350	425	100	
	15.0	400	400	350	425	100	
	15.0	400	400	350	425	100	
	15.0	400	400	350	425	100	
	15.0	400	400	350	425	100	
	15.0	400	400	350	425	100	
	15.0	400	400	350	425	100	
	15.0	400	400	350	425	100	
	15.0	400	400	350	425	100	
	15.0	400	400	350	425	100	
	15.0	400	400	350	425	100	
	15.0	400	400	350	425	100	
	15.0	400	400	350	425	100	
	15.0	400	400	350	425	100	

Grade	Filler	Sp Grav	Shrink, mils/in	Melt flow, g/10 min	Melt temp, °F	Back pres, psi	Drying temp, °F
LS 6080-00			15.0-45.0	8.50			
LS 6090-00		0.96	15.0-45.0	12.50			
LS 6180-00			15.0-45.0	22.00			
LS 6300-00			15.0-45.0	36.00			
LS 6402-00			15.0-45.0	4.20			
LS 6601-00		0.95	15.0-45.0	6.50			
LS 6901-00			15.0-45.0	10.50			
LS 9010-46				1.40			
LT 5704-00				22.00			
LT 5705-00				20.00			
LT 6009-00				0.85			
LT 6109-00				0.85			
LT 6180-00				1.15			
LY 6000-00		0.95		8.00			
LY 6560-00		0.95	15.0-45.0	6.20			
LY 9550-00		0.95		8.80			
HDPE		**RTP Polymers**			**RTP**		
700	GLF	0.96	20.0			50	175
701	GLF	1.02	5.0			50	175
703	GLF	1.10	3.0			50	175
705	GLF	1.18	2.0			50	175
707	GLF	1.28	2.0			50	175
748	MI	1.30	8.0			50	175
HDPE		**Shuman**					
603		0.95		2.00			
HDPE Copolymer		**Alathon**			**OxyChem**		
7230		0.95		6.00 E		50- 100	150-160
H5310		0.95		10.00 E		50- 100	150-160
H5520		0.96		20.00 E		50- 100	150-160
H5618		0.96		18.00 E		50- 100	150-160
M4550		0.95		5.00		50- 100	150-160
M4612		0.95		1.20 E		50- 100	150-160
M4621		0.95		2.10 E		50- 100	150-160
M5350		0.95		5.00 E		50- 100	150-160
M5370		0.95		7.00 E		50- 100	150-160
M5372		0.95		7.20 E		50- 100	150-160
M5390		0.95		9.00 E		50- 100	150-160
M5562		0.96		6.00 E		50- 100	150-160
HIPS		**RTP Polymers**			**RTP**		
ESD-A-480 HI	CF	1.06	2.0				
ESD-C-480 HI	CF	1.08	1.0				
Ionomer		**Formion**			**A. Schulman**		
FI 105		0.97	6.0	1.50 E		145	
FI 107H GF	GLF	1.01	3.0	5.00 E		145	
FI 120		0.97	6.0	1.50 E		145	
FI 128 GF	GLF	1.02	3.0	5.00 E		145	
FI 241E	UNS	0.94	6.0	0.95 E		145	
FI 294G GF	GLF	1.10	2.0	0.95 E		145	
FI 310	UNS	1.10	3.0	1.15 E		145	
FI 317 GF	GLF	0.99	3.0	13.50 E		145	
FI 335 GF	GLF	1.11	2.0	1.50 E		145	

Drying time, hr	Inj time, sec	Front temp, °F	Mid temp, °F	Rear temp, °F	Nozzle temp, °F	Mold temp, °F	Proc temp, °F
	15.0	400	400	350	425	100	
	15.0	400	400	350	425	100	
	15.0	400	400	350	425	100	
	15.0	400	400	350	425	100	
	15.0	400	400	350	425	100	
	15.0	400	400	350	425	100	
	15.0	400	400	350	425	100	
	15.0	400	400	350	425	100	
	15.0	400	400	350	425	100	
	15.0	400	400	350	425	100	
	15.0	400	400	350	425	100	
	15.0	400	400	350	425	100	
	15.0	400	400	350	425	100	
	15.0	400	400	350	425	100	
	15.0	400	400	350	425	100	
	15.0	400	400	350	425	100	
2.0		390-420	370-390	350-370		75-130	
2.0		390-420	370-390	350-370		75-130	
2.0		390-420	370-390	350-370		75-130	
2.0		390-420	370-390	350-370		75-130	
2.0		390-420	370-390	350-370		75-130	
2.0		330-360	320-350	310-340		75-125	
							360-410
0.5- 1.0	15.0	350-400	350-400	325-350	350-425	80-100	400-550
0.5- 1.0	15.0	350-400	350-400	325-350	350-425	80-100	400-550
0.5- 1.0	15.0	350-400	350-400	325-350	350-425	80-100	400-550
0.5- 1.0	15.0	350-400	350-400	325-350	350-425	80-100	400-550
0.5- 1.0	15.0	350-400	350-400	325-350	350-425	80-100	400-550
0.5- 1.0	15.0	350-400	350-400	325-350	350-425	80-100	400-550
0.5- 1.0	15.0	350-400	350-400	325-350	350-425	80-100	400-550
0.5- 1.0	15.0	350-400	350-400	325-350	350-425	80-100	400-550
0.5- 1.0	15.0	350-400	350-400	325-350	350-425	80-100	400-550
0.5- 1.0	15.0	350-400	350-400	325-350	350-425	80-100	400-550
0.5- 1.0	15.0	350-400	350-400	325-350	350-425	80-100	400-550
0.5- 1.0	15.0	350-400	350-400	325-350	350-425	80-100	400-550
		400-550	400-550	400-550		100-160	
		400-550	400-550	400-550		100-160	
	5.0- 8.						450
	5.0- 8.			375		60	450
	5.0- 8.			375		60	450
	5.0- 8.			375		60	450
	5.0- 8.			375		60	450
	5.0- 8.			375		60	450
	5.0- 8.			375		60	450
	5.0- 8.			375		60	450
	5.0- 8.			375		60	450

Grade	Filler	Sp Grav	Shrink, mils/in	Melt flow, g/10 min	Melt temp, °F	Back pres, psi	Drying temp, °F
FI 340 GF	GLF	1.11	2.0	2.50 E		145	
FI 341	UNS	0.95	3.0	4.00 E		145	
FI 342 GF	GLF	1.04	2.0	4.00 E		145	
FI 344 GF	GLF	0.97	4.0	20.00 E		145	
Ionomer		**Surlyn**			**Du Pont**		
7930		0.94		1.80	178-205		
7940		0.94		2.60	178-205		
8020		0.95		1.00	178-205		
8528		0.94		1.30	178-205		
8550		0.94		3.90	178-205		
8660		0.94		10.00	178-205		
8920		0.95		0.90	178-205		
8940		0.95		2.80	178-205		
9020		0.96		1.10	178-205		
9450		0.94		5.50	178-205		
9520		0.95		1.10	178-205		
9650		0.96		5.00	178-205		
9720		0.96		1.00	178-205		
9721		0.96		1.00	178-205		
9730		0.95		1.60	178-205		
9910		0.97		0.70	178-205		
9950		0.96		5.50	178-205		
9970		0.96		5.50	178-205		
LCP		**CTI General**			**CTI**		
LCP-20CF/000	CF 20	1.41					250
LCP		**RTP Polymers**			**RTP**		
3400-3 MS	MOS	1.43	0.2				300
3400-4 MS	MOS	1.43	0.2				300
3403-3	GLF	1.52	0.3				300
3403-4	GLF	1.50	0.3				300
3405-3	GLF	1.61	0.2				300
3405-3 TFE 15	GLF	1.69	0.1				300
3405-4	GLF	1.60	0.2				300
3405-4 TFE 15	GLF	1.69	0.1				300
3407-3	GLF	1.70	0.2				300
3407-4	GLF	1.70	0.2				300
3483-3	CF	1.47	0.1				300
3487-3	CF	1.55					300
3499-3X57404A	GMN	1.61	0.3				300
3499-3X57404B	GMN	1.61	0.5				300
3499-3X57407B	GRP	1.52	2.0				300
3499-3X57408B	GRP	1.52	1.5				300
LCP		**Vectra**			**Hoechst**		
A115	GLF	1.50			536	0- 25	285-300
A130	GLF	1.61	1.0- 2.0		536	0- 25	285-300
A150	GLF	1.79	2.0		536	0- 25	285-300
A230	CF	1.49	1.0- 2.0		536	0- 25	285-300
A410	GMN	1.84	2.0- 3.0		536	0- 25	285-300
A420	GMN	1.89	1.0- 2.0		536	0- 25	285-300
A422	GLF	1.68	1.0- 3.0		536	0- 25	285-300
A430	PTF	1.50			538	0- 25	285-300
A435	GLF	1.62			538	0- 25	285-300

Drying time, hr	Inj time, sec	Front temp, °F	Mid temp, °F	Rear temp, °F	Nozzle temp, °F	Mold temp, °F	Proc temp, °F
	5.0- 8.			375		60	450
	5.0- 8.			375		60	450
	5.0- 8.			375		60	450
	5.0- 8.			375		60	450
		475	475	375	475	40-120	475-500
		475	475	375	475	40-120	475-500
		475	475	375	475	40-120	475-500
		475	475	375	475	40-120	475-500
		475	475	375	475	40-120	475-600
		400	400	350	400	40-120	400-425
		475	475	375	475	40-120	475-500
		475	475	375	475	40-120	475-500
		475	475	375	475	40-120	475-500
		400	400	350	400	40-120	400-425
		475	475	375	475	40-120	475-500
		400	400	350	400	40-120	400-425
		475	475	375	475	40-120	475-500
		475	475	375	475	40-120	475-500
		475	475	375	475	40-120	475-500
		475	475	375	475	40-120	475-500
		400	400	350	400	40-120	400-425
		400	400	350	400	40-120	400-425
						200-250	600
8.0		650-715	640-705	630-695	650-715	150-275	
8.0		650-715	640-705	630-695	650-715	150-275	
8.0		650-715	640-705	630-695	650-715	150-275	
8.0		650-715	640-705	630-695	650-715	150-275	
8.0		650-715	640-705	630-695	650-715	150-275	
8.0		650-715	640-705	630-695	650-715	150-275	
8.0		650-715	640-705	630-695	650-715	150-275	
8.0		650-715	640-705	630-695	650-715	150-275	
8.0		650-715	640-705	630-695	650-715	150-275	
8.0		650-715	640-705	630-695	650-715	150-275	
8.0		650-715	640-705	630-695	650-715	150-275	
8.0		650-715	640-705	630-695	650-715	150-275	
8.0		650-715	640-705	630-695	650-715	150-275	
8.0		650-715	640-705	630-695	650-715	150-275	
8.0		650-715	640-705	630-695	650-715	150-275	
8.0		650-715	640-705	630-695	650-715	150-275	
4.0-24.0		545	545	545	555	85-200	555
4.0-24.0		545	545	545	555	85-200	555
4.0-24.0		545	545	545	555	85-200	555
4.0-24.0		545	545	545	555	85-200	555
4.0-24.0		545	545	545	555	85-200	555
4.0-24.0		545	545	545	555	85-200	555
4.0-24.0		545	545	545	555	85-200	555
4.0-24.0		545	545	545	555	85-200	555
4.0-24.0		545	545	545	555	85-200	555

Grade	Filler	Sp Grav	Shrink, mils/in	Melt flow, g/10 min	Melt temp, °F	Back pres, psi	Drying temp, °F
A515	MN	1.52	3.0- 4.0		536	0- 25	285-300
A530	MN	1.65			538	0- 25	285-300
A540	MN	1.76	2.0- 4.0			0- 25	285-300
A625	GRK	1.54	0.1- 3.0		536	0- 25	285-300
A700	GLF 30	1.61			538	0- 25	285-300
B130	GLF	1.61	0.1		536	0- 25	285-300
B230	CF	1.49				0- 25	285-300
B420	GLF	1.89			536	0- 25	285-300
C115	GLF	1.50	2.0		620	0- 25	285-300
C130	GLF	1.61	1.0- 2.0		620	0- 25	285-300
C150	GLF	1.79	2.0			0- 25	285-300
C550	MN	1.89	3.0- 4.0		620	0- 25	285-300
E130	GLF				670	0- 25	285-300
V140		1.67	1.2- 4.5		580	0- 25	285-300
LCP		**Xydar**			**Amoco Perform.**		
G-330	GLF	1.62					300
G-345	GLF	1.76					300
G-430	GLF	1.64					300
G-445	GLF	1.75					300
G-540	GLF	1.70					300
G-640	GLF	1.70					300
M-350	MN	1.84					300
M-450	MN	1.84					300
MG-350	GMN	1.78					300
MG-450	GMN	1.79					300
LDPE		**Dow LDPE**			**Dow**		
4005M		0.91		5.50 E			
4012M		0.91		12.00 E			
5004IM		0.92		4.00 E			
5004M		0.92		4.00 E			
510M		0.92		2.00 E			
640M		0.92		2.00 E			
722M		0.91		8.00 E			
779I		0.92		7.00 E			
919I		0.92		35.00 E			
955I		0.92		35.00 E			
959I		0.92		55.00 E			
993		0.92		25.00 E			
993I		0.93		25.00 E			
LDPE		**Petrothene**			**Quantum**		
NA 117-000			15.0-45.0	8.50			
NA 143-000				2.00			
NA 143-063				2.00			
NA 145-000				3.50			
NA 145-055				3.50			
NA 145-232				3.50			
NA 150-086				1.40			
NA 154		0.92		2.20			
NA 154-050				2.20			
NA 154-052				2.20			
NA 154-055				2.20			
NA 154-086				2.20			
NA 154-088				2.20			

Drying time, hr	Inj time, sec	Front temp, °F	Mid temp, °F	Rear temp, °F	Nozzle temp, °F	Mold temp, °F	Proc temp, °F
4.0-24.0		545	545	545	555	85-200	555
4.0-24.0		545	545	545	555	85-200	555
4.0-24.0		545	545	545	555	85-200	555
4.0-24.0		545	545	545	555	85-200	555
4.0-24.0		545	545	545	555	85-200	555
4.0		570	570	570	555	85-200	555
4.0-24.0		570	570	570	555	85-200	555
4.0-24.0		570	570	570	555	85-200	555
4.0-24.0		605	605	600	605	85-200	605
4.0-24.0		605	605	600	605	85-200	605
4.0-24.0		605	605	600	605	85-200	605
4.0-24.0		605	605	600	605	85-200	605
4.0-24.0		640	640	635	625	85-200	640
4.0-24.0		580	580	570	580	165	585
8.0						150-200	610-680
8.0						150-200	610-710
8.0						180-220	660-730
8.0						180-220	660-730
6.0- 8.0						150-200	590-645
6.0- 8.0						150-200	590-645
8.0						180-220	660-730
8.0						180-220	660-730
8.0						150-200	610-710
8.0						180-220	660-730
		325-550	325-550	325-550			
		325-550	325-550	325-550			
		325-550	325-550	325-550			
		325-550	325-550	325-550			
		325-550	325-550	325-550			
		325-550	325-550	325-550			
		325-550	325-550	325-550			
		325-550	325-550	325-550			
		325-550	325-550	325-550			
		325-550	325-550	325-550			
		325-550	325-550	325-550			
		325-550	325-550	325-550			
		325-550	325-550	325-550			
	15.0	350	350	325	350	80	
	15.0	350	350	325	350	80	
	15.0	350	350	325	350	80	
	15.0	350	350	325	350	80	
	15.0	350	350	325	350	80	
	15.0	350	350	325	350	80	
	15.0	350	350	325	350	80	
	15.0	350	350	325	350	80	
	15.0	350	350	325	350	80	
	15.0	350	350	325	350	80	
	15.0	350	350	325	350	80	
	15.0	350	350	325	350	80	
	15.0	350	350	325	350	80	

Grade	Filler	Sp Grav	Shrink, mils/in	Melt flow, g/10 min	Melt temp, °F	Back pres, psi	Drying temp, °F
NA 155				2.50			
NA 155-000				2.50			
NA 155-131				2.50			
NA 156-219				3.00			
NA 171-000		0.93		1.35			
NA 202-000		0.92		22.00			
NA 203-000		0.92		8.00			
NA 204-000			15.0-45.0	7.00			
NA 205-000		0.92		3.00			
NA 206-000				13.00			
NA 211-000		0.92		4.50			
NA 212-000		0.91		12.00			
NA 215-000				4.50			
NA 216-000				3.70			
NA 217-000				5.50			
NA 219-000				10.00			
NA 241-000			15.0-45.0	22.00			
NA 249-000		0.93	15.0-45.0	57.00			
NA 254-000		0.93	15.0-45.0	5.00			
NA 270-000		0.91	15.0-45.0	70.00			
NA 279-000		0.92		2.10			
NA 300-000				2.20			
NA 300-340				2.20			
NA 301-000				1.20			
NA 301-060				1.20			
NA 304				2.00			
NA 304-013				2.00			
NA 304-134				2.00			
NA 304-161				2.00			
NA 306-000		0.92	15.0-45.0	2.20			
NA 309-000		0.92	15.0-45.0	0.90			
NA 310-060		0.92		0.26			
NA 314				2.00			
NA 314-013				2.00			
NA 314-160				2.00			
NA 314-161				2.00			
NA 316				1.30			
NA 316-138				1.30			
NA 317-009				2.00			
NA 317-018				2.00			
NA 320		0.93		2.50			
NA 321				2.00			
NA 321-003				2.00			
NA 321-150				2.00			
NA 321-155				2.00			
NA 324-009				3.00			
NA 334				6.00			
NA 334-000				6.00			
NA 336				6.50			
NA 336-149				6.00			
NA 340				1.00			
NA 340-013				1.00			
NA 340-114				1.00			
NA 340-141				1.00			
NA 340-163				1.00			
NA 341				1.20			

Drying time, hr	Inj time, sec	Front temp, °F	Mid temp, °F	Rear temp, °F	Nozzle temp, °F	Mold temp, °F	Proc temp, °F
	15.0	350	350	325	350	80	
	15.0	350	350	325	350	80	
	15.0	350	350	325	350	80	
	15.0	350	350	325	350	80	
	15.0	350	350	325	350	80	
	15.0	350	350	325	350	80	
	15.0	350	350	325	350	80	
	15.0	350	350	325	350	80	
	15.0	350	350	325	350	80	
	15.0	350	350	325	350	80	
	15.0	350	350	325	350	80	
	15.0	350	350	325	350	80	
	15.0	350	350	325	350	80	
	15.0	350	350	325	350	80	
	15.0	350	350	325	350	80	
	15.0	350	350	325	350	80	
	15.0	350	350	325	350	80	
	15.0	350	350	325	350	80	
	15.0	350	350	325	350	80	
	15.0	350	350	325	350	80	
	15.0	350	350	325	350	80	
	15.0	350	350	325	350	80	
	15.0	350	350	325	350	80	
	15.0	350	350	325	350	80	
	15.0	350	350	325	350	80	
	15.0	350	350	325	350	80	
	15.0	350	350	325	350	80	
	15.0	350	350	325	350	80	
	15.0	350	350	325	350	80	
	15.0	350	350	325	350	80	
	15.0	350	350	325	350	80	
	15.0	350	350	325	350	80	
	15.0	350	350	325	350	80	
	15.0	350	350	325	350	80	
	15.0	350	350	325	350	80	
	15.0	350	350	325	350	80	
	15.0	350	350	325	350	80	
	15.0	350	350	325	350	80	
	15.0	350	350	325	350	80	
	15.0	350	350	325	350	80	
	15.0	350	350	325	350	80	
	15.0	350	350	325	350	80	
	15.0	350	350	325	350	80	
	15.0	350	350	325	350	80	
	15.0	350	350	325	350	80	
	15.0	350	350	325	350	80	
	15.0	350	350	325	350	80	
	15.0	350	350	325	350	80	
	15.0	350	350	325	350	80	
	15.0	350	350	325	350	80	
	15.0	350	350	325	350	80	
	15.0	350	350	325	350	80	
	15.0	350	350	325	350	80	
	15.0	350	350	325	350	80	
	15.0	350	350	325	350	80	
	15.0	350	350	325	350	80	

Grade	Filler	Sp Grav	Shrink, mils/in	Melt flow, g/10 min	Melt temp, °F	Back pres, psi	Drying temp, °F
NA 341-124				1.20			
NA 342		0.93		1.00			
NA 343-213				2.00			
NA 345				1.80			
NA 345-000		0.93	15.0-45.0	1.80			
NA 345-009				1.80			
NA 345-013				1.80			
NA 345-166				1.80			
NA 345-195				1.80			
NA 345-196				1.80			
NA 346				3.60			
NA 346-009				3.60			
NA 346-100				3.60			
NA 347-211				2.00			
NA 351				0.30			
NA 353-000				2.00			
NA 355				0.50			
NA 357				0.30			
NA 362				0.40			
NA 362-176				0.50			
NA 420				2.50			
NA 420-000				2.50			
NA 420-127				2.50			
NA 420-142				2.50			
NA 420-234				2.50			
NA 426-225				2.50			
NA 440				2.50			
NA 440-118				2.50			
NA 440-137				2.50			
NA 440-216				2.50			
NA 441				1.00			
NA 441-137				1.00			
NA 480				0.25			
NA 480-145				0.25			
NA 480-158				0.25			
NA 480-177				0.25			
NA 480-178				0.25			
NA 495		0.92		1.20			
NA 496				1.80			
NA 496-013				1.80			
NA 496-195				1.80			
NA 496-196				1.80			
NA 510-000		0.92		2.00			
NA 520-024				0.25			
NA 593-00				22.00 E	219		
NA 594-00				70.00 E	214		
NA 596-00				150.00 E	210		
NA 597-00				250.00 E	225		
NA 598-00				8.60 A	210		
NA 601-00/04				60.00 A	199		
NA 603-04		0.90		60.00 A			
NA 605-04		0.90		138.00 A			
NA 810-000		0.92		1.30			
NA 820-000		0.92	15.0-45.0	2.20			
NA 831-000		0.92	15.0-45.0	9.00			
NA 836-071		0.92		6.00			

Drying time, hr	Inj time, sec	Front temp, °F	Mid temp, °F	Rear temp, °F	Nozzle temp, °F	Mold temp, °F	Proc temp, °F
	15.0	350	350	325	350	80	
	15.0	350	350	325	350	80	
	15.0	350	350	325	350	80	
	15.0	350	350	325	350	80	
	15.0	350	350	325	350	80	
	15.0	350	350	325	350	80	
	15.0	350	350	325	350	80	
	15.0	350	350	325	350	80	
	15.0	350	350	325	350	80	
	15.0	350	350	325	350	80	
	15.0	350	350	325	350	80	
	15.0	350	350	325	350	80	
	15.0	350	350	325	350	80	
	15.0	350	350	325	350	80	
	15.0	350	350	325	350	80	
	15.0	350	350	325	350	80	
	15.0	350	350	325	350	80	
	15.0	350	350	325	350	80	
	15.0	350	350	325	350	80	
	15.0	350	350	325	350	80	
	15.0	350	350	325	350	80	
	15.0	350	350	325	350	80	
	15.0	350	350	325	350	80	
	15.0	350	350	325	350	80	
	15.0	350	350	325	350	80	
	15.0	350	350	325	350	80	
	15.0	350	350	325	350	80	
	15.0	350	350	325	350	80	
	15.0	350	350	325	350	80	
	15.0	350	350	325	350	80	
	15.0	350	350	325	350	80	
	15.0	350	350	325	350	80	
	15.0	350	350	325	350	80	
	15.0	350	350	325	350	80	
	15.0	350	350	325	350	80	
	15.0	350	350	325	350	80	
	15.0	350	350	325	350	80	
	15.0	350	350	325	350	80	
	15.0	350	350	325	350	80	
	15.0	350	350	325	350	80	
	15.0	350	350	325	350	80	
	15.0	350	350	325	350	80	
	15.0	350	350	325	350	80	
	15.0	350	350	325	350	80	
	15.0	350	350	325	350	80	
	15.0	350	350	325	350	80	
	15.0	350	350	325	350	80	
	15.0	350	350	325	350	80	
	15.0	350	350	325	350	80	
	15.0	350	350	325	350	80	
	15.0	350	350	325	350	80	
	15.0	350	350	325	350	80	
	15.0	350	350	325	350	80	
	15.0	350	350	325	350	80	
	15.0	350	350	325	350	80	
	15.0	350	350	325	350	80	

Grade	Filler	Sp Grav	Shrink, mils/in	Melt flow, g/10 min	Melt temp, °F	Back pres, psi	Drying temp, °F
NA 836-154		0.92	15.0-45.0	6.00			
NA 850-000		0.92	15.0-45.0	2.00			
NA 853-000		0.92	15.0-45.0	2.00			
NA 860-000		0.92	15.0-45.0	24.00			
NA 940				0.25			
NA 940-000				0.25			
NA 940-085				0.25			
NA 940-094				0.25			
NA 942-085				0.18			
NA 942-094				0.18			
NA 951				2.16			
NA 951-000				2.30			
NA 951-050				2.30			
NA 951-055				2.30			
NA 951-232				2.30			
NA 952				2.00			
NA 952-000				2.00			
NA 952-094				2.00			
NA 952-095				2.00			
NA 952-096				2.00			
NA 957				2.60			
NA 957-000				2.60			
NA 957-061				2.60			
NA 957-097				2.60			
NA 960				0.90			
NA 960-000				0.90			
NA 960-050				0.90			
NA 960-055				0.90			
NA 960-057				0.90			
NA 960-062				0.90			
NA 960-086				0.90			
NA 960-186				0.90			
NA 961-000				1.20			
NA 961-062				1.20			
NA 961-074				1.20			
NA 962-000				0.93			
NA 962-062				0.93			
NA 962-187				0.93			
NA 963-083				0.70			
NA 964-085				1.10			
NA 966-000				0.85			
NA 980		0.92		0.25			
NA 980-063				0.25			
NA 980-081				0.25			
NA 985		0.92		0.25			
PE 130	CBL 3			0.20			
PE 190	CBL 3			0.25			
PM 92829	CBL 35			0.70			
YR 92866	CBL 3			0.48			

LLDPE Dowlex Dow

Grade	Filler	Sp Grav	Shrink, mils/in	Melt flow, g/10 min	Melt temp, °F	Back pres, psi	Drying temp, °F
2500		0.92		58.00 E			
2503		0.93		105.00 E			
2503.10		0.93		85.00 E			
2505		0.92		85.00 E			
2517		0.91		25.00 E			

Drying time, hr	Inj time, sec	Front temp, °F	Mid temp, °F	Rear temp, °F	Nozzle temp, °F	Mold temp, °F	Proc temp, °F
	15.0	350	350	325	350	80	
	15.0	350	350	325	350	80	
	15.0	350	350	325	350	80	
	15.0	350	350	325	350	80	
	15.0	350	350	325	350	80	
	15.0	350	350	325	350	80	
	15.0	350	350	325	350	80	
	15.0	350	350	325	350	80	
	15.0	350	350	325	350	80	
	15.0	350	350	325	350	80	
	15.0	350	350	325	350	80	
	15.0	350	350	325	350	80	
	15.0	350	350	325	350	80	
	15.0	350	350	325	350	80	
	15.0	350	350	325	350	80	
	15.0	350	350	325	350	80	
	15.0	350	350	325	350	80	
	15.0	350	350	325	350	80	
	15.0	350	350	325	350	80	
	15.0	350	350	325	350	80	
	15.0	350	350	325	350	80	
	15.0	350	350	325	350	80	
	15.0	350	350	325	350	80	
	15.0	350	350	325	350	80	
	15.0	350	350	325	350	80	
	15.0	350	350	325	350	80	
	15.0	350	350	325	350	80	
	15.0	350	350	325	350	80	
	15.0	350	350	325	350	80	
	15.0	350	350	325	350	80	
	15.0	350	350	325	350	80	
	15.0	350	350	325	350	80	
	15.0	350	350	325	350	80	
	15.0	350	350	325	350	80	
	15.0	350	350	325	350	80	
	15.0	350	350	325	350	80	
	15.0	350	350	325	350	80	
	15.0	350	350	325	350	80	
	15.0	350	350	325	350	80	
	15.0	350	350	325	350	80	
	15.0	350	350	325	350	80	
	15.0	350	350	325	350	80	
	15.0	350	350	325	350	80	
	15.0	350	350	325	350	80	
	15.0	350	350	325	350	80	
	15.0	350	350	325	350	80	
	15.0	350	350	325	350	80	
	15.0	350	350	325	350	80	
	15.0	350	350	325	350	80	
		375-550	375-550	375-550			
		375-550	375-550	375-550			
		375-550	375-550	375-550			
		375-550	375-550	375-550			
		375-550	375-550	375-550			

Grade	Filler	Sp Grav	Shrink, mils/in	Melt flow, g/10 min	Melt temp, °F	Back pres, psi	Drying temp, °F
2535		0.92		6.00 E			
2553							

LLDPE Novapol Novacor

Grade	Filler	Sp Grav	Shrink, mils/in	Melt flow, g/10 min	Melt temp, °F	Back pres, psi	Drying temp, °F
GI-2024-A				20.00 E			
GI-5026-A				50.00			

LLDPE Petrothene Quantum

Grade	Filler	Sp Grav	Shrink, mils/in	Melt flow, g/10 min	Melt temp, °F	Back pres, psi	Drying temp, °F
GA 501		0.92		1.00 E			
GA 501-010				1.00 E			
GA 501-011				1.00 E			
GA 501-012				1.00 E			
GA 501-013				1.00 E			
GA 502		0.92		2.00 E			
GA 502-010				2.00 E			
GA 502-012				2.00 E			
GA 502-013				2.00 E			
GA 502-021				2.00 E			
GA 502-024				2.00 E			
GA 502-025				2.00 E			
GA 564			15.0-40.0	20.00 E			
GA 568			15.0-45.0				
GA 568-000			15.0-45.0	32.00			
GA 574			15.0-35.0	50.00 E			
GA 578-000			15.0-35.0	80.00			
GA 584			15.0-45.0	100.00 E			
GA 594		0.93	15.0-45.0	135.00 E			
GA 594-000			15.0-45.0	135.00			
GA 601		0.92		1.00 E			
GA 604		0.91		2.50 E			
GA 605		0.92		0.80 E			
GA 625				5.00 E			
GA 635				6.50 E			
GA 643				3.50 E			
GA 700-750				6.00			
GA 808-090				0.70			
GA 808-093				0.70			
GA 818-070				0.70			
GA 818-073				0.70			
GA TR007				7.00			
GB 501-010				1.00 E			
GB 501-011				1.00 E			
GB 501-012				1.00 E			
GB 501-013				1.00 E			
GB 502-010				2.00 E			
GB 502-012				2.00 E			
GB 502-013				2.00 E			
GB 502-021				2.00 E			
GB 502-024				2.00 E			
GB 502-025				2.00 E			
GB 564			15.0-45.0	20.00 E			
GB 568		0.93	15.0-45.0	35.00 E			
GB 568-000			15.0-45.0	32.00			
GB 574			15.0-45.0	50.00 E			
PA 436				5.50 E			
PA 444				4.00 E			

Drying time, hr	Inj time, sec	Front temp, °F	Mid temp, °F	Rear temp, °F	Nozzle temp, °F	Mold temp, °F	Proc temp, °F
		375-550	375-550	375-550			
		375-550	375-550	375-550			
							320-450
							320-392
	15.0	375	375	350	375	80	
	15.0	375	375	350	375	80	
	15.0	375	375	350	375	80	
	15.0	375	375	350	375	80	
	15.0	375	375	350	375	80	
	15.0	375	375	350	375	80	
	15.0	375	375	350	375	80	
	15.0	375	375	350	375	80	
	15.0	375	375	350	375	80	
	15.0	375	375	350	375	80	
	15.0	375	375	350	375	80	
	15.0	375	375	350	375	80	
	15.0	375	375	350	375	80	
	15.0	375	375	350	375	80	
	15.0	375	375	350	375	80	
	15.0	375	375	350	375	80	
	15.0	375	375	350	375	80	
	15.0	375	375	350	375	80	
	15.0	375	375	350	375	80	
	15.0	375	375	350	375	80	
	15.0	375	375	350	375	80	
	15.0	375	375	350	375	80	
	15.0	375	375	350	375	80	
	15.0	375	375	350	375	80	
	15.0	375	375	350	375	80	
	15.0	375	375	350	375	80	
	15.0	375	375	350	375	80	
	15.0	375	375	350	375	80	
	15.0	375	375	350	375	80	
	15.0	375	375	350	375	80	
	15.0	375	375	350	375	80	
	15.0	375	375	350	375	80	
	15.0	375	375	350	375	80	
	15.0	375	375	350	375	80	
	15.0	375	375	350	375	80	
	15.0	375	375	350	375	80	
	15.0	375	375	350	375	80	
	15.0	375	375	350	375	80	
	15.0	375	375	350	375	80	
	15.0	375	375	350	375	80	
	15.0	375	375	350	375	80	
	15.0	375	375	350	375	80	
	15.0	375	375	350	375	80	
	15.0	375	375	350	375	80	
	15.0	375	375	350	375	80	
	15.0	375	375	350	375	80	
	15.0	375	375	350	375	80	
	15.0	375	375	350	375	80	
	15.0	375	375	350	375	80	
	15.0	375	375	350	375	80	
	15.0	375	375	350	375	80	

Grade	Filler	Sp Grav	Shrink, mils/in	Melt flow, g/10 min	Melt temp, °F	Back pres, psi	Drying temp, °F
PR 92710	CBL 3			0.61			
PR 92735	CBL 3			0.59			
PR 92813				0.63			
PR 92819				0.62			
XL 03512	CBL 25						
MBS Terpolymer		**Network Polymers**					
MS 150		1.05	4.0	3.70 G			180
MS 175		1.02	8.0	5.50 G			180
MBS Terpolymer		**Zylar**			**Novacor**		
90		1.12	4.0	1.50 N			
MBS Terpolymer		**Zylar ST**			**Novacor**		
94-560		1.05	4.0	8.40 N			
94-561		1.05	4.0	9.80 N			
94-562		1.05	4.0	10.80 N			
94-567		1.05	4.0	10.80 N			
94-568		1.05	4.0	6.00 N			
Mel Formaldehyd		**Cymel**			**Cytec**		
1077-1295	AC	1.50	8.0- 9.0				
1077-DXM	AC	1.50	8.0- 9.0				
1077-P	AC	1.50	8.0- 9.0				
1077-S	AC	1.50	8.0- 9.0				
1077-T	AC	1.50	8.0- 9.0				
Mel Formaldehyd		**Perstorp Melamine**			**Perstorp**		
751	AC	1.50	7.0				
752	AC	1.50	7.0				
755	AC	1.50	7.0				
756	AC	1.50	7.0				
791	AC	1.50	7.0				
792	AC	1.50	7.0				
795	AC	1.50	7.0				
796	AC	1.50	7.0				
Mel Formaldehyd		**PMC Melamine**			**Plastics Mfg.**		
MMC		1.49	8.0-13.0				
Mel Phenolic		**Plenco**			**Plenco**		
00755		1.60	6.0- 9.0				
00757	FLK	1.70	6.0-10.0				
Nylon		**Ashlene**			**Ashley**		
870		1.17	5.0			50- 150	150-160
Nylon		**Polyamide**			**Custom Resins**		
401		1.14			430		
404		1.14			430		
406		1.13					
409		1.13					
451		1.14			430		
454		1.14			430		

Drying time, hr	Inj time, sec	Front temp, °F	Mid temp, °F	Rear temp, °F	Nozzle temp, °F	Mold temp, °F	Proc temp, °F
	15.0	375	375	350	375	80	
	15.0	375	375	350	375	80	
	15.0	375	375	350	375	80	
	15.0	375	375	350	375	80	
	15.0	375	375	350	375	80	
2.0						130	430-460
2.0						130	430-460
		460	440	400		130	460
		385	375	365		95-115	410
		385	375	365		95-115	410
		385	375	365		95-115	410
		385	375	365		95-115	410
		385	375	365		95-115	410
	5.0- 12.	180-215		80-170	180-230	300-330	180-240
	5.0- 12.	180-215		80-170	180-230	300-330	180-240
	5.0- 12.	180-215		80-170	180-230	300-330	180-240
	5.0- 12.	180-215		80-170	180-230	300-330	180-240
	5.0- 12.	180-215		80-170	180-230	300-330	180-240
							310-335
							310-335
							310-335
							325-335
							310-335
							310-335
							310-335
							325-335
						300-335	
						300-360	230-245
						300-360	230-245
2.0- 3.0		554	550	536	530-560	150-200	570-610
						180-200	450-470
						180-200	450-470
						180-200	450-470
						180-200	450-470
						180-200	460-480
						180-200	460-480

Grade	Filler	Sp Grav	Shrink, mils/in	Melt flow, g/10 min	Melt temp, °F	Back pres, psi	Drying temp, °F
Nylon		**Electrafil**			**DSM**		
J-7/20/EC	CBL	1.36	2.0				165-220
M-1526/EC	CBL	1.19	13.0				165-220
Nylon		**Gapex AY**			**Ferro**		
RNP23	GLF 23	1.17	3.0			50- 100	200
RNP33	GLF 33	1.25	2.0			50- 100	200
RNP43	GLF 43	1.35	2.0			50- 100	200
Nylon		**Grivory**			**EMS**		
G21 6443		1.18	0.5		275		176
G355NZ		1.09	7.0		275		176
GV-2H	GLF 20	1.29	10.0		507		176
GV-4H	GLF 40	1.47	10.0		500		176
GV-5H	GLF 50	1.58	0.5		500		176
GV-6H	GLF 60	1.72	0.5		500		176
GVN-35H	GLF 35	1.39	10.0		498		176
XE3215		1.10	7.0		262		176
XE3290	GLF 35	1.36					176
XE3291	GLF 40	1.43					176
XE3309	GLF 50	1.56					176
XE3342	GLF 50	1.52					176
XE3343	GLF 60	1.65					176
XS1201	GLF 25	1.36			257		176
Nylon		**NSC**			**Thermofil**		
Esbrid LSG-440A	UNS 60	1.73	4.0			50	220
Esbrid NSG-240A	UNS 50	1.59	4.0			50	220
Esbrid NSG-440A	UNS 60	1.73	4.0			50	220
Esbrid NSG-730A	UNS 65	1.81	4.0			50	220
Nylon		**RTP Polymers**			**RTP**		
200H		1.08	18.0			25- 50	175
200H FR		1.27	22.0			50- 90	175
200H TFE 20	PTF	1.20	15.0			25- 50	175
201E	GLF 10	1.25	6.0			25- 50	
201H	GLF 10	1.15	12.0			25- 50	175
203E	GLF 20	1.32	5.0			25- 50	
203H	GLF 20	1.23	5.0			25- 50	175
205E	GLF 30	1.41	4.0			25- 50	
205H	GLF 30	1.32	4.0			25- 50	175
205H TFE 15	GLF	1.43	2.0			25- 50	175
207E	GLF 40	1.50	2.0			25- 50	
207H	GLF 40	1.41	3.0			25- 50	175
Nylon		**Selar PA**			**Du Pont**		
3426		1.19					175-205
Nylon		**Zytel**			**Du Pont**		
330		1.80	5.0				175
450 HSL BK-152		1.08			491		175
FR-60	MNF	1.65	5.0- 8.0		486		175
Nylon 11		**CTI Nylon**			**CTI**		
NH-20GF	GLF 20	1.19	4.0- 6.0				200

Drying time, hr	Inj time, sec	Front temp, °F	Mid temp, °F	Rear temp, °F	Nozzle temp, °F	Mold temp, °F	Proc temp, °F
2.0-16.0		500-540	520-550	490-520	480-550	140-180	480-530
2.0-16.0		500-540	520-550	490-520	480-550	140-180	480-530
6.0		555-565	550-565	550-565	565-575	180-220	
6.0		555-565	550-565	550-565	565-575	180-220	
6.0		555-565	550-565	550-565	565-575	180-220	
6.0-10.0		536	518	500		176	536-554
6.0-10.0		520	510	500	500	176	518-554
6.0-10.0		572	554	545		176-212	518-590
6.0-10.0		572	554	545		176-212	518-590
6.0-10.0		572	554	545		176-212	518-590
6.0-10.0		572	554	545		176-212	518-590
6.0-10.0		572	554	545		176-212	518-590
6.0-10.0		520	510	500		176	518-554
4.0- 6.0						140-145	500-535
4.0- 6.0						140-145	500-535
4.0- 6.0						140-145	500-535
4.0- 6.0						140-145	500-535
4.0- 6.0						140-145	500-535
4.0- 6.0		510	500	490	500	176	510
3.0		530-550	520-540	500-520	540-560	150-230	
3.0		500-550	490-530	470-510	510-560	150-200	
3.0		500-550	490-530	470-510	510-560	150-200	
3.0		500-550	490-530	470-510	510-560	150-200	
4.0		550-575	535-565	525-550		150-225	
4.0		530-560	520-550	510-540		150-210	
4.0		550-575	535-565	525-550		150-225	
		520-560	510-550	500-540		190-250	
4.0		550-575	535-565	525-550		150-225	
		520-560	510-550	500-540		190-250	
4.0		550-575	535-565	525-550		150-225	
		520-560	510-550	500-540		190-250	
4.0		550-575	535-565	525-550		150-225	
4.0		550-575	535-565	525-550		150-225	
		520-560	510-550	500-540		190-250	
4.0		550-575	535-565	525-550		150-225	
4.0		572	554	536	536	158-200	554-608
2.0-24.0		550	555	535	530-560	150-200	570-610
2.0-24.0		525	535	560	500-570	100-200	550-560
2.0-24.0		540-550	550-560	560-570	560-580	150-210	560-580
						200	490-510

Grade	Filler	Sp Grav	Shrink, mils/in	Melt flow, g/10 min	Melt temp, °F	Back pres, psi	Drying temp, °F
NH-30GF	GLF 30	1.24	3.0- 4.0				200
NH-40GF	GLF 40	1.38	2.0- 3.0				200
Nylon 11		**Rilsan**			**Atochem**		
BMNO P40		1.05			367		150-175
Nylon 11		**RTP Polymers**			**RTP**		
200C		1.04	15.0			25- 50	175
200C TFE 20	GLF	1.17	11.0			25- 50	175
201C	GLF 10	1.11	7.0			25- 50	175
203C	GLF 20	1.18	6.0			25- 50	175
203C TFE 20	GLF	1.35	3.0			25- 50	175
205C	GLF 30	1.28	5.0			25- 50	175
207C	GLF 40	1.38	3.0			25- 50	175
Nylon 12		**CTI Amorphous Nylon**				**CTI**	
AN-20GF	GLF 20	1.26	5.0				240
AN-30CF	CF 30	1.30	0.9- 1.0				150
AN-30GF	GLF 30	1.36	4.0				240
AN-40GF	GLF 30	1.44	3.0				
Nylon 12		**CTI Nylon**			**CTI**		
NJ-30GF	GLF 30	1.24	4.5				200
Nylon 12		**Grilamid**			**EMS**		
L16GM		1.01	7.0		352		158-176
L20GHS		1.01	8.0		352		158-176
L20GNZ/100		1.00	9.0		352		158-176
L20GNZ/300		0.98	9.0		352		158-176
L20GNZ/600		0.97	9.0		352		158-176
LV-15H	GLF 15	1.11	2.0		352		158-176
LV-23H	GLF 23	1.17	1.0		352		158-176
LV-3H	GLF 30	1.22	1.0		352		158-176
LV-43H	GLF 43	1.36	1.0		352		158-176
TR55		1.06	7.0	25.00 J			158-176
TR55FC		1.04	6.0				158-176
TR55LX		1.04	6.0				158-176
TR55LY		1.04	6.0				158-176
TR55UV		1.06	7.0				158-176
Nylon 12		**Rilsan**			**Atochem**		
AMNO		1.02			345		150-175
AZMO 30	GLF 30	1.23			345		150-175
Nylon 12		**RTP Polymers**			**RTP**		
205F	GLF 30	1.24	5.0			25- 50	175
207F	GLF 40	1.34	4.0			25- 50	175
Nylon 12		**UBE Nylon**			**UBE**		
3014U		1.02	10.0-16.0		349-356	1421	
3020LU1		1.02	10.0-16.0		349-356	1421	
3024LU		1.02	10.0-16.0		349-356	1421	
3024U		1.02	10.0-16.0		349-356	1421	
3035JU6		1.03	7.0-18.0		331-338	1421	

Drying time, hr	Inj time, sec	Front temp, °F	Mid temp, °F	Rear temp, °F	Nozzle temp, °F	Mold temp, °F	Proc temp, °F
						200	490-530
						200	490-530
3.0- 4.0		430	410	390	430	80-150	
4.0		435-530	425-520	415-510		100-150	
4.0		435-530	425-520	415-510		100-150	
4.0		435-530	425-520	415-510		100-150	
4.0		435-530	425-520	415-510		100-150	
4.0		435-530	425-520	415-510		100-150	
4.0		435-530	425-520	415-510		100-150	
4.0		435-530	425-520	415-510		100-150	
						225-250	580-600
						150-200	570-630
						225-250	580-600
						225-250	580-600
						200	500-540
6.0-10.0		445	430	410	440	105	455
6.0-10.0		445	430	410	440	105	455
6.0-10.0		465	445	430	445	105	475
6.0-10.0		465	445	430	445	105	475
6.0-10.0		465	445	430	445	105	475
6.0-10.0		520	500	490	500	140	465-570
6.0-10.0		520	500	490	500	175	465-570
6.0-10.0		520	500	490	500	175	465-570
6.0-10.0		530	505	495	500	175	465-570
6.0-10.0		500	480	465	480	175	500-580
6.0-10.0		545	520	480	530	100	500-580
6.0-10.0		480	475	465	475	100	464-535
6.0-10.0		480	475	465	475	100	464-535
6.0-10.0		555	545	520		180	545
3.0- 4.0		430	410	390	430	80-150	
3.0- 4.0		475	450	425	475	80-150	
4.0		440-510	430-500	420-490		150-220	
4.0		440-510	430-500	420-490		150-220	
	15.0	428	428	374	428	158-176	455
	15.0	428	428	374	428	158-176	455
	15.0	428	428	374	428	158-176	455
	15.0	428	428	374	428	158-176	455
	15.0	428	428	374	428	158-176	455

Grade	Filler	Sp Grav	Shrink, mils/in	Melt flow, g/10 min	Melt temp, °F	Back pres, psi	Drying temp, °F
Nylon 12 Elast.		**Grilamid**			**EMS**		
ELY20NZ		0.99	8.0		338		158-176
ELY2702		1.02	5.0		327		158-176
ELY2742		1.01	5.0		307		158-176
ELY60		1.01	4.0		320		158-176
Nylon 12/12		**Zytel**			**Du Pont**		
CFE3535HL		1.02	10.0-12.0		363		
CFE3536HL		1.02	10.0-12.0		363		
CFE5012H	GL 23	1.14			363		
CFE5013H	GL 33	1.24			363		
Nylon 4/6		**Nylatron**			**DSM**		
GSU	MOS	1.15			563		185
GSU-51	GLF 46	1.40			563		185
Nylon 4/6		**Stanyl**			**DSM**		
NCU002		1.21	16.0-20.0		563		220
NCU012	GLF 30	1.42	4.0- 6.0		563		220
TE200F6	GLF 30	1.41	4.0- 6.0		563		220
TE250F3	GLF 15	1.47	6.0- 9.0		563		220
TE250F6	GLF 30	1.68	4.0- 6.0		563		220
TE300		1.18	18.0-20.0		563		220
TE341		1.18	7.0-20.0		563		220
TE350		1.38	18.0-20.0		563		220
TW200F3	GLF 15	1.30	5.0- 9.0		563		220
TW200F6	GLF 30	1.41	4.0- 6.0		563		220
TW200F8	GLF 40	1.51	4.0-19.0		563		220
TW200K8	MN 40	1.55	13.0-20.0		563		220
TW241F10	GLF 50	1.62	2.0-14.0		563		220
TW242FM10	GMN 50	1.60	2.0-16.0		563		220
TW250F6	GLF 30	1.68	2.0-17.0		563		220
TW271F6	GLF 30	1.54	4.0- 6.0		563		220
TW275F6	GLF 30	1.40	4.0- 6.0		563		220
TW300		1.18	18.0-20.0		563		220
TW341		1.18	18.0-20.0		563		220
TW350		1.38	6.0-18.0		563		220
TW363		1.10	18.0-20.0		563		220
TW371		1.31	15.0-20.0		563		220
TW373		1.21	18.0-20.0		563		220
TW400		1.18	15.0-24.0		563		220
TW441		1.18	15.0-24.0		563		220
Nylon 6		**Akulon**			**DSM**		
K222-D		1.13	8.3				220
K224-TG0	GLF 50	1.56	6.0			0- 50	230
K225-KS		1.16	1.1				220
Nylon 6		**Albis Polyamid 6**			**Albis Canada**		
PA45/1		1.13			423-428		170-180
PA45/1GF25	GLF 25	1.32			423-428		170-180
PA45/1GF30	GLF 30	1.36			423-428		170-180
PA45/1GF35	GLF 35	1.41			423-428		170-180
PA45/2GF30	GLF 30	1.36			423-428		170-180
PA45/3FS		1.17			423-428		170-180

Drying time, hr	Inj time, sec	Front temp, °F	Mid temp, °F	Rear temp, °F	Nozzle temp, °F	Mold temp, °F	Proc temp, °F
6.0-10.0		385	375	365		70	420
6.0-10.0		385	375	365		70	420
6.0-10.0		375	365	355		70	410
6.0-10.0		375	365	355		70	410
		430-536	430-536	430-536	430-536	80-120	480-520
		430-536	430-536	430-536	430-536	80-120	480-520
						185	500
						185	500
2.0-12.0		570-600	560-600	540-600	580-590	180-300	580-590
2.0-12.0		570-600	560-600	540-600	580-590	180-300	580-590
2.0- 4.0		570-600	560-600	540-600	580-590	180-300	580-590
2.0- 4.0		570-600	560-600	540-600	580-590	180-300	580-590
2.0- 4.0		570-600	560-600	540-600	580-590	180-300	580-590
2.0- 4.0		570-600	560-600	540-600	580-590	180-300	580-590
2.0- 4.0		570-600	560-600	540-600	580-590	180-300	580-590
2.0- 4.0		570-600	560-600	540-600	580-590	180-300	580-590
2.0- 4.0		570-600	560-600	540-600	580-590	180-300	580-590
2.0- 4.0		570-600	560-600	540-600	580-590	180-300	580-590
2.0- 4.0		570-600	560-600	540-600	580-590	180-300	580-590
2.0- 4.0		570-600	560-600	540-600	580-590	180-300	580-590
2.0- 4.0		570-600	560-600	540-600	580-590	180-300	580-590
2.0- 4.0		570-600	560-600	540-600	580-590	180-300	580-590
2.0- 4.0		570-600	560-600	540-600	580-590	180-300	580-590
2.0- 4.0		570-600	560-600	540-600	530-590	180-300	580-590
2.0- 4.0		570-600	560-600	540-600	580-590	180-300	580-590
2.0- 4.0		570-600	560-600	540-600	580-590	180-300	580-590
2.0- 4.0		570-600	560-600	540-600	580-590	180-300	580-590
2.0- 4.0		570-600	560-600	540-600	580-590	180-300	580-590
2.0- 4.0		570-600	560-600	540-600	580-590	180-300	580-590
2.0- 4.0		570-600	560-600	540-600	580-590	180-300	580-590
2.0- 4.0		570-600	560-600	540-600	580-590	180-300	580-590
2.0- 4.0		570-600	560-600	540-600	580-590	180-300	580-590
2.0- 4.0		570-600	560-600	540-600	580-590	180-300	580-590
2.0- 4.0		570-600	560-600	540-600	580-590	180-300	580-590
2.0- 4.0		570-600	560-600	540-600	580-590	180-300	580-590
2.0- 4.0		460-470	450-460	440-450	430-450	175-190	470-500
3.0		500	490	480	510	175-195	500-555
2.0- 4.0		460-470	450-460	440-450	430-450	175-190	470-500
2.0						160-195	470-520
2.0						175-250	500-535
2.0						175-250	500-535
2.0						175-250	500-535
2.0						175-250	500-535
2.0						160-195	470-520

Grade	Filler	Sp Grav	Shrink, mils/in	Melt flow, g/10 min	Melt temp, °F	Back pres, psi	Drying temp, °F
PA50/1MR30	MN 30	1.35			423-428		170-180
PA50/1MR40	MN 40	1.48			423-428		170-180
PA55 WM		1.11			423-428		170-180
PA55/2		1.13			423-428		170-180
PA55/5TSZ		1.11			423-428		170-180

Nylon 6 Ashlene Ashley

Grade	Filler	Sp Grav	Shrink, mils/in	Melt flow, g/10 min	Melt temp, °F	Back pres, psi	Drying temp, °F
528		1.13	15.0		491	50- 150	150-160
528L		1.13	15.0		491	50- 150	150-160
528LB-2	CBL	1.13	15.0		491	50- 150	150-160
528LS		1.13	15.0		491	50- 150	150-160
630		1.13	14.0		410	50- 150	150-160
630-13G	GLC 13	1.21	5.0		420	50- 150	150-160
630-33G	GLC 33	1.37	4.0		420	50- 150	150-160
630B		1.13	12.0		410	50- 150	150-160
630BU		1.13	13.0		380	50- 150	150-160
630U		1.13	14.0		379	50- 150	150-160
733LD		1.07	1.3		410	50- 150	150-160
734 D		1.10	13.0		410	50- 150	150-160
734LD		1.10	15.0		410	50- 150	150-160
735		1.11	14.0		420	50- 150	150-160
735LD		1.11	14.0		420	50- 150	150-160
736LD		1.09	14.0		420	50- 150	150-160
73M	MN	1.45			415	50- 150	150-160
79MGS	GMN	1.48	4.0		425	50- 150	150-160
830		1.13	12.0		420	50- 150	150-160
830L		1.13	12.0		420	50- 150	150-160
830L-13G	GLF 13	1.23	6.0		420	50- 150	150-160
830L-30G	GLF 30	1.36	3.0		420	50- 150	150-160
830L-33G	GLF 33	1.37	4.0		420	50- 150	150-160
830L-50G	GLF 50	1.56	2.0		420	50- 150	150-160
830LS-30G	GLF 30	1.36	3.0		420	50- 150	150-160
830LW		1.13	12.0		420	50- 150	150-160
835		1.12			420	50- 150	150-160

Nylon 6 Bay Resins

Grade	Filler	Sp Grav	Shrink, mils/in	Melt flow, g/10 min	Melt temp, °F	Back pres, psi	Drying temp, °F
PA-211		1.14	12.0		420	50- 100	180
PA-211CF30	CF 30	1.28	2.0			50- 100	180
PA-211CF30 TF15	CF 30	1.37	2.5			50- 100	180
PA-211G13	GLF 13	1.22			420	50- 100	180
PA-211G33	GLF 33	1.37			420	50- 100	180
PA-211GF30 TF15	GLF 30	1.49	4.0			50- 100	180
PA-211M40P	MN 40	1.51	10.0			50- 100	180
PA-211N40	GLF 15	1.49	6.0			50- 100	180
PA-211TF20		1.26	13.0			50- 100	180
PA-211X012		1.08	13.0			50- 100	180
PA-211X032		1.11	14.0			50- 100	180
PA-212		1.14	9.0		420	50- 100	180
PA-213		1.14	12.0			50- 100	180
PA-213G13	GLF 13	1.22	5.5			50- 100	180
PA-213G33	GLF 33	1.37	3.5			50- 100	180
PA-213M40P	MN 40	1.47			420	50- 100	180
PA-213N40	GLF 15	1.45			420	50- 100	180

Nylon 6 Capron Allied Signal

Grade	Filler	Sp Grav	Shrink, mils/in	Melt flow, g/10 min	Melt temp, °F	Back pres, psi	Drying temp, °F
8200		1.13	12.0		420		180

Drying time, hr	Inj time, sec	Front temp, °F	Mid temp, °F	Rear temp, °F	Nozzle temp, °F	Mold temp, °F	Proc temp, °F
2.0						175-250	500-535
2.0						175-250	500-535
2.0						160-195	470-520
2.0						160-195	470-520
2.0						160-195	470-520
2.0- 3.0		520	525	540	500-570	100-200	535-580
2.0- 3.0		520	525	540	500-570	100-200	535-580
2.0- 3.0		520	525	540	500-570	100-200	535-580
2.0- 3.0		520	525	540	500-570	100-200	535-580
2.0- 3.0		460-520	440-500	430-470	450-500	80-200	460-520
2.0- 3.0		480-540	460-520	440-500	480-540	150-230	490-540
2.0- 3.0		520-580	500-560	480-520	520-580	150-230	520-580
2.0- 3.0		460-520	440-500	430-470	450-500	80-200	460-520
2.0- 3.0		460-520	440-500	430-470	450-500	80-200	460-520
2.0- 3.0		460-520	440-500	430-470	450-500	80-200	460-520
2.0- 3.0		480-540	460-520	440-500	480-540	80-200	480-540
2.0- 3.0		480-540	460-520	440-500	480-540	80-200	480-540
2.0- 3.0		480-540	460-520	440-500	480-540	80-200	480-540
2.0- 3.0		480-540	460-520	440-500	480-540	80-200	480-540
2.0- 3.0		480-540	460-520	440-500	480-540	80-200	480-540
2.0- 3.0		480-540	460-520	440-500	480-540	80-200	480-540
2.0- 3.0		520-580	500-560	480-520	520-580	150-230	520-580
2.0- 3.0		520-560	500-540	480-520	520-560	150-230	520-560
2.0- 3.0		460-520	440-500	430-470	450-500	80-200	460-520
2.0- 3.0		460-520	440-500	430-470	450-500	80-200	460-520
2.0- 3.0		520-580	500-560	480-520	520-580	150-230	520-580
2.0- 3.0		520-580	500-560	480-520	520-580	150-230	520-580
2.0- 3.0		520-580	500-560	480-520	520-580	150-230	520-580
2.0- 3.0		520-580	500-560	480-520	520-580	150-230	520-580
2.0- 3.0		520-580	500-560	480-520	520-580	150-230	520-580
2.0- 3.0		460-520	440-500	430-470	450-500	80-200	460-520
2.0- 3.0		460-520	440-500	430-470	450-500	80-200	460-520
2.0- 4.0		480-525	480-525	480-525		150-200	490-520
2.0- 4.0		480-525	480-525	480-525		150-200	490-520
2.0- 4.0		480-525	480-525	480-525		150-200	490-520
2.0- 4.0		480-525	480-525	480-525		150-200	490-520
2.0- 4.0		480-525	480-525	480-525		150-200	490-520
2.0- 4.0		480-525	480-525	480-525		150-200	490-520
2.0- 4.0		480-525	480-525	480-525		150-200	490-520
2.0- 4.0		480-525	480-525	480-525		150-200	490-520
2.0- 4.0		480-525	480-525	480-525		150-200	490-520
2.0- 4.0		480-525	480-525	480-525		150-200	490-520
2.0- 4.0		480-525	480-525	480-525		150-200	490-520
2.0- 4.0		480-525	480-525	480-525		150-200	490-520
2.0- 4.0		480-525	480-525	480-525		150-200	490-520
2.0- 4.0		480-525	480-525	480-525		150-200	490-520
2.0- 4.0		480-525	480-525	480-525		150-200	490-520
2.0- 4.0		480-525	480-525	480-525		150-200	490-520
2.0- 4.0		480-525	480-525	480-525		150-200	490-520
		480-520	460-500	440-480	450-500	180-200	480-520

Grade	Filler	Sp Grav	Shrink, mils/in	Melt flow, g/10 min	Melt temp, °F	Back pres, psi	Drying temp, °F
8200 BK-102		1.13	12.0		420		180
8200 HS		1.13	12.0		420		180
8200 HS BK-102		1.13	12.0		420		180
8202		1.13	10.0-12.0		420		180
8202 BK-102		1.13	12.0		420		180
8202 BK-106		1.13	12.0		420		180
8202 HS		1.13	12.0		420		180
8202 HS BK-102		1.13	12.0		420		180
8202 HS BK-106		1.13	12.0		420		180
8202 NL		1.13	12.0		420		180
8202C		1.13	9.0		420		180
8202C BK-102		1.13	10.0		420		180
8202C HS		1.13	9.0		420		180
8202C HS BK-102		1.13	10.0		420		180
8202CQ		1.13	10.0		420		180
8202QC		1.13	10.0		420		180
8224 HS		1.13			420		
8230G HS	GLF 6	1.16	8.0		420		180
8230G HS BK-102	GLF 6	1.16	8.0		420		180
8231G HS	GLF 14	1.23	5.0		420		180
8231G HS BK-102	GLF 14	1.23	5.0		420		180
8232G HS	GLF 25	1.49	2.0		420		180
8232G HS FR	GLF 25	1.62	3.0		420		180
8232G HS FR BK-102	GLF 25	1.62	3.0				180
8233G	GLF 33	1.38	3.0		420		180
8233G HS	GLF 33	1.38	3.0		420		180
8233G HS BK-102	GLF 33	1.38	3.0		420		180
8233G HS BK-106	GLF 33	1.38	3.0- 6.0		420		180
8233G HS BK-125	GLF 33	1.38	3.0- 6.0		420		180
8234G HS	GLF 44	1.49	2.0		420		180
8234G HS BK-102	GLF 44	1.49	2.0- 4.0		420		180
8234G HS BK-106	GLF 44	1.49	2.0		420		180
8235G HS	GLF 50	1.56	2.0		420		180
8235G HS BK-102	GLF 50	1.56			420		180
8253		1.09	12.0		420		180
8259		1.09	13.0		420		180
8260	MN 40	1.49	8.0-11.0		420		180
8260 HS	MN 40	1.49	9.0		420		180
8260 HS BK-102	MN 40	1.49	8.0-11.0		420		180
8260 HS BK-104	MN 40	1.49	8.0-11.0		420		180
8266 HS BK-102	GMN 40	1.48	4.0		420		180
8266G HS	GMN 40	1.48	4.0		420		180
8267G HS (Dry)	GMN 40	1.48	4.0		420		180
8267G HS BK-102	GMN 40	1.48	4.0		420		180
8267G HS BK-106	GMN 40	1.48	4.0		420		180
8351		1.07	14.0		420		180
8351 HS		1.07	14.0		420		180
8351 HS BK-102		1.07			420		180
8351 HS BK-106		1.07			420		180
8352 HS		1.07	14.0		420		180
8360	MN 34	1.43	9.0-12.0		420		180
8360 HS	MN 34	1.43	10.0		420		180
8360 HS BK-102	MN 34	1.40	8.0-11.0		420		
8362	MN 34	1.41	9.0-12.0		420		180
8362 HS	MN 34	1.41	10.0		420		180
8362 HS BK-102	MN 34	1.41			420		180

Drying time, hr	Inj time, sec	Front temp, °F	Mid temp, °F	Rear temp, °F	Nozzle temp, °F	Mold temp, °F	Proc temp, °F
		480-520	460-500	440-480	480-520	180	480-520
		480-520	460-500	440-480	480-520	180	480-520
		480-520	460-500	440-480	480-520	180	480-520
		460-520	440-500	430-470	460-520	180-200	460-520
		460-520	440-500	430-470	460-520	180	460-520
		460-520	440-500	430-470	460-520	180	460-520
		460-520	440-500	430-470	460-520	180	460-520
		460-520	440-500	430-470	460-520	180	460-520
		460-520	440-500	430-470	460-520	180	460-520
		460-520	440-500	430-470	460-520	180	460-520
		460-520	440-500	430-470	450-500	180-200	460-520
		460-520	440-500	430-470	460-520	180	460-520
		460-520	440-500	430-470	460-520	180	460-520
		460-520	440-500	430-470	460-520	180	460-520
		460-520	440-500	430-470	460-520	180	460-520
		460-520	440-500	430-470	460-520	180-200	460-520
		480-530	460-510	440-480	480-530	180-220	490-530
		480-530	460-510	440-480	480-530	180-200	490-530
		500-540	480-520	460-500	500-540	180-220	500-540
		500-540	480-520	460-500	500-540	180-200	500-540
		540-580	520-560	500-540	540-580	180-220	540-580
		540-580	520-560	500-540	540-580	180-220	540-580
		540-580	520-560	500-540	540-580	180-200	540-580
		520-560	500-540	480-520	520-560	180-200	520-560
		520-560	500-540	480-520	520-560	180-220	520-560
		520-560	500-540	480-520	520-560	180-200	520-560
		520-560	500-540	480-520	520-560	180-200	520-560
		520-560	500-540	480-520	520-560	180-200	520-560
		520-580	500-560	480-520	520-580	180-220	520-580
		520-560	500-540	480-520	520-560	180-200	520-560
		520-560	500-540	480-520	520-560	180-200	520-560
		520-580	500-560	480-520	520-580	180-200	520-580
		520-580	500-560	480-520	520-580	180-200	520-580
		480-540	460-520	440-500	480-540	180-200	480-540
		480-540	460-520	440-500	480-540	180-200	480-540
		520-570	500-550	460-500	520-570	180-200	520-570
		520-570	500-550	460-500	520-570	180-200	520-570
		520-570	500-550	460-500	520-570	180-200	520-570
		520-570	500-550	460-500	520-570	180-200	520-570
		520-560	500-540	480-520	520-560	180-200	520-560
		520-560	500-540	480-520	520-560	180-220	520-560
		520-560	500-540	480-520	520-560	180-220	520-560
		520-560	500-540	480-520	520-560	180-200	520-560
		520-560	500-540	480-520	520-560	180-200	520-560
		480-540	460-520	440-500	480-540	180-200	480-540
		480-540	460-520	440-500	480-540	180	480-540
		480-540	460-520	440-500	400 510	180	480-540
		480-540	460-520	440-500	480-540	180	480-540
		480-540	460-520	440-500	480-540	180	480-540
		540-580	520-540	500-520	540-580	180-200	540-580
		540-580	520-540	500-520	540-580	180-200	540-580
		540-580	520-540	500-520	540-580	180-200	540-580
		540-580	520-540	500-520	540-580	180-200	540-580
		540-580	520-540	500-520	540-580	180-200	540-580
		540-580	520-540	500-520	540-580	180-200	540-580

Grade	Filler	Sp Grav	Shrink, mils/in	Melt flow, g/10 min	Melt temp, °F	Back pres, psi	Drying temp, °F
D-1426G HS	GMN 55	1.63			420		180
D-8200CQ BLEND		1.13	10.0		420		180
D-8202CQ BK-102		1.13	10.0		420		180
D-8232G HS	GLF 25	1.32			420		180
D-8331G HS BLEND	GLF 14	1.19			420		180
D-8332G	GLF 25	1.36	4.0		420		180
D-8332G BK-5548	GLF 25	1.36	4.0		420		180
D-8333G	GLF 33	1.36			420		180
D-8333G HS	GL 33	1.36			420		180
D-8333G HS BK-102	GLF 33	1.36			420		180
D-8334G HS BK-102	GLF 40	1.47			420		180
D-8352		1.08			420		180
D-8352 HS		1.07			420		180
D-8352 HS BK-102		1.07			420		180
Nylon 6		**Celanese Nylon 6**			**Hoechst**		
2400		1.13			419	0- 50	180
2800		1.13			419	0- 50	180
2800 LU		1.13			419	0- 50	180
3100		1.13			419	0- 50	180
3100 AM		1.13			419	0- 50	180
3100 HSLU		1.13			419	0- 50	180
3100 LU		1.13			419	0- 50	180
3103 MoS2		1.18			419	0- 50	180
3115 FV	GLF 15	1.24			419	0- 50	180
3130 FV	GLF 30	1.34			419	0- 50	180
3140 FV	GLF 40	1.42			419	0- 50	180
Nylon 6		**Celstran**			**Hoechst**		
N6G30-01-4	GLF 30	1.36					175
N6G40-01-4	GLF 40	1.45					175
N6G50-01-4	GLF 50	1.56					175
N6G60-01-4	GLF 60	1.69					175
Nylon 6		**Chem Polymer Nylon**			**Chem Polymer**		
NP 212		1.12	15.0		420		150-180
NP 214GF	GLF 14	1.23	7.0		420		150-180
NP 220GF	GLF 20	1.33	5.0		420		150-180
NP 223		1.12	12.0		420		150-180
NP 225-15MFGF	GLF 15	1.48	6.0		420		150-180
NP 233GF	GLF 30	1.37	3.5		420		150-180
NP 253		1.10	14.0		410		150-180
NP 280		1.08	16.0		428		150-180
Nylon 6		**Comtuf**			**ComAlloy**		
605		1.08	15.0				180
610	UNS	1.46	3.0				180
611	GL 13	1.18	6.0				180
613	GL 33	1.34	4.0				180
615	GL 50	1.51	2.0				180
616	GMN 40	1.41	4.0				180
618	MN	1.27	14.0				180
619	MN 40	1.41	12.0				180

Drying time, hr	Inj time, sec	Front temp, °F	Mid temp, °F	Rear temp, °F	Nozzle temp, °F	Mold temp, °F	Proc temp, °F
		520-580	500-560	480-540	520-580	180-200	520-580
		480-520	460-500	440-480	480-520	180	480-520
		460-520	440-500	430-470	460-520	180	460-520
		520-560	500-540	480-520	520-560	180-200	520-560
		500-540	480-520	460-500	500-540	180-200	500-540
		520-560	500-540	480-520	520-560	180-200	520-560
		520-560	500-540	480-520	520-560	180-200	520-560
		520-560	500-540	480-520	520-560	180-200	520-560
		520-560	500-540	480-520	520-560	180-200	520-560
		520-560	500-540	480-520	520-560	180-200	520-560
		500-540	480-520	460-500	500-540	180-200	520-540
		480-540	460-520	440-500	480-540	180-200	480-540
		480-540	460-520	440-500	480-540	480	480-540
		480-540	460-520	440-500	480-540	180	480-540
3.0- 4.0		460-510	440-490	420-480	460-510	70-200	450-520
3.0- 4.0		460-510	440-490	420-480	460-510	70-200	450-520
3.0- 4.0		460-510	440-490	420-480	460-510	70-200	450-520
3.0- 4.0		460-510	440-490	420-480	460-510	70-200	450-520
3.0- 4.0		460-510	440-490	420-480	460-510	70-200	450-520
3.0- 4.0		460-510	440-490	420-480	460-510	70-200	450-520
3.0- 4.0		460-510	440-490	420-480	460-510	70-200	450-520
3.0- 4.0		460-510	440-490	420-480	460-510	70-200	450-520
3.0- 4.0		500-560	490-540	470-520	500-560	150-200	490-560
3.0- 4.0		500-560	490-540	470-520	500-560	150-200	490-560
3.0- 4.0		500-560	490-540	470-520	500-560	150-200	490-560
4.0		470-490	460-480	450-470	460-480	190-210	470-490
4.0		480-500	470-490	460-480	470-490	190-210	480-500
4.0		490-510	480-500	470-490	480-500	190-210	490-510
4.0		500-520	490-510	480-500	490-510	190-210	500-520
		440	450	470	450	70-200	
		480	510	520	480	140-200	
		480	510	520	480	140-200	
		440	450	470	450	70-200	
		480	510	520	480	140-200	
		480	510	520	480	140-200	
		440	450	470	450	70-200	
		440	450	470	450	70-200	
3.0- 4.0		460-500	470-510	480-520	460-500	100-200	470-510
3.0- 4.0		460-500	470-510	480-520	460-500	100-200	470-510
3.0- 4.0		460-500	470-510	480-520	460-500	100-200	470-510
3.0- 4.0		460-500	470-510	480-520	460-500	100-200	470-510
3.0- 4.0		460-500	470-510	480-520	460-500	100-200	470-510
3.0- 4.0		460-500	470-510	480-520	460-500	100-200	470-510
3.0- 4.0		460-500	470-510	480-520	460-500	100-200	470-510
3.0- 4.0		460-500	470-510	480-520	460-500	100-200	470-510

Grade	Filler	Sp Grav	Shrink, mils/in	Melt flow, g/10 min	Melt temp, °F	Back pres, psi	Drying temp, °F
Nylon 6		**CTI General**			**CTI**		
CTX-312	GLF 10	1.15	4.0- 4.5				180
Nylon 6		**CTI Nylon**			**CTI**		
NY		1.14	14.0				
NY-10GF	GLF 10	1.22	5.0- 7.0				180
NY-20GF	GLF 20	1.29	4.0- 6.0				180
NY-20GF-10GB	GLF 20	1.32	2.0- 4.0				180
NY-30CF	CF 30	1.27	1.0- 3.0				180
NY-30GF	GLF 30	1.38	3.0- 5.0				180
NY-30GF/000	GLF 30	1.36	1.5- 2.0				180
NY-33GF/000	GLF 33	1.39	3.0				180
NY-40GB	GLB 40	1.46	8.0-12.0				180
NY-40GF	GLF 40	1.46	2.0- 4.0				180
NY-50GF	GLF 50	1.56	2.0- 3.0				180
NY-60GF	GLF 50	1.70	2.0				180
Nylon 6		**Dimension**			**Allied Signal**		
D-9130G BK	GLF 30	1.33					230
Nylon 6		**Durethan**			**Miles**		
B 30 S		1.14	11.0-13.0			50- 150	170-180
B 31 SK		1.14	11.0-13.0			50- 150	170-180
B 40 SK		1.14	12.0-13.0			50- 150	170-180
BC 30		1.10	13.0-17.0			50- 150	170-180
BC 303		1.07	16.0			50- 150	176
BC 40		1.10	15.0-19.0			50- 150	170-180
BC 402		1.08	16.0			50- 150	170-180
BG 30 X	GLF	1.35	7.0- 8.0			50- 150	170-180
BKV 115	GLF 15	1.23	3.0- 9.0			50- 150	170-180
BKV 130	GLF 30	1.36	3.0- 9.0			50- 150	170-180
BKV 140	GLF 40	1.46	3.0- 8.0			50- 150	170-180
BKV 15 H	GLF 15	1.23	3.0- 9.0			50- 150	170-180
BKV 30 H	GLF 30	1.36	3.0- 9.0			50- 150	170-180
BKV 35 Z	GLF 35	1.41	3.0- 8.0			50- 150	170-180
BKV 40 H	GLF 40	1.46	3.0- 8.0			50- 150	170-180
BKV 50 H	GLF 50	1.57	3.0- 7.0			50- 150	170-180
BM 230 H	MN 30	1.36	12.0			50- 150	170-180
BM 240 H	MN 40	1.46	10.0			50- 150	170-180
BM 30 X	GMN 30	1.38	5.0- 9.0			50- 150	170-180
BM 40 X	GMN 40	1.46	5.0- 9.0			50- 150	170-180
RM KU 2-2501/30	GLF 30	1.36	3.0- 9.0			50- 150	170-180
RM KU 2-2521/20	GLF 20	1.28	3.0- 9.0			50- 150	170-180
RM KU 2-2521/25	GLF 25	1.32	3.0- 9.0			50- 150	170-180
RM KU 2-2521/30	GLF 30	1.36	3.0- 9.0			50- 150	170-180
RM KU 2-2521/35	GLF 35	1.41	3.0- 9.0			50- 150	170-180
RM KU 2-2561/30	MN 30	1.36	12.0			50- 150	170-180
Nylon 6		**Electrafil**			**DSM**		
J-3/CF/30	CF 30	1.28	1.0				165-220
J-3/CF/40	CF 40	1.33					165-220
NY-3/EC	CBL	1.25	11.0				165-220
Nylon 6		**Ferro Nylon**			**Ferro**		
RNY20LA	GLF 20	1.29	4.0			50- 100	

Drying time, hr	Inj time, sec	Front temp, °F	Mid temp, °F	Rear temp, °F	Nozzle temp, °F	Mold temp, °F	Proc temp, °F
3.0- 4.0		520-540	500-520	480-500	520-530	180-230	
							520-560
3.0- 4.0		520-540	500-520	480-500	520-530	180-230	
3.0- 4.0		520-540	500-520	480-500	520-530	180-230	
3.0- 4.0		520-540	500-520	480-500	520-530	180-230	
3.0- 4.0		520-540	500-520	480-500	520-530	180-230	
3.0- 4.0		520-540	500-520	480-500	520-530	180-230	
3.0- 4.0		520-540	500-520	480-500	520-530	180-230	
3.0- 4.0		520-540	500-520	480-500	520-530	180-230	
3.0- 4.0		520-540	500-520	480-500	520-530	180-230	
3.0- 4.0		520-540	500-520	480-500	520-530	180-230	
3.0- 4.0		520-540	500-520	480-500	520-530	180-230	
3.0- 4.0		520-540	500-520	480-500	520-530	180-230	
		540-580	520-560	500-540	540-580	180-210	540-580
4.0-72.0		500-520	480-500	470-480	520-535	175-250	480-520
4.0-72.0		500-520	480-500	470-480	520-535	175-250	480-520
4.0-72.0		500-520	480-500	470-480	520-535	175-250	480-520
4.0-72.0		520-535	500-520	490-500	520-535	160-195	500-535
4.0- 5.0		510-538	500-518	491-500	518-536	158-194	518-545
4.0-72.0		520-535	500-520	490-500	520-535	160-195	500-535
4.0-72.0		520-535	500-520	490-500	520-535	160-195	500-535
4.0		510-535	480-510	470-480	520-535	160-230	520-535
4.0-72.0		510-535	480-510	470-480	520-535	160-230	520-535
4.0-72.0		510-535	480-510	470-480	520-535	160-230	520-535
4.0-72.0		510-535	480-510	470-480	520-535	160-230	520-535
4.0-72.0		510-535	480-510	470-480	520-535	160-230	520-535
4.0-72.0		510-535	480-510	470-480	520-535	160-230	520-535
4.0-72.0		510-535	480-510	470-480	520-535	160-230	520-535
4.0-72.0		510-535	480-510	470-480	520-535	160-230	520-535
4.0-72.0		510-535	480-510	470-480	520-535	160-230	520-535
4.0-72.0		510-535	480-510	470-480	520-535	160-230	520-535
4.0-72.0		510-535	480-510	470-480	520-535	160-230	520-535
4.0-72.0		510-535	480-510	470-480	520-535	160-230	520-535
4.0-72.0		510-535	480-510	470-480	520-535	160-230	520-535
4.0-72.0		500-520	480-500	465-485	490-510	160-195	510-530
4.0-72.0		500-520	480-500	465-485	490-510	160-195	510-530
4.0-72.0		500-520	480-500	465-485	490-510	160-195	510-530
4.0-72.0		500-520	480-500	465-485	490-510	160-195	510-530
4.0-72.0		500-520	480-500	465-485	490-510	160-195	510-530
4.0		500-520	480 500	465-485	490-510	160-195	510-530
2.0-16.0		510-530	530-550	520-540	500-520	130-180	520-540
2.0-16.0		510-530	530-550	520-540	500-520	130-180	520-540
2.0-16.0		500-540	520-550	490-520	480-550	140-180	480-530
		510-530	510-530	510-530	500-540	150-180	

Grade	Filler	Sp Grav	Shrink, mils/in	Melt flow, g/10 min	Melt temp, °F	Back pres, psi	Drying temp, °F
RNY30LA	GLF 30	1.40	3.0			50- 100	
Nylon 6		**Fiberfil**			**DSM**		
J-3/10	GLF 10	1.23	5.0				165-220
J-3/15/MF/25	GLF 15	1.50	2.0				165-220
J-3/20	GLF 20	1.30	4.0				165-220
J-3/30	GLF 30	1.40	3.0				165-220
J-3/40	GLF 40	1.50	2.0				165-220
NY-17/MF/40	MN 40	1.50	1.0				165-220
Nylon 6		**Fiberfil TN**			**DSM**		
J-7/13	GLF 13	1.18	3.5- 5.0				220
J-7/33	GLF 33	1.33	1.0- 3.0				220
J-7/33/IT	GLF 33	1.33	1.0- 3.0				220
J-7/43	GLF 43	1.43	0.7- 2.0				220
NY-7		1.06	14.0-22.0				220
NY-7/V0		1.26	0.1				165-220
Nylon 6		**Fiberfil V0**			**DSM**		
NY-7/V0		1.26	14.0				220
Nylon 6		**Fiberstran**			**DSM**		
G-1/50	GLF 50	1.57	1.0- 2.0			10- 50	165-180
G-3/10/CF/25	CF 25	1.32	0.5				165-220
G-3/30	GLF 30	1.40	3.0			10- 50	165-180
G-3/30/MS/5	GLF 30	1.40	3.5				165-220
G-3/40	GLF 40	1.47	2.0			10- 50	165-180
G-3/40/MS/5	GLF 40	1.49	3.0				165-220
G-3/50	GLF 50	1.57	1.0- 2.0			10- 50	165-180
Nylon 6		**Grilon**			**EMS**		
A23G 6165		1.14	8.0		428		158-176
A23GM		1.14	8.0		432		158-176
A23GM 6707		1.14	8.0		432		158-176
A28GM		1.14	8.0		432		158-176
A28GM 6707		1.14	8.0		432		158-176
A28NX		1.12	10.0		432		158-176
A28NY		1.08	10.0		428		158-176
A28NZ		1.05			428		176
A28V0		1.16	9.0		432		158-176
BT40Z		1.06	7.0		428		158-176
PMV-4H	GLF 15	1.44	2.0		432		158-176
PV-15H	GLF 15	1.23	2.0		432		158-176
PV-3H	GLM 30	1.35	1.0		432		158-176
PVN-15H	GLF 15	1.18	4.0		419		158-176
PVN-3H	GLF 30	1.32	1.0		432		158-176
PVZ-33H	GLF 33	1.30	1.0		430		176
R40GM		1.14	9.0		432		158-176
R47HW		1.12	10.0		415		158-176
R47NZE		1.08			422		158-176
R70		1.14			428		176
XE3147	GLF 10	1.69	2.0		426		158-176
XS1245	GLF 50	1.55					176
Nylon 6		**Hiloy**			**ComAlloy**		
610	UNS 45	1.49	4.0				180

Drying time, hr	Inj time, sec	Front temp, °F	Mid temp, °F	Rear temp, °F	Nozzle temp, °F	Mold temp, °F	Proc temp, °F
		510-530	510-530	510-530	500-540	150-180	
2.0-16.0		510-530	530-550	520-540	500-520	130-180	520-540
2.0-16.0		510-530	530-550	520-540	500-520	130-180	520-540
2.0-16.0		510-530	530-550	520-540	500-520	130-180	520-540
2.0-16.0		510-530	530-550	520-540	500-520	130-180	520-540
2.0-16.0		510-530	530-550	520-540	500-520	130-180	520-540
2.0-16.0		510-530	530-550	520-540	500-520	130-180	520-540
2.0- 4.0		510-530	530-550	520-540	500-520	130-180	520-540
2.0- 4.0		510-530	530-550	520-540	500-520	130-180	520-540
2.0- 4.0		510-530	530-550	520-540	500-520	130-180	520-540
2.0- 4.0		510-530	530-550	520-540	500-520	130-180	520-540
2.0		460-470	450-460	440-450	430-450	175-190	470-500
2.0-16.0		500-540	520-550	490-520	480-550	140-180	480-530
2.0- 4.0		500-540	520-550	490-520	480-550	140-180	480-530
4.0-16.0		530-550	550-570	540-560	520-540	175-225	540-580
2.0-16.0		510-530	530-550	520-540	500-520	130-180	520-540
4.0-16.0		510-530	530-560	520-540	520-540	175-225	530-570
2.0-16.0		510-530	530-550	520-540	500-520	130-180	520-540
4.0-16.0		510-530	530-560	520-540	520-540	175-225	530-570
2.0-16.0		510-530	530-550	520-540	500-520	130-180	520-540
4.0-16.0		510-530	530-560	520-540	520-540	175-225	530-570
6.0-10.0		465	455	445	465	140	500
6.0-10.0		465	455	445		176	465
6.0-10.0		465	455	445		176	465
6.0-10.0		465	455	445		176	446-500
6.0-10.0		465	455	445		176	446-500
6.0-10.0		490	480	470		176	500
6.0-10.0		490	480	470		176	500
4.0- 6.0						180-200	465-520
6.0-10.0		475	460	445		176	480
6.0-10.0		520	510	480		176	545
6.0-10.0		525	510	490		176	535
6.0-10.0		520	510	490		176	520
6.0-10.0		520	510	490		176	535
6.0-10.0		530	510	490		176	535
6.0-10.0		520	510	490		176	535
2.0		536	528	518		175	550
6.0-10.0		465	455	445		176	475
6.0-10.0		475	465	455		140	500
6.0-10.0		520	520	510		176	500
4.0- 6.0						180-200	465-520
6.0-10.0		555	535	520		176-212	555-570
4.0- 6.0						180-200	465-520
3.0- 4.0		460-500	470-510	480-520	460-500	100-200	470-510

Grade	Filler	Sp Grav	Shrink, mils/in	Melt flow, g/10 min	Melt temp, °F	Back pres, psi	Drying temp, °F
611	GL 13	1.22	6.0				180
612	GL 20	1.28	4.0				180
613	GL 33	1.38	3.0				180
614	GL 43	1.50	2.0				180
615	GL 50	1.56	2.0				180
616	GMN 40	1.48	5.0				180
617	GMN 55	1.63	3.0				180
618	GLF	1.33	12.0				180
619	MN	1.49	10.0				180
Nylon 6		**Kopa**			**Kolon**		
KN111		1.14	12.0-15.0		428	71	167-185
KN131		1.14	12.0-15.0		428	71	167-185
KN1322V0		1.16	14.0-17.0		428	114	167-185
KN133G30	GLF 30	1.36	5.0- 7.0		428	99	167-185
KN133HB	GLF	1.63	4.0- 6.0		428	99	167-185
KN133HR		1.14	12.0-15.0		428	71	167-185
KN133MS		1.16	11.0-12.0		428	114	167-185
KN133MX	MN	1.25	8.0-10.0		428	142	167-185
KN136		1.14	12.0-15.0		428	71	167-185
KN171		1.14	12.0-15.0		428	71	167-185
KN173HI		1.08	14.0-18.0		419-428	71	167-185
KN175HI		1.12	13.0-16.0		419-428	71	167-185
Nylon 6		**Lubrilon**			**ComAlloy**		
605	PTF 20	1.25	14.0				180
613	GL 33	1.50	4.0				180
618	GL 33	1.43	5.0				180
Nylon 6		**MonTor Nylon**			**MonTor**		
CM1001G-15	GLF 15	1.25			436	71- 213	
CM1001G-20	GLF 20	1.29			436	71- 213	
CM1001R	MN 40	1.51			436	71- 213	
CM1003G-R30	GLF 30	1.37			436	71- 213	
CM1007		1.13			436	71- 213	
CM1011G-15	GLF 15	1.25			436	71- 213	
CM1011G-30	GLF 30	1.36			436	71- 213	
CM1011G-45	GLF 45	1.50			436	71- 213	
CM1014-VO		1.18			436	71- 213	
CM1016-K		1.13			436	71- 213	
CM1016G-30	GLF 30	1.36			436	71- 213	
CM1017		1.13			436	71- 213	
CM1017-C		1.13			436	71- 213	
CM1021		1.13			436	71- 213	
CM1023					436	71- 213	
CM1026		1.13			436	71- 213	
UTN121		1.09			436	71- 213	
UTN141		1.06			436	71- 213	
Nylon 6		**Nylafil**			**DSM**		
G-3/30	GLF 30	1.40	3.0				220
G-3/30/MS/5	GLF 30	1.40	3.5				220
G-3/40	GLF 30	1.47	2.0				220
G-3/40/MS/5	GLF 40	1.49	3.0				220
J-3/15/MF/25	GLF 15	1.48	2.0				220
J-3/20	GLF 20	1.30	4.0				220

Drying time, hr	Inj time, sec	Front temp, °F	Mid temp, °F	Rear temp, °F	Nozzle temp, °F	Mold temp, °F	Proc temp, °F
3.0- 4.0		460-500	470-510	480-520	460-500	100-200	470-510
3.0- 4.0		460-500	470-510	480-520	460-500	100-200	470-510
3.0- 4.0		460-500	470-510	480-520	460-500	100-200	470-510
3.0- 4.0		460-500	470-510	480-520	460-500	100-200	470-510
3.0- 4.0		460-500	470-510	480-520	460-500	100-200	470-510
3.0- 4.0		460-500	470-510	480-520	460-500	100-200	470-510
3.0- 4.0		460-500	470-510	480-520	460-500	100-200	470-510
3.0- 4.0		460-500	470-510	480-520	460-500	100-200	470-510
3.0- 4.0		460-500	470-510	480-520	460-500	100-200	470-510
4.0- 6.0	4.0	464	464	428	446	176	496
4.0- 6.0	4.0	464	464	428	446	176	496
4.0- 6.0	4.0	464	464	428	455	176	505
4.0- 6.0	4.0	500	491	464	491	176	541
4.0- 6.0	4.0	500	491	464	491	176	541
4.0- 6.0	4.0	464	464	428	446	176	496
4.0- 6.0	4.0	464	464	428	455	176	505
4.0- 6.0	5.0	491	482	455	491	176	541
4.0- 6.0	4.0	464	464	428	464	176	514
4.0- 6.0	4.0	464	464	428	446	176	496
4.0- 6.0	4.0	473	473	455	482	158	532
4.0- 6.0	4.0	473	473	455	482	158	532
3.0- 4.0		460-500	470-510	480-520	460-500	100-200	470-510
3.0- 4.0		460-500	470-510	480-520	460-500	100-200	470-510
3.0- 4.0		460-500	470-510	480-520	460-500	100-200	470-510
		410-500	410-500	410-500		140-176	428-536
		410-500	410-500	410-500		140-176	428-536
		410-500	410-500	410-500		140-176	428-536
		410-500	410-500	410-500		140-176	428-536
		410-500	410-500	410-500		140-176	428-536
		410-500	410-500	410-500		140-176	428-536
		410-500	410-500	410-500		140-176	428-536
		410-500	410-500	410-500		140-176	428-536
		410-500	410-500	410-500		140-176	428-536
		410-500	410-500	410-500		140-176	428-536
		410-500	410-500	410-500		140-176	428-536
		410-500	410-500	410-500		140-176	428-536
		410-500	410-500	410-500		140-176	428-536
		410-500	410-500	410-500		140-176	428-536
		410-500	410-500	410-500		140-176	428-536
		410-500	410-500	410-500		140-176	428-536
		410 500	410-500	410-500		140-176	428-536
		410-500	410-500	410-500		140-170	420 536
2.0		510-530	530-550	520-540	500-520	130-180	520-540
2.0		510-530	530-550	520-540	500-520	130-180	520-540
2.0		510-530	530-550	520-540	500-520	130-180	520-540
2.0		510-530	530-550	520-540	500-520	130-180	520-540
2.0		510-530	530-550	520-540	500-520	130-180	520-540
2.0		510-530	530-550	520-540	500-520	130-180	520-540

Grade	Filler	Sp Grav	Shrink, mils/in	Melt flow, g/10 min	Melt temp, °F	Back pres, psi	Drying temp, °F
J-3/30	GLF 30	1.40	3.0				220
J-3/40	GLF 40	1.50	2.0				220
J-7/13	GLF 13	1.19	4.0				220
J-7/33	GLF 33	1.33	1.0				220
J-7/43	GLF 43	1.43	0.7				220
Nylon 6		**Nypel**			**Allied Signal**		
2314		1.13	12.0		420		180
2314 FCAT		1.13	9.0		420		180
2314 HS		1.13	12.0		420		180
2314 HS BK		1.13	12.0		420		180
2314C HS		1.13	9.0		420		180
2360	MN 40	1.49	9.0		420		180
2360 HS	MNF	1.49	9.0		420		180
2360 HS BK	MN 40	1.49	9.0		420		180
2365G	GMN 36	1.43	4.0		420		180
2367G HS	GMN 40	1.48	4.0		420		180
2367G HS BK	GMN 40	1.48	4.0		420		180
6015G HS	GL 15	1.24	5.0		420		180
6015G HS BK	GLF 15	1.24			420		180
6030G HS	GLF 30	1.36	3.0		420		180
6030G HS BK	GLF 30	1.36	3.0		420		180
6033G	GLF 33	1.38	3.0		420		180
Nylon 6		**Plaslube**			**DSM**		
G-3/40/MS/5	GLF 40	1.49	3.0				165-220
J-3/30/MS/5	GLF 30	1.40	3.0				165-220
Nylon 6		**PMC Nylon 6**			**PMC**		
60G13	GLF 13	1.22	6.0		420	0- 150	160-180
60G25	GLF 25	1.30	5.0		425	0- 150	160-180
60G33	GLF 33	1.36	4.0		425	0- 150	160-180
60G43	GLF 43	1.43	2.0		425	0- 150	160-180
6253		1.12	17.0		415	0- 800	160-180
Nylon 6		**Radiflam**			**Polymers Intl.**		
S AE		1.18	1.0		428		
S FR		1.15	1.0		428		
S RV300AE		1.40	0.3		428		
Nylon 6		**Radilon**			**Polymers Intl.**		
S		1.13	1.1		428		
S 30E/30EN		1.13	1.1		428		
S 32E/EN		1.13	1.1		428		
S 32ECT		1.13	0.7		423		
S 35E/35EN		1.13	1.0		428		
S 35FL/35FLC		1.13	1.1		428		
S 35FL/FLC		1.13	1.0		428		
S 35HS		1.13	1.1		428		
S 35VHS		1.13	1.2		428		
S 38E/38EN		1.13	1.1		428		
S 40E/EN		1.13	1.1		428		
S 40FL/FLC		1.13	1.1		428		
S 40HS		1.13	1.1		428		
S 40VHS		1.13	1.2		428		
S 42E/42EN		1.13	1.0		428		

Drying time, hr	Inj time, sec	Front temp, °F	Mid temp, °F	Rear temp, °F	Nozzle temp, °F	Mold temp, °F	Proc temp, °F
2.0		510-530	530-550	520-540	500-520	130-180	520-540
2.0		510-530	530-550	520-540	500-520	130-180	520-540
2.0		510-530	530-550	520-540	500-520	130-180	520-540
2.0		510-530	530-550	520-540	500-520	130-180	520-540
2.0		510-530	530-550	520-540	500-520	130-180	520-540
		460-520	440-500	430-470	450-500	180-200	460-520
		460-520	440-500	430-470	450-500	180-200	460-520
		460-520	440-500	430-470	460-520	180-200	460-520
		460-520	440-500	430-470	460-520	180-200	460-520
		460-520	440-500	430-470	460-520	180-200	460-520
		520-580	500-560	480-520	520-580	180-200	520-580
		520-580	500-560	480-520	520-580	180-200	520-580
		520-580	500-560	480-520	520-580	180-200	520-580
		520-560	500-540	480-520	520-560	180-200	520-560
		520-580	500-560	480-520	520-580	180-200	520-580
		520-580	500-560	480-520	520-580	180-200	520-580
		520-580	500-560	480-520	520-580	180-200	520-580
		520-580	500-560	480-520	520-580	180-200	520-580
		520-580	500-560	480-520	520-580	180-200	520-580
		520-580	500-560	480-520	520-580	180-200	520-580
		520-560	500-540	480-520	520-560	180-200	520-560
2.0-16.0		510-530	530-550	520-540	500-520	130-180	520-540
2.0-16.0		510-530	530-550	520-540	500-520	130-200	520-540
		470-510	440-480	410-450	450-500	160-220	
		470-510	440-480	410-450	450-500	160-220	
		470-510	440-480	410-450	450-500	160-220	
		470-510	440-480	410-450	450-500	160-220	
		460-500	430-470	410-440	440-480	130-200	
						68	455
						68	455
						194	482
						68	437
						68	437
						68	446
						68	437
						68	446
						68	437
						68	437
						122	446
						68	446
						68	446
						68	446
						68	437
						122	446
						68	446
						68	446

Grade	Filler	Sp Grav	Shrink, mils/in	Melt flow, g/10 min	Melt temp, °F	Back pres, psi	Drying temp, °F
S 50PL112		1.13	1.1		428		
S 60E/60EN		1.14	0.8		428		
S BHS200/201		1.13	1.4		428		
S CP300		1.37	1.0		428		
S CP400		1.47	0.9		428		
S CV200		1.26	1.1		428		
S CV250		1.31	1.1		428		
S CV300		1.36	1.0		428		
S CV350		1.39	1.0		428		
S CV400		1.43	0.9		428		
S ERV3510		1.26	0.6		428		
S HCT		1.13	0.8		424		
S HS/LHS		1.13	1.2		428		
S HSX		1.12	1.1		428		
S L		1.13	1.2		428		
S LRV250		1.31	0.3		428		
S LRV300		1.36	0.3		428		
S LVHPL80		1.13	1.6		428		
S RCV3015		1.36	0.6		428		
S RV150		1.21	0.4		428		
S RV200		1.26	0.4		428		
S RV250		1.31	0.3		428		
S RV250R		1.31	0.3		428		
S RV300		1.36	0.3		428		
S RV300R		1.36	0.2		428		
S RV330		1.37	0.3		428		
S RV350		1.39	0.2		428		
S RV350R		1.39	0.1		428		
S RV500		1.55	0.1		428		
S USK160		1.09	1.1		428		
S USK200		1.07	1.0		428		
S USK240		1.05	1.0		428		
S USK80		1.11	1.1		428		
S USX160		1.09	1.1		428		
S USX200		1.07	1.0		428		
S USX240		1.05	1.0		428		
S USX80		1.11	1.1		428		
S VHS/LVHS		1.13	1.2		428		
S XCP3303		1.36	1.0		428		
S XRV3510		1.25	0.6		428		
S XRV4010		1.30	0.5		428		
Nylon 6		**RTP Polymers**			**RTP**		
200A		1.13	17.0				
200A MS	MOS	1.17	13.0			25- 50	175
200A TFE 20	GLF	1.26	13.0			25- 50	175
201A	GLF 10	1.21	10.0			25- 50	175
201A M 30	GLF 10	1.48	6.0			25- 50	175
202A M	GLF 15	1.47	5.0			25- 50	175
203A	GLF 20	1.27	7.0			25- 50	175
203A FR	GLF	1.59	4.0			25- 50	175
203A MS	GLF	1.33	5.0			25- 50	175
204A FR	GLF	1.63	3.0			25- 50	175
205A	GLF 30	1.36	5.0			25- 50	175
205A FR	GLF	1.66	2.0			25- 50	175
205A TFE 15	GLF	1.49	4.0			25- 50	175

Drying time, hr	Inj time, sec	Front temp, °F	Mid temp, °F	Rear temp, °F	Nozzle temp, °F	Mold temp, °F	Proc temp, °F
						68	455
						68	446
						122	437
						194	482
						194	482
						194	500
						194	500
						194	500
						194	500
						194	500
						194	500
						68	437
						122	446
						122	437
						68	437
						194	500
						194	500
						68	455
						194	500
						194	491
						194	500
						194	500
						194	500
						194	500
						194	500
						194	500
						194	500
						194	500
						230	518
						68	455
						68	473
						68	473
						68	455
						68	455
						68	473
						68	473
						68	455
						68	446
						158	482
						194	500
						194	500
		450-500	450-500	450-500		150-200	
4.0		455-515	445-505	435-495		150-200	
4.0		455-545	445-535	435-525		150-200	
4.0		455-545	445-535	435-525		150-200	
4.0		455-545	445-535	435-525		150-200	
4.0		455-545	445-535	435-525		150-200	
4.0		455-545	445-535	435-525		150-200	
4.0		470-520	460-510	450-500		150-200	
4.0		455-545	445-535	435-525		150-200	
4.0		470-520	460-510	450-500		150-200	
4.0		455-545	445-535	435-525		150-200	
4.0		470-520	460-510	450-500		150-200	
4.0		455-545	445-535	435-525		150-200	

Grade	Filler	Sp Grav	Shrink, mils/in	Melt flow, g/10 min	Melt temp, °F	Back pres, psi	Drying temp, °F
207A	GLF 40	1.46	3.0			25- 50	175
209A	GLF 50	1.57	3.0			25- 50	175
227A	MN	1.47	8.0			25- 50	175

Nylon 6 — Schulamid 6 — A. Schulman

Grade	Filler	Sp Grav	Shrink, mils/in	Melt flow, g/10 min	Melt temp, °F	Back pres, psi	Drying temp, °F
GB 30	GLB 30	1.36	3.0		420		180
GF 30	GLF 30	1.35	3.0		420		180
MF 40	MN 40	1.47	5.0		420		180
MV 5		1.14	13.0		430		180

Nylon 6 — Stat-Kon — LNP

Grade	Filler	Sp Grav	Shrink, mils/in	Melt flow, g/10 min	Melt temp, °F	Back pres, psi	Drying temp, °F
PDX-P-91519		1.16	12.0-18.0		420	0- 50	180

Nylon 6 — Texalon — Texapol

Grade	Filler	Sp Grav	Shrink, mils/in	Melt flow, g/10 min	Melt temp, °F	Back pres, psi	Drying temp, °F
1000A		1.09	12.0-16.0		420	50	150
1010A		1.11	8.0-12.0		420	50	150
1106A		1.09	14.0-18.0		415	50	150
1108A		1.10	12.0-16.0		415	50	150
1110A PL HS		1.09	12.0-16.0		410	50	150
1110A PL-2 HS		1.09	12.0-16.0	9.00	410-424	50	150
1120		1.13	11.0-15.0	2.00 S	419-428	50	150
1175 BK-18		1.08	11.0-16.0		410	50	150
600A		1.14	10.0-14.0		420	50	150
600A Zip-1		1.14	5.0- 8.0		420	50	150
600A Zip-10		1.14	9.0-12.0	6.50	419-426	50	150
600A Zip-22		1.14	6.0-10.0		420	50	150
604		1.14	6.0-10.0		420	50	150
604 NU		1.14	4.0- 7.0		420	50	150
623 HS BK-13		1.14	7.0-10.0		410	50	150
680A		1.14	10.0-12.0		420	50	150
GF 1000A-14	GLF 14	1.17	4.0- 6.0		420	50	150
GF 600A-13	GLF 13	1.21	3.0- 4.0		420	50	150
GF 600A-33	GLF 33	1.36	1.0- 1.5		420	50	150
GMF 600A-40	GMN 40	1.49	2.0- 3.0		420	50	150
MF 600A-40	MN 40	1.50	4.0- 5.0		420	50	150
MF 604-40 BK-26	MN 40	1.49	5.0- 6.0		420	50	150
XP 2094		1.13	11.0-14.0		420	50	150

Nylon 6 — Thermocomp — LNP

Grade	Filler	Sp Grav	Shrink, mils/in	Melt flow, g/10 min	Melt temp, °F	Back pres, psi	Drying temp, °F
PF-1006 RM	GLF 30	1.24	4.0- 5.0			0- 100	170-180

Nylon 6 — Thermofil Polyamide — Thermofil

Grade	Filler	Sp Grav	Shrink, mils/in	Melt flow, g/10 min	Melt temp, °F	Back pres, psi	Drying temp, °F
N-33FG-0100	GL 33	1.38	2.0			50	170
N-33FG-1100	GL 33	1.38	2.0			50	170
N-40FM-1100	GMN 40	1.47	3.0			50	170
N-40MF-1600	MNF 40	1.48	5.0			50	170
N-43FG-0100	GL 43	1.49	2.0			50	170
N-43FG-1100	GL 43	1.49	2.0			50	170
NSG-240A	CRF 50	1.59	1.0			50	220
NSG-440A	CRF 60	1.73	1.0			50	220
NSG-730A	CRF 65	1.81	2.0			50	220

Nylon 6 — UBE Nylon — UBE

Grade	Filler	Sp Grav	Shrink, mils/in	Melt flow, g/10 min	Melt temp, °F	Back pres, psi	Drying temp, °F
1011FB		1.14	14.0-15.0		419-437	1421	
1011GC4	GLF 20	1.28	4.0- 7.0		419-437	1421	
1013B		1.14	14.0-15.0		419-437	1421	

Drying time, hr	Inj time, sec	Front temp, °F	Mid temp, °F	Rear temp, °F	Nozzle temp, °F	Mold temp, °F	Proc temp, °F
4.0		455-545	445-535	435-525		150-200	
4.0		455-545	445-535	435-525		150-200	
4.0		470-520	485-530	460-510		160-200	
		500	480	465	500	175-210	480-540
		500	480	465	500	175-210	480-540
		500	480	465	500	175-210	480-540
		445	445	440	450	175-210	440-460
4.0- 6.0		420-460	420-460	420-460		50-120	450
4.0		440	435	425	475	125	
4.0		440	435	425	475	125	
4.0		440	435	425	475	125	
4.0		440	435	425	475	125	
4.0		440	435	425	475	125	
4.0		440	435	425	475	125	
4.0		440	435	425	475	125	
4.0		440	435	425	475	125	
4.0		460	450	450	465	125	
4.0		460	450	450	465	125	440-520
4.0		460	450	450	465	125	440-520
4.0		460	450	450	465	125	440-520
4.0		460	450	450	465	125	
4.0		460	450	450	465	125	
4.0		460	450	450	465	125	
4.0		460	450	450	465	125	
4.0		480	470	470	490	125	
4.0		480	470	470	490	125	
4.0		480	470	470	490	125	520-570
4.0		480	470	470	490	125	540-570
4.0		470	460	460	490	125	540-570
4.0		470	460	460	490	125	540-570
4.0		460	450	450	465	125	
4.0- 6.0		480-525	480-525	480-525		150-200	490-520
3.0		500-550	490-530	470-510	510-560	150-200	
3.0		500-550	490-530	470-510	510-560	150-200	
3.0		500-550	490-530	470-510	510-560	150-200	
3.0		500-550	490-530	470-510	510-560	150-200	
3.0		500-550	490-530	470-510	510-560	150-200	
3.0		500-550	490-530	470-510	510-560	150-200	
3.0		500-550	490-530	470-510	510-560	150-200	
3.0		500-550	490-530	470-510	510-560	150-200	
3.0		500-550	490-530	470-510	510-560	150-200	
	12.0	473	446	428	473	158-176	482
	10.0	527	509	464	527	158-176	545
	12.0	473	446	428	473	158-176	482

Grade	Filler	Sp Grav	Shrink, mils/in	Melt flow, g/10 min	Melt temp, °F	Back pres, psi	Drying temp, °F
1013NB		1.14	9.0-12.0		419-437	1421	
1013NH		1.14	12.0-13.0		419-437	1421	
1013NU2		1.14	12.0-13.0		419-437	1421	
1013NW8		1.14	12.0-13.0		419-437	1421	
1013R	MN 30	1.50	7.0- 8.0		419-437	1421	
1013RU1	MN 40	1.50	10.0-12.0		419-437	1421	
1013RW	MN 40	1.50	10.0-11.0		419-437	1421	
1015GC6	GLF 30	1.36	2.0- 7.0		419-437	1421	
1015GC9	GLF 45	1.50	1.0- 5.0		419-437	1421	
1015GI	GLF	1.26	4.0-12.0		419-437	1421	
1018I		1.07	11.0-15.0		419-437	1421	
1022B		1.14	14.0-15.0		419-437	1421	
1022SV2		1.16	9.0-11.0		419-437	1421	
2020GC4	GLF 20	1.28	6.0-14.0		491-509	1421	
2020GC6	GLF 30	1.36	5.0-13.0		491-509	1421	
2020GC9	GLF 45	1.50	4.0-11.0		491-509	1421	
2020GCU	GLF 30	1.60	5.0-13.0		491-509	1421	

Nylon 6 Ultramid BASF

Grade	Filler	Sp Grav	Shrink, mils/in	Melt flow, g/10 min	Melt temp, °F	Back pres, psi	Drying temp, °F
B3		1.13	10.0		428	50- 100	176-230
B35		1.13	12.0		428	50- 100	176-230
B35EG3	GLF 15	1.23	3.0		428	50- 100	176-230
B35K		1.13	12.0		428	50- 100	176-230
B35M		1.13	10.0		420	50- 100	176-230
B35MF01		1.13	10.0		420	50- 100	176-230
B35SK		1.13	7.0		428	50- 100	176-230
B35W		1.13	12.0		428	50- 100	176-230
B35Z		1.08	6.0		428	50- 100	176-230
B35ZGM24	GLF 10	1.34	3.0		428	50- 100	176-230
B36F		1.13	10.0		428	50- 100	176-230
B3EG10	GLF 50	1.55	1.0		428	50- 100	176-230
B3EG3	GLF 15	1.23	3.0		428	50- 100	176-230
B3EG5	GLF 25	1.32	2.0		428	50- 100	176-230
B3EG6	GLF 30	1.36	4.0- 1.5		428	50- 100	176-230
B3EG7	GLF 35	1.41	1.5		428	50- 100	176-230
B3GK24	GLF 10	1.34	4.0		428	50- 100	176-230
B3GM35	MN 25	1.48	4.0		428	50- 100	176-230
B3K		1.13	10.0		428	50- 100	176-230
B3L		1.10	9.0		428	50- 100	176-230
B3M6	MN 30	1.36	8.0		428	50- 100	176-230
B3S		1.13	7.0		428	50- 100	176-230
B3SK		1.13	7.0		428	50- 100	176-230
B3UM6	MN 30	1.45	7.5		428	50- 100	176-230
B3WG10	GLF 50	1.55	1.0		428	50- 100	176-230
B3WG3	GLF 15	1.23	3.0		428	50- 100	176-230
B3WG5	GLF 25	1.32	2.0		428	50- 100	176-230
B3WG6	GLF 30	1.36	4.0- 1.5		428	50- 100	176-230
B3WG7	GLF 35	1.41	1.5		428	50- 100	176-230
B3WM602	MN 30	1.36	5.0		428	50- 100	176-230
B3ZG3	GLF 15	1.22	3.0		428	50- 100	176-230
B3ZG6	GLF 30	1.33	1.2		428	50- 100	176-230
B3ZG8	GLF 40	1.42	2.0		428	50- 100	176-230
B4		1.13	12.0		428	50- 100	176-230
B4K		1.13	12.0		428	50- 100	176-230
B5		1.13	8.0-10.0		428	50- 100	176-230
B5W		1.13	8.0-10.0		428	50- 100	176-230

Drying time, hr	Inj time, sec	Front temp, °F	Mid temp, °F	Rear temp, °F	Nozzle temp, °F	Mold temp, °F	Proc temp, °F
	12.0	473	446	428	473	158-176	482
	12.0	473	446	428	473	158-176	482
	12.0	473	446	428	473	158-176	482
	12.0	473	446	428	473	158-176	482
	10.0	527	509	464	527	158-176	545
	10.0	527	509	464	527	158-176	545
	10.0	527	509	464	527	158-176	545
	10.0	527	509	464	527	158-176	545
	10.0	527	509	464	527	158-176	545
	10.0	527	509	464	527	158-176	545
	12.0	473	446	428	473	158-176	482
	12.0	473	446	428	473	158-176	482
	12.0	473	446	428	473	158-176	482
	10.0	545	536	518	545	158-176	563
	10.0	545	536	518	545	158-176	563
	10.0	545	536	518	545	158-176	563
	10.0	545	536	518	545	158-176	563
4.0-12.0						100-140	500-540
4.0-12.0						100-140	520-550
4.0-12.0						176-194	520-555
4.0-12.0						100-140	520-555
4.0-12.0						100-140	480-520
4.0-12.0						100-140	480-520
4.0-12.0						100-140	520-555
4.0-12.0						104-140	520-555
4.0-12.0						104-140	520-555
4.0-12.0						176-194	520-555
4.0-12.0						100-140	500-540
4.0-12.0						176-194	535-575
4.0-12.0						176-194	520-555
4.0-12.0						176-194	520-555
4.0-12.0						176-194	520-555
4.0-12.0						176-194	520-555
4.0-12.0						176-194	520-555
4.0-12.0						176-194	520-555
4.0-12.0						104-140	480-520
4.0-12.0						104-140	480-520
4.0-12.0						176-194	520-555
4.0-12.0						104-140	480-520
4.0-12.0						100-140	480-520
4.0-12.0						176-194	520-555
4.0-12.0						176-194	535-575
4.0-12.0						176-194	520-555
4.0-12.0						176-194	520-555
4.0-12.0						176-194	520-555
4.0-12.0						176-194	520-555
4.0-12.0						176-194	520-555
4.0-12.0						176-194	520-555
4.0-12.0						176-194	520-555
4.0-12.0						176-194	520-555
4.0-12.0						100-140	480-540
4.0-12.0						104-140	530-555
4.0-12.0						100-140	540-560
4.0-12.0						100-140	540-560

Grade	Filler	Sp Grav	Shrink, mils/in	Melt flow, g/10 min	Melt temp, °F	Back pres, psi	Drying temp, °F
KR4405/1		1.13	10.0		428	50- 100	176-230
KR4430		1.08	9.0		428	50- 100	176-230
Nylon 6		**Verton**			**LNP**		
PF-700-10	GLF 50	1.57	3.0- 4.0			0- 50	180-200
PF-7007	GLF 35	1.42	4.0- 5.0			0- 50	180-200
Nylon 6		**Voloy**			**ComAlloy**		
612	GL	1.54	4.0				180
618	MN 25	1.54	10.0				180
652	GL 25	1.46	4.0				180
658	MN 25	1.46	12.0				180
Nylon 6		**Wellamid Nylon**			**Wellman**		
420-N		1.12	13.0-17.0		410-430		150-180
420-XE-N		1.12	14.0-18.0		420		150-180
42L-N		1.12	13.0-17.0		410-430		150-180
42L-XE-N		1.12	14.0-18.0		420		150-180
42LH-N		1.12	13.0-17.0		410-430		150-180
42LH-XE-N		1.12	14.0-18.0		420		150-180
42LN2-N		1.12	8.0-13.0		410-430		150-180
42LN2-XE-N		1.12	8.0-14.0		420		150-180
GF30-60 42LH-N	GLF	1.32	2.0- 6.0		420		150-170
GF30-60 XE-N	GLF	1.32	2.0- 6.0		420		150-170
GS40-60 42L-N	GLB	1.37	13.0-17.0		420	0- 100	150-180
GS40-60 42LH-N	GLB	1.37	13.0-17.0		420	0- 100	150-180
MRGF25/15 42H-N	GLF	1.42	3.0- 6.0		420	0- 100	150-170
MRGF30/10 42H-N	GLF	1.42	4.0- 8.0		420	0- 100	150-170
Nylon 6 Alloy		**Akuloy RM**			**DSM**		
J-75/30	GLF 30	1.26	3.0				160-170
J-75/30/HI	GLF 30	1.25	3.0				160-170
M-1914		1.03	7.0				160-170
M-1915	GLF 30	1.26	3.0				160-170
NY-75		1.03	9.0				160-170
NY-75/MF/40	MN 40	1.38	7.0				160-170
Nylon 6 Alloy		**Dimension**			**Allied Signal**		
D-9000		1.08					230
D-9130G	GLF 30	1.33					230
Nylon 6 Copoly.		**Capron**			**Allied Signal**		
8253 HS		1.09	12.0	4.00	420		180
8253 HS BK-102		1.09	12.0	4.00	420		180
8253 HS BK-106		1.09	12.0	4.00	420		180
8255 HS		1.08			420		180
Nylon 6 Elast.		**Grilon**			**EMS**		
ELX2112		1.06	12.0		403		176
ELX23NZ		1.03	12.0		405		176
XE3106		1.04	12.0		419		176
Nylon 6/10		**CTI Nylon**			**CTI**		
68CNI-6GF	CBI 68	2.84	5.0				180
NI		1.08	14.0				180
NI-20GF	GLF 20	1.22	5.0				180

Drying time, hr	Inj time, sec	Front temp, °F	Mid temp, °F	Rear temp, °F	Nozzle temp, °F	Mold temp, °F	Proc temp, °F
4.0-12.0						100-140	480-520
4.0-12.0						100-140	520-555
4.0		510-550	510-550	510-550		150-200	
4.0		510-550	510-550	510-550		150-200	
3.0- 4.0		460-500	470-510	480-520	460-500	100-200	470-510
3.0- 4.0		460-500	470-510	480-520	460-500	100-200	470-510
3.0- 4.0		460-500	470-510	480-520	460-500	100-200	470-510
3.0- 4.0		460-500	470-510	480-520	460-500	100-200	470-510
		440	450	470	450	70-200	440-550
		440	450	470	450	70-200	440-550
		440	450	470	450	70-200	440-550
		440	450	470	450	70-200	440-550
		440	450	470	450	70-200	440-550
		440	450	470	450	70-200	440-550
		440	450	470	450	70-200	440-550
		440	450	470	450	70-200	440-550
		480-520	490-530	500-540	480-520	140-200	490-540
		480-520	490-530	500-540	480-520	140-200	490-540
		500-540	500-540	500-540	480-520	70-200	480-530
		500-540	500-540	500-540	480-520	70-200	480-530
		500-550	500-550	500-550	500-550	180-210	500-550
		500-550	500-550	500-550	500-550	180-210	500-550
2.0-16.0	0.5- 1.	460-480	470-490	450-470	470-490	100-180	450-490
2.0-16.0	0.5- 1.	460-480	470-490	450-470	470-490	100-180	450-490
2.0-16.0	0.5- 1.	460-480	470-490	450-470	470-490	100-180	450-490
2.0-16.0	0.5- 1.	460-480	470-490	450-470	470-490	100-180	450-490
2.0-16.0	0.5- 1.	460-480	470-490	450-470	470-490	100-180	450-490
2.0-16.0	0.5- 1.	460-480	470-490	450-470	470-490	100-180	450-490
		540-580	520-560	500-540	540-580	180-210	540-580
		540-580	520-560	500-540	540-580	180-210	540-580
		480-520	460-520	440-500	480-520	180	480-540
		480-540	460-520	440-500	480-540	180	480-540
		480-540	460-520	440-500	480-540	180	480-540
		480-540	460-520	440-500	480-540	180	480-540
6.0-10.0		430	410	390		120	445
6.0-10.0		430	410	390		120	445
6.0-10.0		435	420	400		120	445
2.0		520-540	510-530	490-510	490-530	175-250	
2.0		520	510	490	490	175	
2.0		520-540	510-530	490-510	490-530	175-250	

Grade	Filler	Sp Grav	Shrink, mils/in	Melt flow, g/10 min	Melt temp, °F	Back pres, psi	Drying temp, °F
NI-30GF	GLF 30	1.31	4.0				180
NI-30GF/15T	GLF 30	1.44	3.0				180
NI-40GF	GLF 40	1.40	3.0				180
Nylon 6/10		**Fiberfil**			**DSM**		
J-2/30	GLF 30	1.32	2.0				165-220
J-2/40	GLF 40	1.43	2.0				165-220
Nylon 6/10		**Fiberfil V0**			**DSM**		
J-2/30/V0	GLF 30	1.55	2.0				220
J-3/30/V0	GLF 30	1.62	1.0				220
Nylon 6/10		**Fiberstran**			**DSM**		
G-2/30	GLF 30	1.33	1.5			10- 50	165-180
G-2/40	GLF 40	1.39	1.3			10- 50	165-180
Nylon 6/10		**RTP Polymers**			**RTP**		
200B		1.08	20.0			25- 50	175
201B	GLF 10	1.15	7.0			25- 50	175
203B	GLF 20	1.22	5.0			25- 50	175
205B	GLF 30	1.31	4.0			25- 50	175
205B TFE 15	GLF	1.44	3.0			25- 50	175
207B	GLF 40	1.40	4.0			25- 50	175
209B	GLF 50	1.52	35.0			25- 50	175
Nylon 6/10		**Texalon**			**Texapol**		
1600A		1.08	12.0-14.0		415	50	150
GF 1600A-33	GL 33	1.32	2.0- 4.0		410	50	150
Nylon 6/10		**Ultramid**			**BASF**		
S2		1.07	12.0	481.00	425	50- 100	176-230
S3		1.07	12.0	64.00	425	50- 100	176-230
S3K		1.07	12.0		425	50- 100	176-230
S4		1.07	12.0		425	50- 100	176-230
Nylon 6/10		**Verton**			**LNP**		
QF-700-10 (D)	GLF 50	1.53	3.0- 4.0		424	0- 50	180-200
QF-7007	GLF 35	1.35	3.0- 4.0			0- 50	180-200
Nylon 6/12		**Aqualoy**			**ComAlloy**		
640	UNS	1.46	3.0				180
Nylon 6/12		**Ashlene**			**Ashley**		
980L		1.06	11.0		415	50- 150	150-160
980LS		1.06	11.0		415	50- 150	150-160
980LS-33G	GLC 33	1.32	3.0		413	50- 150	150-160
Nylon 6/12		**Comtuf**			**ComAlloy**		
608		1.00	15.0				180
640	UNS	1.42	2.0				180
643	GL 33	1.25	3.0				180
644	GL 43	1.38	2.0				180
649	MN 40	1.32	10.0				180
Nylon 6/12		**CTI Nylon**			**CTI**		
NL		1.08	12.0				190

Drying time, hr	Inj time, sec	Front temp, °F	Mid temp, °F	Rear temp, °F	Nozzle temp, °F	Mold temp, °F	Proc temp, °F
2.0		520-540	510-530	490-510	490-530	175-250	
2.0		520-540	510-530	490-510	490-530	175-250	
2.0		520-540	510-530	490-510	490-530	175-250	
2.0-16.0		510-530	530-550	520-540	500-520	130-180	520-540
2.0-16.0		510-530	530-550	520-540	500-520	130-180	520-540
2.0- 4.0		500-540	520-550	490-520	480-550	140-180	480-530
2.0- 4.0		500-540	520-550	490-520	480-550	140-180	480-530
4.0-16.0		530-550	550-570	540-560	520-540	175-225	540-580
4.0-16.0		530-550	550-570	540-560	520-540	175-225	540-580
4.0		520-550	510-540	500-530		100-175	
4.0		520-550	510-540	500-530		100-175	
4.0		520-550	510-540	500-530		100-175	
4.0		520-550	510-540	500-530		100-175	
4.0		520-550	510-540	500-530		100-175	
4.0		520-550	510-540	500-530		100-175	
4.0		520-550	510-540	500-530		100-175	
4.0		460	450	450	465	125	
4.0		480	470	470	490	125	
4.0-12.0						100-140	510-530
4.0-12.0						100-140	510-530
4.0-12.0						104-140	500-535
4.0-12.0						100-140	510-540
4.0		510-550	510-550	510-550		150-200	
4.0		510-550	510-550	510-550		150-200	
3.0- 4.0		510-550	530-570	540-580	530-540	100-250	540-580
2.0- 3.0		440	445	460	450	100-200	450-550
2.0- 3.0		440	445	460	450	100-200	450-550
2.0- 3.0		510-520	520-530	540-560	530-560	150-250	540-580
3.0- 4.0		510-550	530-570	540-580	530-540	100-250	540-580
3.0- 4.0		510-550	530-570	540-580	530-540	100-250	540-580
3.0- 4.0		510-550	530-570	540-580	530-540	100-250	540-580
3.0- 4.0		510-550	530-570	540-580	530-540	100-250	540-580
3.0- 4.0		510-550	530-570	540-580	530-540	100-250	540-580
2.0- 3.0		510	520	540	520	190	

Grade	Filler	Sp Grav	Shrink, mils/in	Melt flow, g/10 min	Melt temp, °F	Back pres, psi	Drying temp, °F
NL-10GF	GLF 10	1.14	5.0				190
NL-20CF	CF 20	1.15	2.0- 3.0				190
NL-20GF	GLF 20	1.20	4.0				190
NL-30CF	CF 30	1.20	1.0- 2.0				190
NL-30GF	GLF 30	1.31	3.0- 4.0				190
NL-30GF/000 FR	CF 30	1.58	3.0				190

Nylon 6/12 Electrafil DSM

Grade	Filler	Sp Grav	Shrink, mils/in	Melt flow, g/10 min	Melt temp, °F	Back pres, psi	Drying temp, °F
J-4/CF/30	CF 30	1.22	1.0				165-220
J-4/CF/30/EG	CF 30	1.22	1.0				165-220
J-4/CF/30/TF	CF 40	1.28	1.0				165-220
J-4/CF/50/EG	CF 50	1.34	1.0				165-220
J-4/CN/40	CFN 40	1.31	1.0				165-220

Nylon 6/12 Fiberfil DSM

Grade	Filler	Sp Grav	Shrink, mils/in	Melt flow, g/10 min	Melt temp, °F	Back pres, psi	Drying temp, °F
J-4/35	GLF 35	1.34	2.0				165-220
J-4/45	GLF 45	1.45	1.0				165-220

Nylon 6/12 Fiberfil TN DSM

Grade	Filler	Sp Grav	Shrink, mils/in	Melt flow, g/10 min	Melt temp, °F	Back pres, psi	Drying temp, °F
NY-12		1.03					220

Nylon 6/12 Fiberfil V0 DSM

Grade	Filler	Sp Grav	Shrink, mils/in	Melt flow, g/10 min	Melt temp, °F	Back pres, psi	Drying temp, °F
J-4/15/V0 (Black)	GLF 15	1.41	4.0- 5.0			25- 50	220
J-4/15/V0	GLF 15	1.41	4.0- 5.0			25- 50	220
J-4/15/V0 (Red)	GLF 15	1.41	4.0- 5.0			25- 50	220
J-4/30/V0	GLF 30	1.55	2.0- 3.0			25- 50	220

Nylon 6/12 Fiberstran DSM

Grade	Filler	Sp Grav	Shrink, mils/in	Melt flow, g/10 min	Melt temp, °F	Back pres, psi	Drying temp, °F
G-4/35	GLF 35	1.34	2.0			10- 50	165-180
G-4/40/MS/5	GLF 40	1.46	0.5				165-220
G-4/45	GLF 45	1.45	1.0- 2.0			10- 50	165-180
G-4/SS/15	STS 15	1.22	10.0				165-220

Nylon 6/12 Hiloy ComAlloy

Grade	Filler	Sp Grav	Shrink, mils/in	Melt flow, g/10 min	Melt temp, °F	Back pres, psi	Drying temp, °F
640	UNS	1.46	2.0				180
641	GL 13	1.13	6.0				180
642	GL 20	1.21	4.0				180
643	GL 33	1.32	2.0				180
644	GL 43	1.42	15.0				180
645	GL 50	1.50	1.0				180
646	GMN 40	1.42	4.0				180
648	UNS	1.26	10.0				180
649	MN	1.39	8.0				180

Nylon 6/12 Lubrilon ComAlloy

Grade	Filler	Sp Grav	Shrink, mils/in	Melt flow, g/10 min	Melt temp, °F	Back pres, psi	Drying temp, °F
643	GL 33	1.43	4.0				180

Nylon 6/12 Nybex Ferro

Grade	Filler	Sp Grav	Shrink, mils/in	Melt flow, g/10 min	Melt temp, °F	Back pres, psi	Drying temp, °F
30003 NA	GLF 30	1.30	4.0				200

Nylon 6/12 Nylafil DSM

Grade	Filler	Sp Grav	Shrink, mils/in	Melt flow, g/10 min	Melt temp, °F	Back pres, psi	Drying temp, °F
G-4/35	GLF 35	1.34	2.0				220
G-4/45	GLF 45	1.45	1.0				220
J-4/35	GLF 35	1.34	2.0				220
J-4/45	GLF 45	1.45	1.0				220

Drying time, hr	Inj time, sec	Front temp, °F	Mid temp, °F	Rear temp, °F	Nozzle temp, °F	Mold temp, °F	Proc temp, °F
2.0- 3.0		510-520	520-530	540-560	520-530	190-250	
2.0- 3.0		510-520	520-530	540-560	520-530	190-250	
2.0- 3.0		510-520	520-530	540-560	520-530	190-250	
2.0- 3.0		510	520	540	520	190	
2.0- 3.0		510	520	540	520	190	
2.0- 3.0		510	520	540	520	190	
2.0-16.0		510-530	530-550	520-540	500-520	130-180	520-540
2.0-16.0		510-530	530-550	520-540	500-520	130-180	520-540
2.0-16.0		510-530	530-550	520-540	500-520	130-180	520-540
2.0-16.0		510-530	530-550	520-540	500-520	130-180	520-540
2.0-16.0		510-530	530-550	520-540	500-520	130-180	520-540
2.0-16.0		510-530	530-550	520-540	500-520	130-180	520-540
2.0-16.0		510-530	530-550	520-540	500-520	130-180	520-540
2.0		460-470	450-460	440-450	430-450	175-190	470-500
2.0- 4.0		500-540	520-550	490-520	480-550	140-180	480-530
2.0- 4.0		500-540	520-550	490-520	480-550	140-180	480-530
2.0- 4.0		500-540	520-550	490-520	480-550	140-180	480-530
2.0- 4.0		500-540	520-550	490-520	480-550	140-180	480-530
4.0-16.0		510-530	530-560	520-540	520-540	175-225	530-570
2.0-16.0		510-530	530-550	520-540	500-520	130-180	520-540
4.0-16.0		510-530	530-560	520-540	520-540	175-225	530-570
2.0-16.0		510-530	530-550	520-540	500-520	130-180	520-540
3.0- 4.0		510-550	530-570	540-580	530-540	100-250	540-580
3.0- 4.0		510-550	530-570	540-580	530-540	100-250	540-580
3.0- 4.0		510-550	530-570	540-580	530-540	100-250	540-580
3.0- 4.0		510-550	530-570	540-580	530-540	100-250	540-580
3.0- 4.0		510-550	530-570	540-580	530-540	100-250	540-580
3.0- 4.0		510-550	530-570	540-580	530-540	100-250	540-580
3.0- 4.0		510-550	530-570	540-580	530-540	100-250	540-580
3.0- 4.0		510-550	530-570	540-580	530-540	100-250	540-580
3.0- 4.0		510-550	530-570	540-580	530-540	100-250	540-580
3.0- 4.0		510-550	530-570	540-580	530-540	100-250	540-580
						200	480-550
2.0		510-530	530-550	520-540	500-520	130-180	520-540
2.0		510-530	530-550	520-540	500-520	130-180	520-540
2.0		510-530	530-550	520-540	500-520	130-180	520-540
2.0		510-530	530-550	520-540	500-520	130-180	520-540

Grade	Filler	Sp Grav	Shrink, mils/in	Melt flow, g/10 min	Melt temp, °F	Back pres, psi	Drying temp, °F
Nylon 6/12		**Plaslube**			**DSM**		
J-4/30/TF/15	GLF 30	1.45	2.0				165-220
J-4/CF/30/TF/10	CF 30	1.30	1.3				165-200
J-4/CF/30/TF/13	CF 20	1.30	1.0				165-220
Nylon 6/12		**RTP Polymers**			**RTP**		
200D		1.06	15.0			25- 50	175
200D TFE 20	PTF	1.19	11.0			25- 50	175
201D	GLF 10	1.14	6.0			25- 50	175
203D	GLF 20	1.20	5.0			25- 50	175
204D FR	GLF	1.56	4.0			25- 50	175
204D TFE 15 FR	GLF	1.54	3.0			25- 50	175
205D	GLF 30	1.31	4.0			25- 50	175
205D FR	GLF	1.60	3.0			25- 50	175
205D TFE 15	GLF	1.42	3.0			25- 50	175
207D	GLF 40	1.39	4.0			25- 50	175
299DX50026B		1.15	10.0-20.0			25- 50	175
Nylon 6/12		**Texalon**			**Texapol**		
1800A		1.08	12.0-13.0		410	50	150
GF 1800A-33	GL 33	1.32	2.0- 4.0		419	50	150
Nylon 6/12		**Voloy**			**ComAlloy**		
642	GL 25	1.46	4.0				180
648	MN 25	1.46	90.0				180
Nylon 6/12		**Zytel**			**Du Pont**		
151 L		1.06	11.0		414		175
153 HSL		1.06			414		175
157 HS-L BK-10	CBL	1.06	11.0		414		175
158 L		1.06	11.0		414		175
77G 33L	GLC	1.32	2.0		414		175
77G 43L	GLC	1.46	1.0		414		175
Nylon 6/6		**Akulon**			**DSM**		
S223-D		1.14				0- 50	230
S223-EH		1.14	10.0			0- 50	230
S225-KS		1.16	12.5				220
Nylon 6/6		**Albis Polyamid 66**			**Albis Canada**		
PA140/1GF30	GLF 30	1.36			482-491		170-180
PA140/1GF33	GLF 33	1.41			482-491		170-180
PA145/1MR40	MN 40	1.50			482-491		170-180
PA150/1		1.13			482-491		170-180
Nylon 6/6		**Aqualoy**			**ComAlloy**		
623	UNS	1.28	5.0				180
624	UNS	1.36	3.0				180
Nylon 6/6		**Ashlene**			**Ashley**		
520		1.13	15.0		480	50- 150	150-160
520-13G	GLC 13	1.23	7.0		485	50- 150	150-160
520-25GU	GLC 25	1.30			485	50- 150	150-160
520-33G	GLC 33	1.37	2.0		485	50- 150	150-160
520-33GU	GLC 33	1.37	2.0		485	50- 150	150-160

Drying time, hr	Inj time, sec	Front temp, °F	Mid temp, °F	Rear temp, °F	Nozzle temp, °F	Mold temp, °F	Proc temp, °F
2.0-16.0		510-530	530-550	520-540	500-520	130-200	520-540
2.0-16.0		510-530	530-550	520-540	500-520	130-200	520-540
2.0-16.0		510-530	530-550	520-540	500-520	130-200	520-540
4.0		500-535	495-525	485-515		140-200	
4.0		500-535	495-525	485-515		140-200	
4.0		500-535	495-525	485-515		140-200	
4.0		500-535	495-525	485-515		140-200	
4.0		500-520	490-510	480-500		140-200	
4.0		500-520	490-510	480-500		140-200	
4.0		500-535	495-525	485-515		140-200	
4.0		500-520	490-510	480-500		140-200	
4.0		500-535	495-525	485-515		140-200	
4.0		500-535	495-525	485-515		140-200	
4.0		500-535	495-525	485-515		140-200	
4.0		460	450	450	465	125	
4.0		480	470	470	490	125	
3.0- 4.0		510-550	530-570	540-580	530-540	100-250	540-580
3.0- 4.0		510-550	530-570	540-580	530-540	100-250	540-580
2.0-24.0		440	445	460	450	100-200	450-550
2.0-24.0		440	445	460	450	100-200	450-550
2.0-24.0		440	445	460	450	100-200	450-550
2.0-24.0		440	445	460	450	100-200	450-550
2.0-24.0	0.5- 3.	520-530	510-520	530-560	540-560	150-250	540-580
2.0-24.0	0.5- 3.	520-530	510-520	530-560	540-560	150-250	540-580
3.0		520	535	535	520	175-195	520-555
3.0		520	535	535	525	175-195	520-555
2.0- 4.0		530-540	540-550	540-550	520-530	175-190	520-560
2.0						175-250	520-570
2.0						175-250	520-570
2.0						175-250	520-570
2.0						160-195	500-555
3.0- 4.0		470-510	480-520	490-530	470-510	100-250	480-520
3.0- 4.0		470-510	480-520	490-530	470-510	100-250	480-520
2.0- 3.0		520	525	540	500-570	100-200	535-580
2.0- 3.0		500-510	500-510	520-540	520-530	150-250	520-540
2.0- 3.0		500-510	500-510	520-540	520-530	150-250	520-540
2.0- 3.0		500-510	500-510	520-540	520-530	150-250	520-540
2.0- 3.0		500-510	500-510	520-540	520-530	150-250	520-540

Grade	Filler	Sp Grav	Shrink, mils/in	Melt flow, g/10 min	Melt temp, °F	Back pres, psi	Drying temp, °F
520-50G	GLC 50	1.57	2.0		485	50- 150	150-160
520B		1.13	15.0		480	50- 150	150-160
520BU		1.13			430	50- 150	150-160
520MS	MOS	1.17	12.0		480	50- 150	150-160
521	TIO	1.13	16.0		430	50- 150	150-160
522	TIO 1	1.13	16.0		430	50- 150	150-160
525		1.08	16.0		480	50- 150	150-160
525-13G	GLF 13	1.17			485	50- 150	150-160
525-33G	GLF 33	1.35			485	50- 150	150-160
525LD		1.08	15.0		491	50- 150	150-160
525LD-13G	GLF 13	1.17			489	50- 150	150-160
525LD-33G	GLF 33	1.35			485	50- 150	150-160
526LD		1.11	16.0		491	50- 150	150-160
527		1.07	16.0		480	50- 150	150-160
527-13G	GLC 13	1.19	8.0		485	50- 150	150-160
527-33G	GLC 33	1.34	3.0		485	50- 150	150-160
527LD-13G	GLF 13	1.19	8.0		490	50- 150	150-160
527LD-14G	GL 14				485	50- 150	150-160
527LDS-B2	CBL	1.07	15.0		485	50- 150	150-160
527LDW		1.07	16.0		490	50- 150	150-160
528BR-WO		1.25	15.0		490	50- 150	150-160
528L-13G	GLF 13	1.21	5.0		485	50- 150	150-160
528L-25G	GLF 25	1.31	5.0		485	50- 150	150-160
528L-30G	GLF 30	1.36			485	50- 150	150-160
528L-33G	GLF 33	1.33	2.0		490	50- 150	150-160
528L-33G/MS	GLF 33	1.43	2.0		485	50- 150	150-160
528L-5G	GLF 5	1.16	6.0		485	50- 150	150-160
528L2		1.13	15.0		485	50- 150	150-160
528LW		1.13	15.0		491	50- 150	150-160
528MS	MOS	1.17	12.0		480	50- 150	150-160
528SMS-30GB	GLF 30	1.41	3.4		485	50- 150	150-160
528TF		1.13	15.0		475	50- 150	150-160
60M	MN 40	1.50	13.0		480-510	50- 150	150-160
62M	MN	1.49	9.0		493	50- 150	150-160
63M	MN	1.40	10.0		490	50- 150	150-160

Nylon 6/6 AVP Polymerland

Grade	Filler	Sp Grav	Shrink, mils/in	Melt flow, g/10 min	Melt temp, °F	Back pres, psi	Drying temp, °F
GYYHM		1.14	5.0-15.0	45.00 I		10- 50	165
RYY33	GL 33	1.41	4.0- 6.0	50.00 I		10- 50	165

Nylon 6/6 Bay Resins

Grade	Filler	Sp Grav	Shrink, mils/in	Melt flow, g/10 min	Melt temp, °F	Back pres, psi	Drying temp, °F
PA-111CF30	CF 30	1.28	2.0			50- 100	200
PA-111CF30 TF15	CF 30	1.38	2.5			50- 100	200
PA-111GF30 TF15	GLF 30	1.49	4.0			50- 100	200
PA-111M40	MN 40	1.48	9.0			50- 100	200
PA-111N40	GLF 15	1.49	6.0			50- 100	200
PA-111TF20		1.26	15.0			50- 100	200
PA-113		1.14	17.0			50- 100	200
PA-113CF30	CF 30	1.28	2.0			50- 100	200
PA-113CF30 TF15	CF 30	1.38	2.5			50- 100	200
PA-113G13	GLF 13	1.21	6.0			50- 100	200
PA-113G20	GLF 20	1.28	5.0			50- 100	200
PA-113G33	GLF 33	1.37	4.0			50- 100	200
PA-113G43	GLF 43	1.47	4.0			50- 100	200
PA-113GF30 TF15	GLF 30	1.49	4.0			50- 100	200

Drying time, hr	Inj time, sec	Front temp, °F	Mid temp, °F	Rear temp, °F	Nozzle temp, °F	Mold temp, °F	Proc temp, °F
2.0- 3.0		500-510	500-510	520-540	520-530	150-250	520-540
2.0- 3.0		520	525	540	500-570	100-200	535-580
2.0- 3.0		520	525	540	500-570	100-200	535-580
2.0- 3.0		520	525	540	500-570	100-200	535-580
2.0- 3.0		520	525	540	500-570	100-200	535-580
2.0- 3.0		520	525	540	500-570	100-200	535-580
2.0- 3.0		525	535	560	500-570	100-200	550-560
2.0- 3.0		520-530	530-540	550-570	540-560	150-250	550-590
2.0- 3.0		520-530	530-540	550-570	540-560	150-250	550-590
2.0- 3.0		525	535	560	500-570	100-200	550-560
2.0- 3.0		520-530	530-540	550-570	540-560	150-250	550-590
2.0- 3.0		520-530	530-540	550-570	540-560	150-250	550-590
2.0- 3.0		525	535	560	500-570	100-200	550-560
2.0- 3.0		525	535	560	500-570	100-200	550-560
2.0- 3.0		520-530	530-540	550-570	540-560	150-250	550-590
2.0- 3.0		520-530	530-540	550-570	540-560	150-250	550-590
2.0- 3.0		520-530	530-540	550-570	540-560	150-250	550-590
2.0- 3.0		520-530	530-540	550-570	540-560	150-250	550-590
2.0- 3.0		525	535	560	500-570	100-200	550-560
2.0- 3.0		525	535	560	500-570	100-200	550-560
2.0- 3.0		520	525	540	500-570	100-200	535-580
2.0- 3.0		500-510	500-510	520-540	520-530	150-250	520-540
2.0- 3.0		500-510	500-510	520-540	520-530	150-250	520-540
2.0- 3.0		500-510	500-510	520-540	520-530	150-250	520-540
2.0- 3.0		500-510	500-510	520-540	520-530	150-250	520-540
2.0- 3.0		500-510	500-510	520-540	520-530	150-250	520-540
2.0- 3.0		500-510	500-510	520-540	520-530	150-250	520-540
2.0- 3.0		520	525	540	500-570	100-200	535-580
2.0- 3.0		520	525	540	500-570	100-200	535-580
2.0- 3.0		520	525	540	500-570	100-200	535-580
2.0- 3.0		520	525	540	500-570	100-200	535-580
2.0- 3.0		520	525	540	500-570	100-200	535-580
2.0- 3.0		530-540	530-550	540-560	540-550	150-230	540-570
2.0- 3.0		530-540	530-550	540-560	540-550	150-230	540-570
2.0- 3.0		530-540	530-550	540-560	540-550	150-230	540-570
8.0						140-200	500-550
8.0						160-220	500-550
2.0- 4.0		510-560	510-560	510-560		150-200	510-550
2.0- 4.0		510-560	510-560	510-560		150-200	510-550
2.0- 4.0		510-560	510-560	510-560		150-200	510-550
2.0- 4.0		510-560	510-560	510-560		150-200	510-550
2.0- 4.0		510-560	510-560	510-560		150-200	510-550
2.0- 4.0		510-560	510-560	510-560		150-200	510-550
2.0- 4.0		510-560	510-560	510-560		150-200	510-550
2.0- 4.0		510-560	510-560	510-560		150-200	510-550
2.0- 4.0		510-560	510-560	510-560		150-200	510-550
2.0- 4.0		510-560	510-560	510-560		150-200	510-550
2.0- 4.0		510-560	510-560	510-560		150-200	510-550
2.0- 4.0		510-560	510-560	510-560		150-200	510-550
2.0- 4.0		510-560	510-560	510-560		150-200	510-550
2.0- 4.0		510-560	510-560	510-560		150-200	510-550

Grade	Filler	Sp Grav	Shrink, mils/in	Melt flow, g/10 min	Melt temp, °F	Back pres, psi	Drying temp, °F
PA-113M40	MN 40	1.48	9.0			50- 100	200
PA-113TF20		1.26	15.0			50- 100	200
PA111		1.13			491	50- 100	200
PA111G13	GLF 13	1.20			491	50- 100	200
PA111G20	GLF 20	1.28			491	50- 100	200
PA111G33	GLF 33	1.36			491	50- 100	200
PA111G43	GLF 43	1.48			491	50- 100	200
PA113M40W	MN 40	1.47			492	50- 100	200

Nylon 6/6 Capron Allied Signal

Grade	Filler	Sp Grav	Shrink, mils/in	Melt flow, g/10 min	Melt temp, °F	Back pres, psi	Drying temp, °F
D-5233G HS BK-102	GL 33	1.38			491		180

Nylon 6/6 Celanese Nylon 6/6 Hoechst

Grade	Filler	Sp Grav	Shrink, mils/in	Melt flow, g/10 min	Melt temp, °F	Back pres, psi	Drying temp, °F
1000		1.14	16.0		495	0- 50	140
1000 UV		1.14				0- 50	140
1003		1.14	16.0		495	0- 50	140
1310		1.14	14.0		495	0- 50	140
1500	GLF 33	1.38	4.0		495	0- 50	140
1500 FDA	GLF 33	1.38	4.0		495	0- 50	140
1500 UV	GLF 33	1.38				0- 50	140
1503	GLF 33	1.38	4.0		495	0- 50	140
1503 FDA	GLF 33	1.38	4.0		495	0- 50	140
6020		1.11	14.0		495	0- 50	140
6023		1.11	14.0		495	0- 50	140
6030		1.09	13.0		495	0- 50	140
6030 UV		1.09				0- 50	140
6033		1.09	13.0		495	0- 50	140
7010		1.10	16.0		495	0- 50	140
7013		1.10	16.0		495	0- 50	140
7020		1.09	16.0		495	0- 50	140
7023		1.09	16.0		495	0- 50	140
7030		1.08	16.0		495	0- 50	140
7033		1.08	16.0		495	0- 50	140
7420	GLF 13	1.17	5.0		495	0- 50	140
7423	GLF 13	1.17	5.0		495	0- 50	140
7520	GLF 33	1.33	4.0		495	0- 50	140
7523	GLF 33	1.33	4.0		495	0- 50	140

Nylon 6/6 Celstran Hoechst

Grade	Filler	Sp Grav	Shrink, mils/in	Melt flow, g/10 min	Melt temp, °F	Back pres, psi	Drying temp, °F
N66C40-01-4	CF 40	1.33					175
N66G30-01-4	GLF 30	1.36					175
N66G30-02-4	GLF 30	1.36					175
N66G40-01-4	GLF 40	1.45					175
N66G40-02-4	GLF 40	1.45					175
N66G50-01-4	GLF 50	1.56					175
N66G50-02-4	GLF 50	1.56					175
N66G60-01-4	GLF 60	1.69					175
N66G60-02-4	GLF 60	1.69					175
N66K35-02-4	KEV 35	1.22					175

Nylon 6/6 Celstran S Hoechst

Grade	Filler	Sp Grav	Shrink, mils/in	Melt flow, g/10 min	Melt temp, °F	Back pres, psi	Drying temp, °F
N66S10-01-4	STS 10	1.24	3.0- 4.0				175
N66S10-02-4	STS 10	1.24	3.0- 4.0				175
N66S10-03-4	STS 10	1.48	3.0- 5.0				175
N66S6-01-4	STS 6	1.19	4.0- 5.0				175
N66S6-02-4	STS 6	1.19	4.0- 5.0				175

Drying time, hr	Inj time, sec	Front temp, °F	Mid temp, °F	Rear temp, °F	Nozzle temp, °F	Mold temp, °F	Proc temp, °F
2.0- 4.0		510-560	510-560	510-560		150-200	510-550
2.0- 4.0		510-560	510-560	510-560		150-200	510-550
2.0- 4.0		510-560	510-560	510-560		150-200	510-550
2.0- 4.0		510-560	510-560	510-560		150-200	510-550
2.0- 4.0		510-560	510-560	510-560		150-200	510-550
2.0- 4.0		510-560	510-560	510-560		150-200	510-550
2.0- 4.0		510-560	510-560	510-560		150-200	510-550
2.0- 4.0		510-560	510-560	510-560		150-200	510-550
		540-580	520-560	500-540	540-580	180-200	540-580
1.0- 2.0	1.0- 5.	530	520	510	540	140-212	530-580
1.0- 2.0	1.0- 5.	530	520	510	540	140-212	530-580
1.0- 2.0	1.0- 5.	530	520	510	540	140-212	530-580
1.0- 2.0	1.0- 5.	530	520	510	540	140-212	530-580
1.0- 2.0	1.0- 5.	530	520	510	540	140-212	530-580
1.0- 2.0	1.0- 5.	530	520	510	540	140-212	530-580
1.0- 2.0	1.0- 5.	530	520	510	540	140-212	530-580
1.0- 2.0	1.0- 5.	530	520	510	540	140-212	530-580
1.0- 2.0	1.0- 5.	530	520	510	540	140-212	530-580
1.0- 2.0	1.0- 5.	530	520	510	540	140-212	530-580
1.0- 2.0	1.0- 5.	530	520	510	540	140-212	530-580
1.0- 2.0	1.0- 5.	530	520	510	540	140-212	530-580
1.0- 2.0	1.0- 5.	530	520	510	540	140-212	530-580
1.0- 2.0	1.0- 5.	530	520	510	540	140-212	530-580
1.0- 2.0	1.0- 5.	530	520	510	540	140-212	530-580
1.0- 2.0	1.0- 5.	530	520	510	540	140-212	530-580
1.0- 2.0	1.0- 5.	530	520	510	540	140-212	530-580
1.0- 2.0	1.0- 5.	530	520	510	540	140-212	530-580
1.0- 2.0	1.0- 5.	530	520	510	540	140-212	530-580
1.0- 2.0	1.0- 5.	530	520	510	540	140-212	530-580
1.0- 2.0	1.0- 5.	530	520	510	540	140-212	530-580
1.0- 2.0	1.0- 5.	530	520	510	540	140-212	530-580
1.0- 2.0	1.0- 5.	530	520	510	540	140-212	530-580
1.0- 2.0	1.0- 5.	530	520	510	540	140-212	530-580
4.0		580-600	570-590	560-580	580-600	190-210	580-600
4.0		550	550	540	560	200	
4.0		540-560	530-550	520-540	530-550	190-210	540-560
4.0		560	560	550	570	200	
4.0		550-570	540-560	530-550	540-560	190-210	550-570
4.0		570	570	560	580	200	
4.0		560-580	550-570	540-560	550-570	190-210	560-580
4.0		585	580	570	590	200	
4.0		570-590	560-580	550-570	560-580	190-210	570-590
4.0		590-610	580-600	550-570	580-600	190-210	590-610
4.0		490	480	470	500	200	
4.0		550	540	540	550	200	
4.0		540	520	510	540	200	
4.0		490	480	470	500	200	
4.0		550	540	540	550	200	

Grade	Filler	Sp Grav	Shrink, mils/in	Melt flow, g/10 min	Melt temp, °F	Back pres, psi	Drying temp, °F
N66S6-03-4	STS 6	1.44	4.0- 6.0				175
Nylon 6/6		**Chem Polymer Nylon**			**Chem Polyme**		
NP 100		1.14	15.0		490		150-180
NP 100MD		1.16	15.0		486		150-180
NP 102		1.14	11.0		490		150-180
NP 104		1.09	15.0		490		150-180
NP 104-13	GL 13	1.17	5.0		490		
NP 109		1.07	18.0		485		150-180
NP 109-33	GL 33	1.34	3.0		490		150-180
NP 113GF	GLF 13	1.22	6.0		490		150-180
NP 125-15MFGF	GLF 15	1.42	5.0		490		150-180
NP 125GF	GLF 25	1.33	4.0		490		150-180
NP 133GF	GLF 33	1.38	3.0		490		150-180
NP 140MF	MN 40	1.50	8.0		490		150-180
NP 143GF	GLF 43	1.50	2.0		490		150-180
Nylon 6/6		**Comtuf**			**ComAlloy**		
607		1.08	15.0				180
620	UNS	1.46	3.0				180
621	GL 13	1.18	6.0				180
623	GL 33	1.34	4.0				180
626	GMN 40	1.41	4.0				180
629	MN 40	1.41	12.0				180
630	UNS	1.46	3.0				180
631	UNS 13	1.18	6.0				180
633	GL	1.34	4.0				180
634	GL 43	1.43	3.0				180
636	GMN 40	1.41	4.0				180
639	MN 40	1.41					180
Nylon 6/6		**CTI General**			**CTI**		
CTC-3300	GLF 33	1.40	3.0				180
CTX-323	GLF 13	1.18	7.0- 7.5				180
Nylon 6/6		**CTI Nylon**			**CTI**		
NN		1.14	16.0				180
NN-10GF	GLF 10	1.22	7.0				180
NN-10GF/000FR	GLF 10	1.37	9.0-10.0				180
NN-10GF/20GB	GLF 10	1.37	6.0				180
NN-10KV	KEV 10	1.17	8.0				180
NN-15GF/000FR	GLF 15	1.42	8.5- 9.0				180
NN-20CF	CF 20	1.22	2.0- 4.0				180
NN-20GF	GLF 20	1.29	6.0				180
NN-29/17T-2S	BR 39	1.96	8.5- 9.5				180
NN-30CF	CF 30	1.27	2.0				180
NN-30GF	GLF 30	1.37	4.0- 5.0				180
NN-40CF	CF 40	1.28	1.0- 3.0				180
NN-40GB	GLB 40	1.46	12.0				180
NN-40GF	GLF 40	1.46					180
NN-40GM	GL 40	1.46	6.0- 8.0				180
NN-40MN	MN 40	1.47	12.0				180
NN-50GF	GLF 50	1.57					180
NN-60CF	CF 60	1.43	1.0- 3.0				180
NN-60GF	GLF 60	1.71					180

Drying time, hr	Inj time, sec	Front temp, °F	Mid temp, °F	Rear temp, °F	Nozzle temp, °F	Mold temp, °F	Proc temp, °F
4.0		540	520	510	540	200	
		500	520	540	500	70-200	
		500	520	540	500	70-200	
		500	520	540	500	70-200	
		500	520	540	500	70-200	
		530	540	560	530	140-200	
		500	520	540	500	70-200	
		530	540	560	530	140-200	
		530	540	560	530	140-200	
		530	540	560	530	140-200	
		530	540	560	530	140-200	
		530	540	560	530	140-200	
		530	540	560	530	140-200	
		530	540	560	530	140-200	
3.0- 4.0		470-510	480-520	490-530	470-510	100-250	480-520
3.0- 4.0		470-510	480-520	490-530	470-510	100-250	480-520
3.0- 4.0		470-510	480-520	490-530	470-510	100-250	480-520
3.0- 4.0		470-510	480-520	490-530	470-510	100-250	480-520
3.0- 4.0		470-510	480-520	490-530	470-510	100-250	480-520
3.0- 4.0		470-510	480-520	490-530	470-510	100-250	480-520
3.0- 4.0		470-510	480-520	490-530	470-510	100-250	480-520
3.0- 4.0		470-510	480-520	490-530	470-510	100-250	480-520
3.0- 4.0		470-510	480-520	490-530	470-510	100-250	480-520
3.0- 4.0		470-510	480-520	490-530	470-510	100-250	480-520
3.0- 4.0		470-510	480-520	490-530	470-510	100-250	480-520
3.0- 4.0		470-510	480-520	490-530	470-510	100-250	480-520
3.0- 4.0		520-530	530-540	550-570	530-550	200-250	
3.0- 4.0		520	530	550	530	200	
3.0- 4.0		520	530	550	530	200	
3.0- 4.0		520	530	550	530	200	
3.0- 4.0		520	530	550	530	200	
3.0- 4.0		520	530	550	530	200	
3.0- 4.0		520	530	550	530	200	
3.0- 4.0		520	530	550	530	200	
3.0- 4.0		520	530	550	530	200	
3.0- 4.0		520	530	550	530	200	
3.0- 4.0		520-530	530-540	550-570	530-550	200-250	
3.0- 4.0		520-530	530-540	550-570	530-550	200-250	
3.0- 4.0		520-530	530-540	550-570	530-550	200-250	
3.0- 4.0		520-530	530-540	550-570	530-550	200-250	
3.0- 4.0		520-530	530-540	550-570	530-550	200-250	
3.0- 4.0		520-530	530-540	550-570	530-550	200-250	
3.0- 4.0		520-530	530-540	550-570	530-550	200-250	
3.0- 4.0		520-530	530-540	550-570	530-550	200-250	
3.0- 4.0		520-530	530-540	550-570	530-550	200-250	
3.0- 4.0		520-530	530-540	550-570	530-550	200-250	
3.0- 4.0		520-530	530-540	550-570	530-550	200-250	

Grade	Filler	Sp Grav	Shrink, mils/in	Melt flow, g/10 min	Melt temp, °F	Back pres, psi	Drying temp, °F
Nylon 6/6		**CTI Super Tough Nylon**				**CTI**	
NT-15GF/000	GLF 15	1.18	8.5- 9.0				180
NT-30GF/000	GLF 30	1.31	2.0				180
NT-33GF/000 FR	GLF 33	1.58	2.0				180
NT-40MN/000	MN 40	1.42	8.0				180
Nylon 6/6		**Electrafil**			**DSM**		
J-1/30/CF/7/H	CF 7	1.40	1.0				165-220
J-1/CF/10	CF 10	1.18	3.0				165-220
J-1/CF/15/TF/20	CF 15	1.33	2.5				165-220
J-1/CF/20	CF 20	1.23	2.0				165-220
J-1/CF/30	CF 30	1.28	1.0				165-220
J-1/CF/30/TF/13	CF 30	1.36	1.0				165-220
J-1/CF/30/TF/15	CF 30	1.38	1.0				165-220
J-1/CF/40	CF 40	1.33	0.5				165-220
J-1/CF/50/EG	CF 50	1.38	0.5				165-220
J-17/CF/15/EG/VO	CF 15	1.42	1.0			50	220
Nylon 6/6		**Ferro Nylon**			**Ferro**		
RNY20MA	GLF 20	1.28	4.0			50- 100	
RNY33MA	GLF 33	1.39	3.0			50- 100	
Nylon 6/6		**Fiberfil**			**DSM**		
J-1/10	GLF 10	1.23	5.0				165-220
J-1/15/MF/25	GLF 15	1.50	2.0				165-220
J-1/20	GLF 20	1.30	4.0				165-220
J-1/30	GLF 30	1.39	2.0				165-220
J-1/40	GLF 40	1.50	1.0				165-220
NY-16/MF/40	MN 40	1.50	1.0				165-220
Nylon 6/6		**Fiberfil TN**			**DSM**		
J-8/13	GLF 13	1.18	4.0- 7.0				220
J-8/33	GLF 33	1.32	2.0- 4.0				220
J-8/33/IT	GLF 33	1.38	1.0- 3.0				220
J-8/43	GLF 43	1.42	1.5- 3.5				220
NY-8		1.08	20.0-28.0				220
NY-8/MF/36	MN 36	1.41	17.0-22.0				220
Nylon 6/6		**Fiberfil VO**			**DSM**		
J-1/20/FR	GLF 20	1.51	4.0- 5.0				220
J-17/20/VO	GLF 20	1.53	2.0- 3.0				220
J-17/30/VO	GLF 30	1.62	2.0- 2.0				220
Xylon		1.42	10.0-16.0				220
Nylon 6/6		**Fiberstran**			**DSM**		
G-1/30	GLF 30	1.40	3.0			10- 50	165-180
G-1/30/MS/5	GLF 30	1.44					165-220
G-1/30/SI/2	GLF 30	1.40	3.0				165-220
G-1/30/TF/15	GLF 30	1.52	2.0				165-220
G-1/40	GLF 40	1.47	2.0			10- 50	165-180
G-1/40/MS/5	GLF 40	1.51	3.0				165-220
G-1/SS/5	STS 5	1.22	15.0				165-220
G-8/40	GLF 40	1.40					165-220

Drying time, hr	Inj time, sec	Front temp, °F	Mid temp, °F	Rear temp, °F	Nozzle temp, °F	Mold temp, °F	Proc temp, °F
3.0- 4.0		520	530	550	530	200	
3.0- 4.0		520-530	530-540	550-570	530-550	200-250	
3.0- 4.0		520-530	530-540	550-570	530-550	200-250	
3.0- 4.0		520-530	530-540	550-570	530-550	200-250	
2.0-16.0		530-550	550-570	540-560	520-540	130-200	540-580
2.0-16.0		530-550	550-570	540-560	520-540	130-200	540-580
2.0-16.0		530-550	550-570	540-560	520-540	130-200	540-580
2.0-16.0		530-550	550-570	540-560	520-540	130-200	540-580
2.0-16.0		530-550	550-570	540-560	520-540	130-200	540-580
2.0-16.0		530-550	550-570	540-560	520-540	130-200	540-580
2.0-16.0		530-550	550-570	540-560	520-540	130-200	540-580
2.0-16.0		530-550	550-570	540-560	520-540	130-200	540-580
2.0-16.0		530-550	550-570	540-560	520-540	130-200	540-580
2.0		510	530	520	500	140	520
		510-530	510-530	510-530	500-540	150-180	
		510-530	510-530	510-530	500-540	150-180	
2.0-16.0		530-550	550-570	540-560	540-560	130-200	540-580
2.0-16.0		530-550	550-570	540-560	540-560	130-200	540-580
2.0-16.0		530-550	550-570	540-560	540-560	130-200	540-580
2.0-16.0		530-550	550-570	540-560	540-560	130-200	540-580
2.0-16.0		530-550	550-570	540-560	540-560	130-200	540-580
2.0-16.0		530-550	550-570	540-560	540-560	130-200	540-580
2.0- 4.0		530-550	550-570	540-560	540-560	130-200	540-580
2.0- 4.0		530-550	550-570	540-560	540-560	130-200	540-580
2.0- 4.0		530-550	550-570	540-560	540-560	130-200	540-580
2.0- 4.0		530-550	550-570	540-560	540-560	130-200	540-580
2.0- 4.0		530-550	550-570	540-560	540-560	130-200	540-580
2.0- 4.0		530-550	550-570	540-560	540-560	130-200	540-580
2.0- 4.0		500-540	520-550	490-520	480-550	140-180	480-530
2.0- 4.0		500-540	520-550	490-520	480-550	140-180	480-530
2.0- 4.0		500-540	520-550	490-520	480-550	140-180	480-530
2.0- 4.0		500-540	520-550	490-520	480-550	140-180	480-530
4.0-16.0		530-550	550-570	540-560	520-540	175-225	540-580
2.0-16.0		530-550	550-570	540-560	520-540	130-200	540-580
2.0-16.0		530-550	550-570	540-560	520-540	130-200	540-580
2.0-16.0		530-550	550-570	540-560	520-540	130-200	540-580
4.0-16.0		530-550	550-570	540-560	520-540	175-225	540-580
2.0-16.0		530-550	550-570	540-560	520-540	130-200	540-580
2.0-16.0		530-550	550-570	540-560	520-540	130-200	540-580
2.0-16.0		530-550	550-570	540-560	520-540	130-200	540-580

Grade	Filler	Sp Grav	Shrink, mils/in	Melt flow, g/10 min	Melt temp, °F	Back pres, psi	Drying temp, °F
Nylon 6/6		**Grilon**			**EMS**		
T300FC		1.14	6.0-10.0		493		176
T300GM		1.14	7.0-12.0		493		176
T300GMH		1.13	7.0-12.0		500		176
T302VO		1.14	11.0-16.0		493		176
TV-15H	GLF 15	1.24	4.0-16.0		493		176
TV-35H	GLF 35	1.39	2.0-16.0		493		176
TV-3H	GLF 30	1.34	2.0-20.0		493		176
Nylon 6/6		**Hiloy**			**ComAlloy**		
620	GLF	1.49	4.0				180
621	GL 13	1.22	6.0				180
622	GL 20	1.28	4.0				180
623	GL 33	1.38	3.0				180
624	GL 43	1.50	2.0				180
625	GL 50	1.56	2.0				180
626	GMN 40	1.48	5.0				180
629	MN 40	1.49	10.0				180
630	UNS	1.51	3.0				180
Nylon 6/6		**Kopa**			**Kolon**		
KN331		1.14	10.0-12.0		491	142	158-176
KN3321G15VO	GLF 15	1.43	5.0-10.0		491	213	158-176
KN3322VO		1.16	13.0-16.0		491	142	158-176
KN333G30	GLF 30	1.37	3.0- 7.0		491	213	158-176
KN333G30VO	GLF 30	1.60	3.0- 7.0		491	213	158-176
KN333HB	GLF	1.42	3.0- 7.0		491	213	158-176
KN333HI		1.07	13.0-18.0		491	142	158-176
KN333HR		1.14	10.0-12.0		491	142	158-176
KN333MS		1.16	10.0-12.0		491	142	158-176
KN333MT30	MN 30	1.37	3.0- 8.0		491	213	158-176
Nylon 6/6		**Lubricomp**			**LNP**		
RAL-4022 FR	KEV 10	1.23	8.0-11.0			0- 75	160
RFL-4036	GLF 30	1.49	4.0- 5.0		500	50- 100	160
Nylon 6/6		**Lubrilon**			**ComAlloy**		
602	MOS 4	1.17	15.0				180
606	PTF 20	1.25	14.0				180
623	GL 33	1.50	4.0				180
628	GL 33	1.43	5.0				180
629	GL 33	1.51	4.0				180
Nylon 6/6		**Lupon**			**Lucky**		
GP-2251A-F	GLF 25	1.56	4.0- 8.0			0- 568	194
GP-2330A	GLF 33	1.38	2.0- 6.0			0- 568	194
GP-2337A	GLF 33	1.38	2.0- 6.0			0- 568	194
GP-2430A	GLF 43	1.51	2.0- 4.0			0- 568	194
HI-1002A		1.08	10.0-18.0			0- 568	194
HI-2332A	GLF 33	1.34	3.0- 8.0			0- 568	194
LW-3400A	MN 40	1.51	8.0-10.0			0- 568	194
LW-3402A	MN 40	1.42	2.0- 4.0			0- 568	194
LW-4400A	GMN 40	1.43	2.0- 3.0			0- 568	194
SL-2339A	GLF 33	1.45	2.0- 6.0			0- 568	194

Drying time, hr	Inj time, sec	Front temp, °F	Mid temp, °F	Rear temp, °F	Nozzle temp, °F	Mold temp, °F	Proc temp, °F
2.0		510	500	490		175	520
2.0		510	500	490		175	520
2.0		510	500	490		175	520
2.0		510	500	490		175	520
2.0		530	520	510		175-250	550
2.0		530	520	510		175-250	550
2.0		530	520	510		175-250	550
3.0- 4.0		470-510	480-520	490-530	470-510	100-250	480-520
3.0- 4.0		470-510	480-520	490-530	470-510	100-250	480-520
3.0- 4.0		470-510	480-520	490-530	470-510	100-250	480-520
3.0- 4.0		470-510	480-520	490-530	470-510	100-250	480-520
3.0- 4.0		470-510	480-520	490-530	470-510	100-250	480-520
3.0- 4.0		470-510	480-520	490-530	470-510	100-250	480-520
3.0- 4.0		470-510	480-520	490-530	470-510	100-250	480-520
3.0- 4.0		470-510	480-520	490-530	470-510	100-250	480-520
3.0- 4.0		470-510	480-520	490-530	470-510	100-250	480-520
5.0	5.0	536	518	500	518	176	
5.0	4.0	545	536	500	527	176	
5.0	5.0	527	518	500	518	176	
5.0	4.0	554	545	509	536	176	
5.0	4.0	545	536	500	527	176	
5.0	4.0	554	545	509	536	176	
5.0	5.0	527	518	500	509	158	
5.0	5.0	536	518	500	518	176	
5.0	4.0	536	518	491	518	176	
5.0	4.0	554	545	509	536	176	
4.0- 6.0		510-555	510-555	510-555		125-200	525-550
4.0		530-580	530-580	530-580		150-225	540-575
3.0- 4.0		470-510	480-520	490-530	470-510	100-250	480-520
3.0- 4.0		470-510	480-520	490-530	470-510	100-250	480-520
3.0- 4.0		470-510	480-520	490-530	470-510	100-250	480-520
3.0- 4.0		470-510	480-520	490-530	470-510	100-250	480-520
3.0- 4.0		470-510	480-520	490-530	470-510	100-250	480-520
2.0- 4.0		536-554	527-554	536-572	536-554	158-212	536-572
2.0- 4.0		536-554	527-554	536-572	536-554	158-212	536-572
2.0- 4.0		536-554	527-554	536-572	536-554	158-212	536-572
2.0- 4.0		536-554	527-554	536-572	536-554	158-212	536-572
2.0- 4.0		536-554	527-554	536-572	536-554	158-212	536-572
2.0- 4.0		536-554	527-554	536-572	536-554	158-212	536-572
2.0- 4.0		536-554	527-554	536-572	536-554	158-212	536-572
2.0- 4.0		536-554	527-554	536-572	536-554	158-212	536-572
2.0- 4.0		536-554	527-554	536-572	536-554	158-212	536-572
2.0- 4.0		536-554	527-554	536-572	536-554	158-212	536-572

Grade	Filler	Sp Grav	Shrink, mils/in	Melt flow, g/10 min	Melt temp, °F	Back pres, psi	Drying temp, °F
Nylon 6/6		**Minlon**			**Du Pont**		
10B40	MN	1.51	8.0-10.0		504	50	175
11C40	MN	1.48	9.0-13.0		498	50	175
12T	MN	1.42	10.0-12.0		498	50	175
20B	GMN	1.42	3.0-10.0		504	50	175
22C	GMN	1.45	5.0- 9.5		504	50	175
Nylon 6/6		**MonTor Nylon**			**MonTor**		
CM3001-N		1.13			508	71- 213	
CM3001G-15	GLF 15	1.26			508	71- 213	
CM3001G-30	GLF 30	1.37			508	71- 213	
CM3001G-30B1	GLF 30	1.37			508	71- 213	
CM3001G-45	GLF 45	1.50			508	71- 213	
CM3003					508	71- 213	
CM3003G-R30	GLF 30	1.38			508	71- 213	
CM3004-VO		1.18			508	71- 213	
CM3004G-15	GLF 15	1.47			508	71- 213	
CM3004G-30	GLF 30	1.59			508	71- 213	
CM3006		1.13			508	71- 213	
CM3006-E		1.13			508	71- 213	
CM3006G-30	GLF 30	1.37			508	71- 213	
CM3007		1.13			508	71- 213	
UTN320		1.09			490	71- 213	
UTN325		1.07				71- 213	
Nylon 6/6		**Nylafil**			**DSM**		
G-1/30	GLF 30	1.40	3.0				220
G-1/30/MS/5	GLF 30	1.44	3.0				220
G-1/30/TF/15	GLF 30	1.52	2.0				220
G-1/40	GLF 40	1.47	2.0				220
G-1/40/MS/5	GLF 40	1.51	3.0				220
G-8/40	GLF 40	1.40	1.5				220
J-1/15/MF/25	GLF 15	1.49	3.0				220
J-1/20	GLF 20	1.30	4.0				220
J-1/30	GLF 30	1.39	2.0				220
J-1/30/MS/5	GLF 30	1.44	3.0				220
J-1/30/TF/15	GLF 30	1.52	2.0				220
J-1/43	GLF 43	1.50	1.0				220
J-8/13	GLF 13	1.19	5.0				220
J-8/33	GLF 13	1.34	2.5				220
J-8/43	GLF 13	1.42	1.5				220
NY-8/MF/36	GLF 40	1.41	17.0				220
Nylon 6/6		**Nylatron**			**DSM**		
1018 HS	GLF 33	1.41	3.0		490		
1024 HS		1.14	15.0		490		
1025	MN 40	1.51	3.0		490		
1027	GLF 30	1.41	3.0		490		
1040 HS	GLF 40	1.48	4.0		490		
GS	MOS 2	1.16	7.0		480-498		185
GS HS	MOS 2	1.16	7.0		480-498		185
GS-21	MOS	1.16	7.0-17.0		480-498		
GS-51	GLF 30	1.41	2.0		480-498		185
GS-51-13	GLF 40	1.46	1.0		480-498		
NSB-90		1.18	15.0-25.0		490		

Drying time, hr	Inj time, sec	Front temp, °F	Mid temp, °F	Rear temp, °F	Nozzle temp, °F	Mold temp, °F	Proc temp, °F
24.0		540-550	550-560	560-570	560-580	150-240	560-580
24.0		530-540	530-550	540-570	540-550	150-240	540-570
24.0		530-540	530-550	540-570	540-550	150-240	540-570
24.0		540-550	550-560	560-570	560-580	150-240	560-580
24.0		540-550	550-560	560-570	560-580	150-240	560-580
		464-572	464-572	464-572		140-176	500-590
		464-572	464-572	464-572		140-176	500-590
		464-572	464-572	464-572		140-176	500-590
		464-572	464-572	464-572		140-176	500-590
		464-572	464-572	464-572		140-176	500-590
		464-572	464-572	464-572		140-176	500-590
		464-572	464-572	464-572		140-176	500-590
		464-572	464-572	464-572		140-176	500-590
		464-572	464-572	464-572		140-176	500-590
		464-572	464-572	464-572		140-176	500-590
		464-572	464-572	464-572		140-176	500-590
		464-572	464-572	464-572		140-176	500-590
		464-572	464-572	464-572		140-176	500-590
		464-572	464-572	464-572		140-176	500-590
		464-572	464-572	464-572		140-176	500-590
		464-572	464-572	464-572		140-176	500-590
2.0		530-550	550-570	540-560	520-540	130-200	540-580
2.0		530-550	550-570	540-560	520-540	130-200	540-580
2.0		530-550	550-570	540-560	520-540	130-200	540-580
2.0		530-550	550-570	540-560	520-540	130-200	540-580
2.0		530-550	550-570	540-560	520-540	130-200	540-580
2.0		530-550	550-570	540-560	520-540	130-200	540-580
2.0		530-550	550-570	540-560	520-540	130-200	540-580
2.0		530-550	550-570	540-560	520-540	130-200	540-580
2.0		530-550	550-570	540-560	520-540	130-200	540-580
2.0		530-550	550-570	540-560	520-540	130-200	540-580
2.0		530-550	550-570	540-560	520-540	130-200	540-580
2.0		530-550	550-570	540-560	520-540	130-200	540-580
2.0		530-550	550-570	540-560	520-540	130-200	540-580
2.0		530-550	550-570	540-560	520-540	130-200	540-580
2.0		530-550	550-570	540-560	520-540	130-200	540-580
2.0		530-550	550-570	540-560	520-540	130-200	540-580
		520	520	530		180	525-570
		520	530	540		180	530-560
		520	530	540		180	525-575
		520	530	540		180	525-575
		540	540	550		180	540
2.0- 6.0		520-540	510-530	500-520	500	180	520-560
2.0- 6.0		520-540	510-530	500-520	500	180	520-560
		540	525	510		175	530
2.0- 6.0		520-540	510-530	500-520	500	180	520-560
		550	530	515		180	540
		530	520	510		180	530

Grade	Filler	Sp Grav	Shrink, mils/in	Melt flow, g/10 min	Melt temp, °F	Back pres, psi	Drying temp, °F
Nylon 6/6		**Nyloy**			**Nytex Compos.**		
M-4005	MS	1.18	20.0				
MG-0077N-VO	GLF 33	1.58	3.0				
MS-0100		1.08	22.0				
Nylon 6/6		**Nytron**			**Nytex Compos.**		
LMC-0045	CF 45	1.33	0.1				
LMC-4030	CF 30	1.30	0.3				
LMC-5022	CF 10	1.20	1.5				
LMC-5024	CF 10	1.32					
LMG-0030 (Carbon)	CF 30	1.27	0.3				
LMG-0030 (Glass)	GLF 30	1.40	2.0				
LMG-0050	GLF 50	1.54	1.0				
LMG-1045	GLF 45	1.47	2.0				
Nylon 6/6		**Plaslube**			**DSM**		
G-1/30/MS/5	GLF 30	1.44	3.0				165-220
G-1/30/SI/2	GLF 30	1.40	3.0				165-220
G-1/30/TF/15	GLF 30	1.52	2.0				165-220
G-1/40/MS/5	GLF 40	1.51	3.0				165-220
G-3/30/MS/5	GLF 30	1.40	3.5				165-220
J-1/30/MS/5	GLF 30	1.44	3.0				165-220
J-1/30/SI/3	GLF 30	1.40	2.0				165-220
J-1/30/TF/15	GLF 30	1.52	2.0				165-220
J-1/33/TF/13/SI/2	GLF 33	1.49	2.0				165-220
J-1/CF/15/TF/20	CF 15	1.33	2.5				165-220
J-1/CF/30/TF/13	CF 30	1.36	1.0				165-220
J-1/CF/30/TF/15	CF 30	1.38	1.0				165-220
Nylon 6/6		**PMC Nylon 6/6**			**PMC**		
6601 BK-09		1.13			498	0- 800	160-180
6601-FR		1.14			500	0- 800	160-180
6601-L		1.13			498	0- 800	160-180
6601-LN		1.14			508	0- 800	160-180
6602 BK-02		1.13			498	0- 800	160-180
6602-NNL		1.14			498	0- 800	160-180
6603-HSL		1.14			498	0- 800	160-180
6604-L		1.09			508	0- 800	160-180
6604G13	GLF 13	1.18			500	0- 150	160-180
6604G33	GLF 33	1.34			500	0- 150	160-180
6604G43	GLF 43	1.49			500	0- 150	160-180
6605 BK-10	CBL				498	0- 150	160-180
6608 BLACK					498	0- 800	160-180
6608G33 BLACK	GLF 33	1.40			501	0- 150	160-180
6618G15	GLF 15				501	0- 150	160-180
6642M	MOS				501	0- 800	160-180
66G13	GLF 13	1.22			501	0- 150	160-180
66G25	GLF 25	1.35			501	0- 150	160-180
66G25-FR	GLF 25	1.32			500	0- 150	160-180
66G25M15	GLF 25				501	0- 150	160-180
66G33	GLF 33	1.38			501	0- 150	160-180
66G33M	GLF 33				501	0- 150	160-180
66G43	GLF 43	1.51			501	0- 150	160-180
66M40	MNF 40				501	0- 150	160-180

Drying time, hr	Inj time, sec	Front temp, °F	Mid temp, °F	Rear temp, °F	Nozzle temp, °F	Mold temp, °F	Proc temp, °F
							510
							527
							500
							600
							590
							600
							600
							590
							580
							590
							580
2.0-16.0		530-550	550-570	540-560	520-540	130-200	540-580
2.0-16.0		530-550	550-570	540-560	520-540	130-200	540-580
2.0-16.0		530-550	550-570	540-560	520-540	130-200	540-580
2.0-16.0		530-550	550-570	540-560	520-540	130-200	540-580
2.0-16.0		530-550	550-570	540-560	520-540	130-200	540-580
2.0-16.0		530-550	550-570	540-560	520-540	130-200	540-580
2.0-16.0		530-550	550-570	540-560	520-540	130-200	540-580
2.0-16.0		530-550	550-570	540-560	520-540	130-200	540-580
2.0-16.0		530-550	550-570	540-560	520-540	130-200	540-580
2.0-16.0		530-550	550-570	540-560	520-540	130-200	540-580
2.0-16.0		530-550	550-570	540-560	520-540	130-200	540-580
2.0-16.0		530-550	550-570	540-560	520-540	130-200	540-580
		540-560	520-550	480-510	510-550	130-200	
		540-560	520-550	480-510	510-550	130-200	
		540-560	520-550	480-510	510-550	130-200	
		540-560	520-550	480-510	510-550	130-200	
		540-560	520-550	480-510	510-550	130-200	
		540-560	520-550	480-510	510-550	130-200	
		540-560	520-550	480-510	510-550	130-200	
		540-560	520-550	480-510	510-550	130-200	
		540-570	520-550	480-520	510-550	160-220	
		540-570	520-550	480-520	510-550	160-220	
		540-570	520-550	480-520	510-550	160-220	
		540-570	520-550	480-520	510-550	160-220	
		540-560	520-550	480-510	510-550	130-200	
		540-570	520-550	480-520	510-550	160-220	
		540-570	520-550	480-520	510-550	160-220	
		540-560	520-550	480-510	510-550	130-200	
		540-570	520-550	480-520	510-550	160-220	
		540-570	520-550	480-520	510-550	160-220	
		540-570	520-550	480-520	510-550	160-220	
		540-570	520-550	480-520	510-550	160-220	
		540-570	520-550	480-520	510-550	160-220	
		540-570	520-550	480-520	510-550	160-220	
		540-570	520-550	480-520	510-550	160-220	
		540-570	520-550	480-520	510-550	160-220	

Grade	Filler	Sp Grav	Shrink, mils/in	Melt flow, g/10 min	Melt temp, °F	Back pres, psi	Drying temp, °F
Nylon 6/6		**Radiflam**			**Polymers Intl.**		
A AE		1.18	1.4		495		
A RV250AF		1.31	0.4		495		
A RV300AE		1.40	0.4		495		
A RV300AF		1.39	0.4		495		
Nylon 6/6		**Radilon**			**Polymers Intl.**		
A		1.14	1.8		495		
A 32E/32EN		1.14	1.5		495		
A 32HS		1.14	1.9		495		
A 32VHS		1.14	1.9		495		
A 35E/35EN		1.14	1.4		495		
A 38E/38EN		1.14	1.4		495		
A 38HS		1.14	1.8		495		
A 38VHS		1.14	1.9		495		
A CP300		1.37	1.0		495		
A CP400		1.47	0.9		495		
A CV200		1.26	1.3		495		
A CV250		1.31	1.2		495		
A CV300		1.34	1.1		495		
A ECP4212		1.31	0.9		496		
A ERV2310		1.19	0.5		495		
A HS/LHS		1.14	1.9		495		
A HSX		1.13	2.5		495		
A L		1.14	1.8		495		
A RV150		1.20	0.5		495		
A RV200		1.26	0.5		495		
A RV250		1.31	0.4		495		
A RV250R		1.31	0.4		495		
A RV300		1.36	0.4		495		
A RV300R		1.36	0.4		495		
A RV350		1.40	0.3		495		
A RV350R		1.40	0.3		495		
A RV500		1.50	0.1		495		
A USX160		1.09	1.3		495		
A USX200		1.07	1.2		495		
A USX80		1.11	1.3		495		
A VHM243		1.12	1.9		495		
A VHS/LVHS		1.14	2.0		495		
A XCP3303		1.36	1.0		495		
A XRV2010		1.18	0.5		495		
Nylon 6/6		**RTP Polymers**			**RTP**		
200		1.14	18.0				
200 FR		1.36	16.0			25- 50	175
200 GB 10	GLB 10	1.21	27.0			25- 50	175
200 GB 20	GLB 20	1.28	24.0			25- 50	175
200 GB 30	GLB 30	1.36	21.0			25- 50	175
200 GB 40	GLB 40	1.45	20.0			25- 50	175
200 GB 50	GLB 50	1.56	20.0			25- 50	175
200 MS	GLF	1.19	11.0			25- 50	175
200 TFE 10	PTF	1.20	18.0			25- 50	175
200 TFE 20	PTF	1.27	14.0			25- 50	175
200 TFE 20 SI	PTF	1.27	14.0			25- 50	175
201	GLF 10	1.21	12.0			25- 50	175

Drying time, hr	Inj time, sec	Front temp, °F	Mid temp, °F	Rear temp, °F	Nozzle temp, °F	Mold temp, °F	Proc temp, °F
						68	500
						221	509
						203	522
						221	522
						122	518
						122	518
						122	522
						122	522
						122	531
						122	536
						122	527
						122	522
						212	518
						212	518
						194	522
						194	522
						194	522
						176	509
						176	509
						122	518
						95	509
						122	513
						194	513
						194	513
						194	513
						203	513
						221	527
						221	518
						221	522
						221	522
						221	536
						68	509
						68	509
						68	509
						68	513
						122	522
						176	518
						176	509
		525-550	525-550	525-550		150-225	
4.0		515-535	505-525	495-515		150-225	
4.0		500-545	490-535	480-525		150-225	
4.0		500-545	490-535	480-525		150-225	
4.0		500-545	490-535	480-525		150-225	
4.0		500-545	490-535	480-525		150-225	
4.0		500-545	490-535	480-525		150-225	
4.0		500-545	490-535	480-525		150-225	
4.0		535-565	525-555	515-545		150-225	
4.0		535-565	525-555	515-545		150-225	
4.0		535-565	525-555	515-545		150-225	
4.0		500-545	490-535	480-525		150-225	

Grade	Filler	Sp Grav	Shrink, mils/in	Melt flow, g/10 min	Melt temp, °F	Back pres, psi	Drying temp, °F
201 FR	GLF	1.50	5.0			25- 50	175
201 FR SP	GLF	1.50	5.0			25- 50	175
201 TFE 5 SI	GLF	1.22	6.0			25- 50	175
201.3	GLF 13	1.23	12.0			25- 50	175
202	GLF 15	1.25	8.0			25- 50	175
202 FR	GLF	1.54	5.0			25- 50	175
202 M	GLF 10	1.47	5.0			25- 50	175
203	GLF 20	1.28	6.0			25- 50	175
203 FR	GLF	1.62	4.0			25- 50	175
203 GB 20	GLB 20	1.46	7.0			25- 50	175
203 M GB 20	GLF 15	1.46	0.5			25- 50	175
203 MG GB 20	GLF 20	1.46	12.0			25- 50	175
203 MS	GLF	1.34	5.0			25- 50	175
203 TFE 10	GLF	1.36	5.0			25- 50	175
203 TFE 15	GLF	1.40	5.0			25- 50	175
203 TFE 20	GLF	1.44	4.0			25- 50	175
203 TFE FR 15	GLF	1.60	4.0			25- 50	175
204 FR	GLF	1.65	4.0			25- 50	175
204 GB FR	GLB	1.60	11.0			25- 50	175
205	GLF 30	1.36	6.0			25- 50	175
205 FR	GLF	1.68	3.0			25- 50	175
205 FR SP	GLF	1.68	3.0			25- 50	175
205 HS	GLF 30	1.36	6.0			25- 50	175
205 MS	GLF	1.44	4.0			25- 50	175
205 TFE 13 SI2	GLF	1.49	3.0			25- 50	175
205 TFE 15	GLF	1.49	3.0			25- 50	175
205 TFE 15 HB	GLF	1.49	3.0			25- 50	175
205 TFE 15 SI	GLF	1.49	3.0			25- 50	175
205 TFE 15 Z	GLF	1.50	3.0			25- 50	175
205 TFE 20	GLF	1.54	3.0			25- 50	175
205 TFE 5	GLF	1.41	4.0			25- 50	175
205 TFE FR 15	GLF	1.75	3.0			25- 50	175
205.3	GLF 33	1.39	6.0			25- 50	175
205.3 HS SI	GLF	1.40	2.0			25- 50	175
205H FR	GLF	1.56	2.0			25- 50	175
207	GLF 40	1.46	5.0			25- 50	175
207 MS	GLF	1.54	3.0			25- 50	175
207.3	GLF 43	1.49	5.0			25- 50	175
209	GLF 50	1.57	4.0			25- 50	175
211	GLF 60	1.68	3.0			25- 50	175
225	TAL 20	1.30	12.0			25- 50	175
227	TAL 40	1.50	9.0			25- 50	175
281 TFE 20	CF	1.32	4.0			25- 50	175
281H	CF	1.14	2.0			25- 50	175
282	CF	1.18	2.0			25- 50	175
282 FR	CF	1.47	1.0			25- 50	175
283	CF	1.23	2.0			25- 50	175
283 TFE 10	CF	1.29	2.0			25- 50	175
283H	CF	1.17	2.0			25- 50	175
285	CF	1.28	1.0			25- 50	175
285 TFE 13 SI2	CF	1.37	1.0			25- 50	175
285 TFE 15	CF	1.38	1.0			25- 50	175
285H	CF	1.22	1.0			25- 50	175
285P	CF	1.28	3.0			25- 50	175
287	CF	1.33	1.0			25- 50	175
287 TFE 10	CF	1.39	1.0			25- 50	175

Drying time, hr	Inj time, sec	Front temp, °F	Mid temp, °F	Rear temp, °F	Nozzle temp, °F	Mold temp, °F	Proc temp, °F
4.0		515-535	505-525	495-515		150-225	
4.0		515-535	505-525	495-515		150-225	
4.0		500-545	490-535	480-525		150-225	
4.0		500-545	490-535	480-525		150-225	
4.0		500-545	490-535	480-525		150-225	
4.0		515-535	505-525	495-515		150-225	
4.0		500-545	490-535	480-525		150-225	
4.0		500-545	490-535	480-525		150-225	
4.0		515-535	505-525	495-515		150-225	
4.0		500-545	490-535	480-525		150-225	
4.0		500-545	490-535	480-525		150-225	
4.0		500-545	490-535	480-525		150-225	
4.0		500-545	490-535	480-525		150-225	
4.0		500-545	490-535	480-525		150-225	
4.0		500-545	490-535	480-525		150-225	
4.0		500-545	490-535	480-525		150-225	
4.0		515-535	505-525	495-515		150-225	
4.0		515-535	505-525	495-515		150-225	
4.0		515-535	505-525	495-515		150-225	
4.0		500-545	490-535	480-525		150-225	
4.0		515-535	505-525	495-515		150-225	
4.0		515-535	505-525	495-515		150-225	
4.0		500-545	490-535	480-525		150-225	
4.0		500-545	490-535	480-525		150-225	
4.0		500-545	490-535	480-525		150-225	
4.0		500-545	490-535	480-525		150-225	
4.0		515-535	505-525	495-515		150-225	
4.0		500-545	490-535	480-525		150-225	
4.0		500-545	490-535	480-525		150-225	
4.0		500-545	490-535	480-525		150-225	
4.0		500-545	490-535	480-525		150-225	
4.0		515-535	505-525	495-515		150-225	
4.0		500-545	490-535	480-525		150-225	
4.0		500-545	490-535	480-525		150-225	
4.0		515-535	505-525	495-515		150-225	
4.0		500-545	490-535	480-525		150-225	
4.0		500-545	490-535	480-525		150-225	
4.0		500-545	490-535	480-525		150-225	
4.0		500-545	490-535	480-525		150-225	
4.0		500-545	490-535	480-525		150-225	
4.0		495-530	485-520	475-510		180-220	
4.0		495-530	485-520	475-510		180-220	
4.0		525-575	515-565	500-550		150-200	
4.0		525-575	515-565	500-550		150-200	
4.0		525-575	515-565	500-550		150-200	
4.0		515-535	505-525	495-515		150-225	
4.0		525-575	515-565	500-550		150-200	
4.0		525-575	515-565	500-550		150-200	
4.0		525-575	515-565	500-550		150-200	
4.0		525-575	515-565	500-550		150-200	
4.0		525-575	515-565	500-550		150-200	
4.0		525-575	515-565	500-550		150-200	
4.0		525-575	515-565	500-550		150-200	
4.0		525-575	515-565	500-550		150-200	
4.0		525-575	515-565	500-550		150-200	
4.0		525-575	515-565	500-550		150-200	

Grade	Filler	Sp Grav	Shrink, mils/in	Melt flow, g/10 min	Melt temp, °F	Back pres, psi	Drying temp, °F
287H	CF	1.27	1.0			25- 50	175
287P	CF	1.35	2.0			25- 50	175
299X50446D	GLF	1.35	4.0			25- 50	175
299X51265F	MN	2.13	14.0			25- 50	175
EMI-261H	STS	1.20	12.0-16.0				
EMI-281H	GRN	1.21	2.0				
ESD-A-200H		1.13	16.0				
ESD-A-202H	GLF	1.24	6.0				
ESD-A-280H	CF	1.12	3.0-20.0				
ESD-C-200H		1.15	16.0				
ESD-C-202H	GLF	1.26	6.0				
ESD-C-260H	STS	1.14	15.0-20.0				
ESD-C-280H	CF	1.14	2.0-20.0				

Nylon 6/6 — Schulamid 6.6 — A. Schulman

Grade	Filler	Sp Grav	Shrink, mils/in	Melt flow, g/10 min	Melt temp, °F	Back pres, psi	Drying temp, °F
GF 13	GLF 13	1.19	8.0		491		180
GF 15	GLF 15	1.20	8.0		500		180
GF 30	GLF 30	1.35	3.0		500		180
GF 45	GLF 45	1.50	4.0		500		180
MF 40	MN 40	1.47	8.0		500		180
MV 3		1.14	15.0		495		180
MV 3 HI-IMP		1.07	15.0		482		180
MWG 40	GLF 15	1.49	5.0		500		180
N 68		1.08	15.0		500		180

Nylon 6/6 — Texalon — Texapol

Grade	Filler	Sp Grav	Shrink, mils/in	Melt flow, g/10 min	Melt temp, °F	Back pres, psi	Drying temp, °F
1200 A		1.14	12.0-16.0		491-496	50	150
1200 AL		1.14	12.0-17.0		496	50	150
1200 FPL 20-BK-13	PTF 20	1.14	16.0-18.0		496	50	150
1200A HR-2 BK-16		1.14	11.0-15.0		491-496	50	150
1200A Zip-1		1.14	8.0-12.0		496	50	150
1200A ZIP-10		1.14	14.0-18.0		491-496	50	150
1200A Zip-25		1.14	8.0-12.0		496	50	150
1200A-BK-11	CBL 2	1.14	10.0-13.0		496	50	150
1200A-FR		1.30	10.0-15.0		496	50	150
1200A-Zip 17		1.14	12.0-17.0		496	50	150
1204		1.14	12.0-16.0		496	50	150
1204 HS BK-13		1.14	10.0-14.0		496	50	150
1280A		1.14	15.0-18.0		496	50	150
1308A		1.10	15.0-20.0		496	50	150
1310		1.12	14.0-18.0		493	50	150
2000A		1.09	17.0-20.0		493	50	150
2004		1.09	17.0-20.0		493	50	150
900A ZIP-10		1.14	12.0-16.0		446-464	50	150
920A		1.14	15.0-20.0		491	50	150
GF 1200A-13	GL 13	1.22	5.0- 6.0		500	50	150
GF 1200A-33	GL 33	1.37	2.0- 3.0		500	50	150
GF 1308A-13	GL 13	1.18	5.0- 7.0		496	50	150
GF 2000A-14	GL 14	1.17	4.0- 6.0		493	50	150
GF 2000A-33	GL 33	1.33	2.0		493-498	50	150
GF 900A-33	GL 33	1.38	1.0	7.80	446-464	50	150
GF 920A-33	GL 33	1.35	2.0- 3.0		460	50	150
GMF 1200A-40	GMN 40	1.47	3.0- 4.0		500	50	150
MF 1200A-40	MN 40	1.50	5.0- 8.0		500	50	150
MF 1204-40 BK	MN 40	1.50	5.0- 8.0		500	50	150
MS 1204-2A	MOS 2	1.14	4.0- 6.0		496	50	150

Drying time, hr	Inj time, sec	Front temp, °F	Mid temp, °F	Rear temp, °F	Nozzle temp, °F	Mold temp, °F	Proc temp, °F
4.0		525-575	515-565	500-550		150-200	
4.0		525-575	515-565	500-550		150-200	
4.0		500-545	490-535	480-525		150-225	
4.0		515-535	505-525	495-515		150-225	
		500-550	500-550	500-550		175-250	
		500-525	500-525	500-525		175-250	
		525-550	525-550	525-550		175-250	
		525-550	525-550	525-550		175-250	
		500-550	500-550	500-550		175-250	
		525-550	525-550	525-550		175-250	
		525-550	525-550	525-550		175-250	
		500-550	500-550	500-550		175-250	
		500-550	500-550	500-550		175-250	
		530	530	540	540	175-210	540-560
		530	530	540	540	175-210	540-560
		530	530	540	540	175-210	540-560
		530	530	540	540	175-210	540-560
		530	530	540	540	175-210	540-560
		510	520	530	510	175-210	520-560
		510	520	530	510	175-210	520-560
		530	530	540	540	175-210	540-560
		510	520	530	510	175-210	520-560
4.0		510	500	500	530	125	520-580
4.0		510	500	500	530	125	520-580
4.0		510	500	500	530	125	
4.0		510	500	500	530	125	520-580
4.0		510	500	500	530	125	520-580
4.0		510	500	500	530	125	520-580
4.0		510	500	500	530	125	520-580
4.0		510	500	500	530	125	
4.0		500	500	490	520	125	
4.0		510	500	500	530	125	520-580
4.0		510	500	500	530	125	
4.0		510	500	500	530	125	
4.0		510	500	500	530	125	
4.0		530	520	520	540	125	
4.0		530	520	520	540	125	
4.0		530	520	520	540	125	
4.0		530	520	520	540	125	
4.0		510	500	500	530	125	
4.0		510	500	500	530	125	
4.0		550	540	540	545	125	
4.0		550	540	540	545	125	
4.0		550	540	540	545	125	
4.0		550	540	540	545	125	
4.0		550	540	540	545	125	
4.0		550	540	540	545	125	
4.0		550	540	540	545	125	
4.0		540	530	530	530	125	
4.0		540	530	530	530	125	
4.0		510	500	500	530	125	
4.0		510	500	500	530	125	

Grade	Filler	Sp Grav	Shrink, mils/in	Melt flow, g/10 min	Melt temp, °F	Back pres, psi	Drying temp, °F
Nylon 6/6		**Thermocomp**			**LNP**		
RF-1006 RM	GLF 30	1.25	4.0- 5.0		500	0- 100	170-180
Nylon 6/6		**Thermofil Polyamide**				**Thermofil**	
LSG-440A	CRF 60	1.73	2.0			50	220
N3-13FG-0100	GL 13	1.23	7.0			50	170
N3-13FG-0127	GL 13	1.23	7.0			50	170
N3-13FG-0626	GL 13	1.23	7.0			50	170
N3-13FG-4100	GL 13	1.23	7.0			50	170
N3-13FG-4127	GL 13	1.23	7.0			50	170
N3-13FG-4626	GL 13	1.23	7.0			50	170
N3-33FG-0100	GL 33	1.40	3.0			50	170
N3-33FG-0127	GL 33	1.40	3.0			50	170
N3-33FG-0626	GL 33	1.40	3.0			50	170
N3-33FG-4100	GL 33	1.40	3.0			50	170
N3-33FG-4127	GL 33	1.40	3.0			50	170
N3-33FG-4626	GL 33	1.40	3.0			50	170
N3-40FM-1600	GMN 40	1.48	3.0			50	170
N3-40MF-1600	MNF 40	1.49	7.0			50	170
N3-9900-0231	MOS	1.19	22.0			50	170
Nylon 6/6		**UBE Nylon**			**UBE**		
2015B		1.14	17.0-22.0		491-509	1421	
2015SV		1.16	12.0-17.0		491-509	1421	
2020B		1.14	17.0-22.0		491-509	1421	
2020H		1.14	17.0-22.0		491-509	1421	
2020U		1.14	17.0-22.0		491-509	1421	
2020UW1		1.14	17.0-22.0		491-509	1421	
Nylon 6/6		**Ultramid**			**BASF**		
A3		1.13	13.0		500	50- 100	176-230
A3EG10	GLF 50	1.55	1.0		500	50- 100	176-230
A3EG3	GLF 15	1.24	4.0		500	50- 100	176-230
A3EG5	GLF 25	1.30	3.0		500	50- 100	176-230
A3EG6	GLF 30	1.35	2.0		500	50- 100	176-230
A3EG7	GLF 35	1.40	2.0		500	50- 100	176-230
A3HG10	GLF 50	1.55	1.0		500	50- 100	176-230
A3HG3	GLF 15	1.24	4.0		500	50- 100	176-230
A3HG5	GLF 25	1.30	3.0		500	50- 100	176-230
A3HG7	GLF 35	1.40	2.0		500	50- 100	176-230
A3K		1.13	13.0		500	50- 100	176-230
A3K Q602		1.13	13.0		500	50- 100	176-230
A3K-HF		1.13			500	50- 100	176-230
A3M8	MN 40	1.44	6.5		500	50- 100	176-230
A3R	PE	1.10	8.0		500	50- 100	176-230
A3SK		1.14	7.0		500	50- 100	176-230
A3W		1.13	13.0		500	50- 100	176-230
A3WC4	CF 20	1.22	0.7		500	50- 100	176-230
A3WG10	GLF 50	1.55	1.0		500	50- 100	176-230
A3WG3	GLF 15	1.24	4.0		500	50- 100	176-230
A3WG5	GLF 25	1.30	3.0		500	50- 100	176-230
A3WG6	GLF 30	1.35	2.0		500	50- 100	176-230
A3WG7	GLF 35	1.40	2.0		500	50- 100	176-230
A3WGM35	GLF 15	1.39			500	50- 100	176-230
A3X1G10	GLF 50	1.60	1.5		500	50- 100	176-230

Drying time, hr	Inj time, sec	Front temp, °F	Mid temp, °F	Rear temp, °F	Nozzle temp, °F	Mold temp, °F	Proc temp, °F
4.0- 6.0		510-540	510-540	510-540		150-200	525-535
3.0		530-550	520-540	500-520	540-560	150-230	
3.0		530-550	520-540	500-520	540-560	150-230	
3.0		530-550	520-540	500-520	540-560	150-230	
3.0		530-550	520-540	500-520	540-560	150-230	
3.0		530-550	520-540	500-520	540-560	150-230	
3.0		530-550	520-540	500-520	540-560	150-230	
3.0		530-550	520-540	500-520	540-560	150-230	
3.0		530-550	520-540	500-520	540-560	150-230	
3.0		530-550	520-540	500-520	540-560	150-230	
3.0		530-550	520-540	500-520	540-560	150-230	
3.0		530-550	520-540	500-520	540-560	150-230	
3.0		530-550	520-540	500-520	540-560	150-230	
3.0		530-550	520-540	500-520	540-560	150-230	
3.0		530-550	520-540	500-520	540-560	150-230	
3.0		530-550	520-540	500-520	540-560	150-230	
3.0		530-550	520-540	500-520	540-560	150-230	
	12.0	536	527	509	536	158-176	554
	12.0	536	527	509	536	158-176	554
	12.0	536	527	509	536	158-176	554
	12.0	536	527	509	536	158-176	554
	12.0	536	527	509	536	158-176	554
	12.0	536	527	509	536	158-176	554
4.0-12.0						100-140	530-570
4.0-12.0						176-194	555-590
4.0-12.0						176-194	535-575
4.0-12.0						176-194	535-575
4.0-12.0						176-194	535-575
4.0-12.0						176-194	535-575
4.0-12.0						176-194	555-590
4.0-12.0						176-194	535-575
4.0-12.0						176-194	535-575
4.0-12.0						176-194	535-575
4.0-12.0						104-140	535-575
4.0-12.0						104-140	535-575
4.0-12.0						100-140	535-575
4.0-12.0						176-194	520-555
4.0-12.0						100-140	535-575
4.0-12.0						104-140	535-575
4.0-12.0						100-140	535-575
4.0-12.0						176-194	535-575
4.0-12.0						176-194	555-590
4.0-12.0						176-194	535-575
4.0-12.0						176-194	535-575
4.0-12.0						176-194	535-575
4.0-12.0						176-194	535-575
4.0-12.0						176-194	535-575
4.0-12.0						176-194	535-575

Grade	Filler	Sp Grav	Shrink, mils/in	Melt flow, g/10 min	Melt temp, °F	Back pres, psi	Drying temp, °F
A3X1G5	GLF 25	1.33	3.0		500	50- 100	176-230
A3X1G7	GLF 35	1.45	2.5		500	50- 100	176-230
A3X2G5	GLF 25	1.33	3.0- 4.0		500	50- 100	176-230
A3X2G7	GLF 35	1.45	2.5- 3.5		500	50- 100	176-230
A3XG5	GLF 25	1.33	3.0		500	50- 100	176-230
A3XG7	GLF 35	1.45	2.5		500	50- 100	176-230
A3XZG5	GLF 25	1.32	2.5		500	50- 100	176-230
A3Z		1.07	16.0		500	50- 100	176-230
A3ZG6	GLF 30	1.33	2.0		500	50- 100	176-230
A3ZM4	MN 20	1.20	11.0		500	50- 100	176-230
A4		1.13	14.0		500	50- 100	176-230
A4 Type 24		1.13	14.0		500	50- 100	176-230
A4H		1.13	14.0		500	50- 100	176-230
A4K		1.13	14.0		500	50- 100	176-230
A5		1.13	14.0		500	50- 100	176-230
KR4206		1.13	13.0		500	50- 100	176-230
Nylon 6/6		**Verton**			**LNP**		
RF-100-10 (C)	GLF 50		3.0- 4.0			0- 50	200
RF-100-10 (D)	GLF 50	1.57	3.0- 4.0			0- 50	200
RF-100-12 (C)	GLF 60		3.0- 4.0			0- 50	200
RF-100-12 (D)	GLF 60	1.70	3.0- 4.0			0- 50	200
RF-1007 (C)	SGL 35		4.0- 5.0			0- 50	200
RF-1007 (D)	SGL 35	1.41	4.0- 5.0			0- 50	200
RF-700-10 EM HS (C)	GLF 50		3.0- 4.0			0- 50	200
RF-700-10 EM HS (D)	GLF 50	1.57	3.0- 4.0		460	0- 50	200
RF-700-10 HS (C)	GLF 50		3.0- 4.0			0- 50	200
RF-700-10 HS (D)	GLF 50	1.57	3.0- 4.0			0- 50	200
RF-700-12 EM HS (C)	GLF		3.0- 4.0		500	0- 50	200
RF-700-12 HS (C)	GLF 60		3.0- 4.0			0- 50	200
RF-700-12 HS (D)	GLF 60	1.70	3.0- 4.0			0- 50	200
RF-7007 EM HS (C)	GLF		4.0- 5.0		464	0- 50	200
RF-7007 EM HS (D)	GLF	1.41	4.0- 5.0		460	0- 50	200
RF-7007 HS (C)	GLF 35		4.0- 5.0			0- 50	200
RF-7007 HS (D)	GLF 35	1.41	4.0- 5.0		500	0- 50	180
RFL-8019	GLF 45	1.57	3.0- 4.0			0- 50	180-200
RFL-8028	GLF 40	1.56	3.0- 4.0			0- 50	180-200
Nylon 6/6		**Voloy**			**ComAlloy**		
622	GL 25	1.54	4.0				180
628	MN 25	1.54	10.0				180
632	GL 25	1.52	4.0				180
638	MN	1.52	10.0				180
672	GL 25	1.46	4.0				180
678	MN 25	1.46	12.0				180
681	MN	1.55	5.0				200-220
682	GL	1.55	2.0				200-220
683	UNS	1.60	3.0				200-220
684	UNS	1.60	3.0				200-220
685	UNS	1.61	3.0				200-220
686	UNS	1.62	2.0				200-220
Nylon 6/6		**Vydyne**			**Monsanto**		
10V		1.14	15.0-20.0		500	0- 150	150-170
20M		1.14	15.0-20.0		500	0- 150	150-170
20N		1.14	8.0-12.0		500	0- 150	150-170

Drying time, hr	Inj time, sec	Front temp, °F	Mid temp, °F	Rear temp, °F	Nozzle temp, °F	Mold temp, °F	Proc temp, °F
4.0-12.0						176-194	535-575
4.0-12.0						176-194	535-575
4.0-12.0						176-194	535-575
4.0-12.0						176-194	536-572
4.0-12.0						176-194	535-575
4.0-12.0						176-194	535-575
4.0-12.0						176-194	535-575
4.0-12.0						104-140	535-575
4.0-12.0						176-194	535-575
4.0-12.0						176-194	535-575
4.0-12.0						100-140	555-575
4.0-12.0						100-140	555-575
4.0-12.0						104-140	555-575
4.0-12.0						104-140	555-575
4.0-12.0						100-140	530-550
4.0-12.0						100-140	535-575
4.0		565-575	565-575	565-575		175-225	
4.0		565-575	565-575	565-575		175-225	
4.0		565-575	565-575	565-575		175-225	
4.0		565-575	565-575	565-575		175-225	
4.0		565-575	565-575	565-575		175-225	
4.0		565-575	565-575	565-575		175-225	
4.0		565-575	565-575	565-575		175-225	
4.0		565-575	565-575	565-575		175-225	
4.0		565-575	565-575	565-575		175-225	
4.0		565-575	565-575	565-575		175-225	
4.0		565-575	565-575	565-575		175-225	
4.0		565-575	565-575	565-575		175-225	
4.0		565-575	565-575	565-575		175-225	
4.0		565-575	565-575	565-575		175-225	
4.0		565-575	565-575	565-575		175-225	
4.0		565-575	565-575	565-575		175-225	
2.0- 4.0		565-575	565-575	565-575		175-225	
4.0		565-575	565-575	565-575		150-225	
4.0		565-575	565-575	565-575		150-225	
3.0- 4.0		470-510	480-520	490-530	470-510	100-250	480-520
3.0- 4.0		470-510	480-520	490-530	470-510	100-250	480-520
3.0- 4.0		470-510	480-520	490-530	470-510	100-250	480-520
3.0- 4.0		470-510	480-520	490-530	470-510	100-250	480-520
3.0- 4.0		470-510	480-520	490-530	470-510	100-250	480-520
3.0- 4.0		470-510	480-520	490-530	470-510	100-250	480-520
3.0- 4.0		560-580	550-570	540-560	580	230	570-590
3.0- 4.0		560-580	550-570	540-560	580	230	570-590
3.0- 4.0		560-580	550-570	540-560	580	230	570-590
3.0- 4.0		560-580	550-570	540-560	580	230	570-590
3.0- 4.0		560-580	550-570	540-560	580	230	570-590
3.0- 4.0		560-580	550-570	540-560	580	230	570-590
1.0- 4.0		540-580	520-550	500-540	500-550	100-200	520-560
1.0- 4.0		540-580	520-550	500-540	500-550	100-200	520-560
1.0- 4.0		540-580	520-550	500-540	490-500	100-200	520-560

Grade	Filler	Sp Grav	Shrink, mils/in	Melt flow, g/10 min	Melt temp, °F	Back pres, psi	Drying temp, °F
21		1.14	15.0-20.0		500	0- 150	150-180
21SP		1.14	15.0-20.0		500	0- 150	150-180
21X		1.14	15.0-20.0		500	0- 150	150-180
22H		1.14	15.0-20.0		500	25- 150	160
22HSP		1.14	15.0-20.0		500	25- 150	160
24NSP		1.14	8.0-12.0		500	0- 150	140
25T	CBL 2	1.14	19.0		496	25- 150	160
25W	CBL 2	1.14	16.0-20.0		496	25- 150	160
25WSP	CBL 2	1.14	16.0-20.0		500	25- 150	160
66B		1.14	15.0-20.0		500	0- 150	
66J		1.14	15.0-20.0		500	0- 150	
900	GLF 20	1.56	2.0- 6.0		490	0- 150	160
909	GLF 25	1.47	2.0- 6.0		482	0- 150	160
909-01	GLF 25	1.47	2.0- 6.0		482	0- 150	160
M-340		1.24	15.0-20.0		480	0- 150	160
M-344		1.27	15.0-20.0		480	0- 150	160
M-344-01		1.27	15.0-20.0		480	0- 150	160
M-345		1.31	14.0-20.0		460	0- 150	160
R-100	MN 40	1.47	12.0-22.0		500		160
R-108	MN 40	1.47	12.0-22.0		500		160
R-200	MN 40	1.47	12.0-22.0		500		160
R-208	MN 40	1.47	12.0-22.0		500		160
R-220	MN 40	1.46	12.0-22.0		480		160
R-228	MN 40	1.46	12.0-22.0		480		160
R-240	MN 32	1.39	10.0-15.0		480		160
R-250	MN 40	1.46	10.0-22.0		450		160
R-250-01	MN 40	1.46	10.0-22.0		450		160
R-400G	GMN	1.42	5.0-10.0		471	50- 75	160
R-400G-01	GMN	1.42	5.0-10.0		471	50- 75	160
R-513	GLF 13	1.22	5.0		500		160-180
R-513-01	GLF 13	1.22	5.0		500		160-180
R-513H	GLF 13	1.22	5.0		500		160-180
R-513H-01	GLF 13	1.22	5.0		500		160-180
R-525H-02	GLF 25	1.32	4.0				160
R-533	GLF 33	1.40	4.0		500	0- 150	160-180
R-533-01	GLF 33	1.40	4.0		500	0- 150	160-180
R-533H	GLF 33	1.40	4.0		500	0- 150	160-180
R-533H-01	GLF 33	1.40	4.0		500	0- 150	160-180
R-538H-02	GLF 33	1.40	4.0		500	0- 150	160-180
R-543	GLF 43	1.50	3.0		500	0- 150	160-180
R-543-01	GLF 43	1.50	3.0		500	0- 150	160-180
R-543H (Dry)	GLF 43	1.50	3.0		500	0- 150	160-180
R-543H-01	GLF 43	1.50	3.0		500	0- 150	160-180
Nylon 6/6		**Wellamid Nylon**			**Wellman**		
210-N		1.13	14.0-18.0		500		150-180
21L-N		1.13	14.0-18.0		500		150-180
21LH-N		1.13	14.0-18.0		500		150-180
21LN2-NNT		1.13	11.0-15.0		500		150-180
21U-NBK1		1.11	13.0-18.0		500		150-180
220-N		1.13	15.0-20.0		500		150-180
220-XE-N1		1.13	15.0-20.0		500		150-180
22L-N		1.13	15.0-20.0		500		150-180
22L-XE-N1		1.13	15.0-20.0		500		150-180
22LH-N		1.13	15.0-20.0		500		150-180
22LH-XE-N		1.13	15.0-20.0		500		150-180

Drying time, hr	Inj time, sec	Front temp, °F	Mid temp, °F	Rear temp, °F	Nozzle temp, °F	Mold temp, °F	Proc temp, °F
1.0- 3.0		520-560	530-550	480-520	530-550	100-190	520-560
1.0- 3.0		520-560	530-550	480-520	530-550	100-190	520-560
1.0- 3.0		520-560	530-550	480-520	530-550	100-190	520-560
3.0		520-560	520-550	500-540	500-550	100-200	520-560
3.0		520-560	520-550	500-540	500-550	100-200	520-560
3.0		540-560	530-550	480-520	530-550	100-200	520-560
3.0		540-560	530-550	480-520	530-550	100-200	540-560
3.0		540-560	530-550	480-520	530-550	100-200	540-560
3.0		540-560	530-550	480-520	530-550	100-200	540-560
		520-560	520-560	480-500	490-500	100-200	520-560
		520-560	520-560	480-500	490-500	100-200	520-560
3.0		490-560	475-540	470-530	480-550	100-200	490-560
3.0		480-500	475-500	470-500	480-500	100-200	490-510
3.0		480-500	475-500	470-500	480-500	100-200	490-510
3.0		480-520	480-520	460-500	470-490	70-200	490-510
3.0		480-520	480-520	460-500	470-490	70-200	490-510
3.0		480-520	480-520	460-500	470-490	70-200	490-510
3.0		465-495	455-485	450-480	460-480	70-200	465-500
1.0- 4.0	1.0- 2.	540-570	540-570	530-560	530-560	180-220	550-580
1.0- 4.0	1.0- 2.	540-570	540-570	530-560	530-560	180-220	550-580
1.0- 4.0	1.0- 2.	540-570	540-570	530-560	530-560	180-220	550-580
1.0- 4.0	1.0- 2.	540-570	540-570	530-560	530-560	180-220	550-580
1.0- 4.0	1.0- 2.	510-540	510-540	500-530	500-530	180-220	520-550
1.0- 4.0	1.0- 2.	510-540	510-540	500-530	500-530	180-220	520-550
1.0- 4.0	1.0- 2.	510-540	510-540	500-530	500-530	180-220	520-550
1.0- 4.0	1.0- 2.	510-540	510-540	500-530	500-530	180-220	520-550
1.0- 4.0	1.0- 2.	510-540	510-540	500-530	500-530	180-220	520-550
1.0- 4.0	1.0- 2.	510-540	510-540	500-530	500-530	150-200	520-550
1.0- 4.0	1.0- 2.	510-540	510-540	500-530	500-530	150-200	520-550
1.0- 4.0		545-572	545-572	518-536	545-572	150-200	530-570
1.0- 4.0		545-572	545-572	518-536	545-572	150-200	530-570
1.0- 4.0		545-572	545-572	518-536	545-572	150-200	530-570
1.0- 4.0		545-572	545-572	518-536	545-572	150-200	530-570
2.0		545-572	545-572	518-536	545-572	150-200	530-570
1.0- 4.0		545-572	545-572	518-536	545-572	150-200	530-570
1.0- 4.0		545-572	545-572	518-536	545-572	150-200	530-570
1.0- 4.0		545-572	545-572	518-536	545-572	150-200	530-570
1.0- 4.0		545-572	545-572	518-536	545-572	150-200	530-570
1.0- 4.0		545-572	545-572	518-536	545-572	150-200	530-570
1.0- 4.0		545-572	545-572	518-536	545-572	150-200	530-570
1.0- 4.0		545-572	545-572	518-536	545-572	150-200	530-570
1.0- 4.0		545-572	545-572	518-536	545-572	150-200	530-570
1.0- 4.0		545-572	545-572	518-536	545-572	150-200	530-570
		500	520	540	500	70-200	520-580
		500	520	540	500	70-200	520-580
		500	520	540	500	70-200	520-580
		500	520	540	500	70-200	520-580
		500	520	540	500	70-200	520-580
		500	520	540	500	70-200	520-580
		500	520	540	500	70-200	520-580
		500	520	540	500	70-200	520-580
		500	520	540	500	70-200	520-580
		500	520	540	500	70-200	520-580
		500	520	540	500	70-200	520-580

Grade	Filler	Sp Grav	Shrink, mils/in	Melt flow, g/10 min	Melt temp, °F	Back pres, psi	Drying temp, °F
22LH13-XE-N		1.08	16.0-20.0		500		150-170
22LH14-XE-N		1.06	18.0-22.0		500		150-170
22LH15-XE-N		1.04	18.0-22.0		500		150-170
22LH16-XE-N		1.06	16.0-20.0		500		150-170
22LN2-N		1.13	8.0-12.0		500		150-180
FR22F-N		1.28	14.0-18.0		500		150-180
FRGF25-66N	GLF 25	1.48	2.0- 6.0		500		150-180
FRGS25-66N	GLB 25	1.48	2.0- 6.0		500		150-180
GF13-66 22LH-N	GLF	1.20	3.0- 7.0		500		150-170
GF13-66 XE-N	GLF	1.20	3.0- 7.0		500		150-170
GF33-66 22LH-N	GLF	1.32	2.0- 6.0		500		150-170
GF33-66 XE-N	GLF	1.32	2.0- 6.0		500		150-170
GF33-66 XE-N1	GLF	1.32	2.0- 6.0		500		150-170
GF43-66 XE-N	GLF	1.48	2.0- 4.0		500		150-170
GFT13-66 XE-N	GLF	1.14	4.0- 8.0		500		150-170
GS25-66 22L-N	GLB	1.26	13.0-17.0		500	0- 100	150-180
GS25-66 22LH-N	GLB	1.26	13.0-17.0		500	0- 100	150-180
GS40-66 22L-N	GLB	1.39	13.0-17.0		500	0- 100	150-180
GSF25/15-66 22L-N	GLB	1.39	13.0-17.0		500	0- 100	150-180
Nylon 6/6		**Zytel**			**Du Pont**		
101		1.14	15.0		491		175
101 F		1.14	15.0		491		175
101 L		1.14	15.0		491		175
103 HS-L		1.14	15.0		491		175
105 BK-10A	CBL	1.15	15.0		491		175
122 L		1.14	15.0		491		175
133 L		1.14	6.0		493		175
3189		1.11	21.0-24.0		491		175
408		1.09	15.0		491		175
408 HS		1.09	15.0		491		175
408 L		1.09	15.0		491		175
70G 13 HS1-L	GLC	1.22	5.0		491		175
70G 13L	GLC	1.22	5.0		491		175
70G 33 HR-L	GLC	1.38	2.0		491		175
70G 33 HS1-L	GLC	1.38	2.0		491		175
70G 33L	GLC	1.38	2.0		491		175
70G 43L	GLC	1.51	2.0		491		175
71G 13L	GLC	1.18	6.0		491		175
71G 33L	GLC	1.35	3.0		491		175
8018 NC10	GLF 14	1.19			504		175
8018HS NC10	GLF 14	1.19			504		175
80G 33HS1-L	GLC	1.34	3.0-11.0				175
FE3071		1.14			491		175
FE3574		1.14			491		175
FR-10		1.24	12.0		450		175
FR-50	GLC	1.56	4.0- 8.0		491		175
ST 801		1.08	15.0		491		175
ST 801 BK-10	CBL	1.08			491		175
ST 801 HS		1.08	15.0		491		175
Nylon 6/6 Alloy		**Akuloy RM**			**DSM**		
J-70/15	GLF 15	1.15					160-170
J-70/30	GLF 30	1.26	3.0				160-170

Drying time, hr	Inj time, sec	Front temp, °F	Mid temp, °F	Rear temp, °F	Nozzle temp, °F	Mold temp, °F	Proc temp, °F
		500	520	520	500	70-200	520-580
		500	520	520	500	70-200	520-580
		500	520	520	500	70-200	520-580
		500	520	520	500	70-200	520-580
		500	520	540	500	70-200	520-580
		500-520	510-530	520-540	520-540	160-200	510-530
		500-520	510-530	520-540	520-540	160-200	510-530
		500-520	510-530	520-540	520-540	160-200	510-530
		520-540	520-550	540-580	520-540	140-200	530-580
		520-540	520-550	540-580	520-540	140-200	530-580
		520-540	520-550	540-580	520-540	140-200	530-580
		520-540	520-550	540-580	520-540	140-200	530-580
		520-540	520-550	540-580	520-540	140-200	530-580
		520-540	520-550	540-580	520-540	140-200	530-580
		520-540	520-550	540-580	520-540	140-200	530-580
		500-530	520-550	540-580	500-530	70-200	530-570
		500-530	520-550	540-580	500-530	70-200	530-570
		500-530	520-550	540-580	500-530	70-200	530-570
		500-530	520-550	540-580	500-530	70-200	530-570
2.0-24.0		520	525	540	500-570	100-200	535-580
2.0-24.0		520	525	540	500-570	100-200	535-580
2.0-24.0		520	525	540	500-570	100-200	535-580
2.0-24.0		520	525	540	500-570	100-200	535-580
2.0-24.0		520	525	540	500-570	100-200	535-580
2.0-24.0		520	525	540	500-570	100-200	535-580
2.0-24.0		520	525	540	500-570	100-200	535-580
2.0-24.0		525	535	560	500-570	100-200	550-560
2.0-24.0		525	535	560	500-570	100-200	550-560
2.0-24.0		525	535	560	500-570	100-200	550-560
2.0-24.0		525	535	560	500-570	100-200	550-560
2.0-24.0	0.5- 3.	530-540	520-530	540-560	550-570	150-250	550-590
2.0-24.0	0.5- 3.	530-540	520-530	540-560	550-570	150-250	550-590
2.0-24.0	0.5- 3.	530-540	520-530	540-560	550-570	150-250	550-590
2.0-24.0	0.5- 3.	530-540	520-530	540-560	550-570	150-250	550-590
2.0-24.0	0.5- 3.	530-540	520-530	540-560	550-570	150-250	550-590
2.0-24.0	0.5- 3.	530-540	520-530	540-560	550-570	150-250	550-590
2.0-24.0	0.5- 3.	530-540	520-530	540-560	550-570	150-250	550-590
2.0-24.0	0.5- 3.	530-540	520-530	540-560	550-570	150-250	550-590
2.0-20.0		520-530	530-540	550-570	530-560	150-250	550-590
2.0-20.0		520-530	530-540	550-570	530-560	150-250	550-590
2.0-24.0	0.5- 3.	530-540	520-530	530-560	550-570	150-250	550-590
3.0-48.0		530	540	550	530	160	545-575
3.0-48.0		530	540	550	530	160	545-575
2.0-24.0		450-460	460-470	470-480	430-480	100-150	480-525
2.0-24.0		520-530	530-540	550-570	540-560	150-250	550-580
2.0-24.0		525	535	560	500-570	100-200	550-560
2.0-24.0		525	535	560	500-570	100-200	550-560
2.0-24.0		525	535	560	500-570	100-200	550-560
2.0-16.0	0.5- 1.	460-480	470-490	450-470	470-490	100-180	450-490
2.0-16.0	0.5- 1.	460-480	470-490	450-470	470-490	100-180	450-490

Grade	Filler	Sp Grav	Shrink, mils/in	Melt flow, g/10 min	Melt temp, °F	Back pres, psi	Drying temp, °F
Nylon 6/6 Alloy		**Hiloy**			**ComAlloy**		
631	GL 13	1.22	6.0				180
632	GL 20	1.26	4.0				180
633	GL 33	1.36	3.0				180
634	GL 43	1.47	2.0				180
635	GL 50	1.56	2.0				180
636	GMN 40	1.46	6.0				180
638	GLF	1.31	12.0				180
639	MN	1.47	10.0				180
691	GL 13	1.18	5.0				180
693	GL 33	1.37	3.0				180
Nylon 6/6 Alloy		**Nylatron**			**DSM**		
1028 SL		1.14	10.0		490		
Nylon 6/6 Alloy		**Nyloy**			**Nytex Compos.**		
MG-0077N-Vo	GLF 33		1.5- 2.5				
Nylon 6/66 Cop.		**Ultramid**			**BASF**		
C35		1.13	9.5		385	50- 100	176-230
C3EG6	GLF 30	1.35	1.2		469	50- 100	176-230
C3K		1.13	13.0		470	50- 100	176-230
C3M8	MN 40	1.50	8.5		469	50- 100	176-230
C3U		1.16	8.0		470	50- 100	176-230
C3ZM3	MN 15	1.21	11.0		469	50- 100	176-230
C3ZM6	MN 30	1.37	9.5		469	50- 100	176-230
KR4205		1.16	8.0		470	50- 100	176-230
Nylon 6/66 Cop.		**Vydyne**			**Monsanto**		
80X		1.14	15.0-20.0		470-475	0-1000	
Nylon 6/66 Cop.		**Wellamid Nylon**			**Wellman**		
MR259 22LH-N	MN	1.28	14.0-18.0		500		150-170
MR409 22H-N	MN	1.41	14.0-18.0		500		150-170
MR410 22H-N	MN	1.44	7.0-11.0		500		150-170
Nylon 6/6T		**Ultramid T**			**BASF**		
KR4350		1.18	7.0		563		176-230
KR4351		1.18	7.0		563		176-230
KR4352		1.18	7.0		563		176-230
KR4355 G5	GLF 25	1.36			563		176-230
KR4355 G7	GLF 35	1.44	4.0		563		176-230
KR4357 G6	GLF 30	1.39	5.0		563		176-230
KR4360 M6	MN 30	1.41			563		176-230
KR4370 C6	CF 30	1.32			563		176-230
Nylon 6/9		**Vydyne**			**Monsanto**		
602M		1.08	10.0-14.0		410	0- 150	
Nylon MXD6		**Reny**			**Mitsubishi Gas**		
6002		1.21	14.1		469		
6301		1.17	15.0		469		
PAEK		**RTP Polymers**			**RTP**		
3801	GLF	1.37	4.0			25- 100	300

Drying time, hr	Inj time, sec	Front temp, °F	Mid temp, °F	Rear temp, °F	Nozzle temp, °F	Mold temp, °F	Proc temp, °F
3.0- 4.0		470-510	480-520	490-530	470-510	100-250	480-520
3.0- 4.0		470-510	480-520	490-530	470-510	100-250	480-520
3.0- 4.0		470-510	480-520	490-530	470-510	100-250	480-520
3.0- 4.0		470-510	480-520	490-530	470-510	100-250	480-520
3.0- 4.0		470-510	480-520	490-530	470-510	100-250	480-520
3.0- 4.0		470-510	480-520	490-530	470-510	100-250	480-520
3.0- 4.0		470-510	480-520	490-530	470-510	100-250	480-520
3.0- 4.0		470-510	480-520	490-530	470-510	100-250	480-520
3.0- 4.0		470-510	480-520	490-530	470-510	100-250	480-520
3.0- 4.0		470-510	480-520	490-530	470-510	100-250	480-520
		520	530	540		180	530-560
							527
4.0-12.0						100-140	450-480
4.0-12.0						176-194	520-555
4.0-12.0						104-140	520-555
4.0-12.0						176-194	520-555
4.0-12.0						100-140	500-535
4.0-12.0						176-194	520-555
4.0-12.0						176-194	520-555
4.0-12.0						104-140	500-535
		460-540	460-540	470-500	500-540	100-150	500-540
	1.0- 5.	570	570	560	560	180-220	570-590
	1.0- 5.	570	570	560	560	180-220	570-590
	1.0- 5.	570	570	560	560	180-220	570-590
		644-644	608-644	572-644	644-644	140-176	590-644
		644-644	608-644	572-644	644-644	140-176	590-644
		644-644	608-644	572-644	644-644	140-176	590-644
						158-194	608-662
						158-194	608-662
						158-194	608-662
						158-194	608-662
						158-194	608-662
		490-520	490-520	450-475	490-520	70-200	480-520
						248-284	482-518
						248-284	482-518
3.0		690-750	670-730	650-710		300-425	

Grade	Filler	Sp Grav	Shrink, mils/in	Melt flow, g/10 min	Melt temp, °F	Back pres, psi	Drying temp, °F
3803	GLF	1.44	3.0			25- 100	300
3805	GLF 30	1.52	2.0			25- 100	300
3807	GLF 40	1.62	1.0			25- 100	300
3881	CF	1.33	2.0			25- 100	300
3883	CF	1.37	1.0			25- 100	300
3885	CF	1.41	0.5			25- 100	300
3887	CF	1.45	0.5			25- 100	300
PAEK		**Ultrapek**			**BASF**		
KR4176		1.30	14.0-14.5		718		320
KR4176 G4	GLF 20	1.45					
KR4176 G6	GLF 30	1.53					
KR4177		1.30	14.0-15.3		718		320
KR4177 G4	GLF 20	1.45	7.4-11.0		718		320
KR4177 G6	GLF 30	1.53	5.0- 7.0		718		320
KR4178		1.30	14.0-15.5		718		320
PBT		**Arnite**			**DSM**		
T06 200		1.30	20.0		433	45- 85	212-230
T06 202		1.30	20.0		428		180-275
T06 203V		1.45	20.0		433	43- 87	212-230
T08 200		1.30	20.0		433	45- 85	212-230
T08 201T		1.25	20.0		432	43- 87	212-230
TM4 250	MN 25	1.49	1.3- 1.5		433	43- 87	212-230
TV4 260S	GLF 30	1.69	2.5-15.0		433	43- 87	212-230
TV4 260SN	GLF 30	1.67	2.5-15.0		433	43- 87	212-230
TV4 261	UNS 30	1.52	2.5-15.0		433		
TV6 240	UNS 20	1.45	25.0-15.0		433		212-230
TV6 241 S	GLF 20	1.60	2.5-15.0		433	87- 145	212-230
TV6 241SN	UNS 20	1.59	2.5-15.0		433	87- 145	212-230
TV8 260SY	GLF 30	1.66	2.5-15.0		433	87- 145	212-230
TZ6 260SY	GLF	1.66	9.0-17.0		433		212-230
PBT		**AVP**			**Polymerland**		
RVV30	GL 30	1.54	4.0- 6.0	9.00 I		25- 50	250
PBT		**Bay Resins PBT**			**Bay Resins**		
PBT-1100		1.31	18.0		428-437	50- 100	160
PBT-1100G15	GLF 15	1.42	9.0		428-437	50- 100	160
PBT-1100G20	GLF 25	1.45	7.0		428-437	50- 100	160
PBT-1100G30	GLF 30	1.53	5.0		428-437	50- 100	160
PBT-1100G30TF15	GLF 30	1.66	5.0		428-437	50- 100	160
PBT-1300		1.30	8.0		425-435	50- 100	160
PBT-1300G25	GLF 25	1.46	6.0		425-435	50- 100	160
PBT-1700		1.40	18.0		428-437	50- 100	160
PBT-1700G15	GLF 15	1.53	9.0		428-437	50- 100	160
PBT-1700G30	GLF 30	1.66	5.0		428-437	50- 100	160
PBT		**Celanex**			**Hoechst**		
1300A		1.31	18.0-20.0			0- 50	250-280
1400A		1.31	18.0-20.0			0- 50	250-280
1462Z	GLF 30	1.52	3.0- 5.0			0- 50	250-280
1612Z	GLF 8	1.35	8.0-10.0			0- 50	250-280
1700A		1.31	18.0-20.0			0- 50	250-280
2000		1.31	18.0-20.0			0- 50	250-280
2000K		1.31	18.0-20.0			0- 50	250-280

Drying time, hr	Inj time, sec	Front temp, °F	Mid temp, °F	Rear temp, °F	Nozzle temp, °F	Mold temp, °F	Proc temp, °F
3.0		690-750	670-730	650-710		300-425	
3.0		690-750	670-730	650-710		300-425	
3.0		690-750	670-730	650-710		300-425	
3.0		725-800	690-785	675-750		300-425	
3.0		725-800	690-785	675-750		300-425	
3.0		725-800	690-785	675-750		300-425	
3.0		725-800	690-785	675-750		300-425	
4.0-15.0		779-788	752-788	716-788	788-788	356-410	734-788
						355-410	750-800
						355-410	750-800
4.0-15.0		779-788	752-788	716-788	788-788	356-410	734-788
4.0-15.0		779-788	752-788	716-788	788-788	356-410	752-806
4.0-15.0		779-788	752-788	716-788	788-788	356-410	752-806
4.0-15.0		779-788	752-788	716-788	788-788	356-410	752-806
6.0-12.0		446-473	437-464	428-455	455-482	140-212	446-518
2.0-16.0		480-500	450-460	440-450	490-510	80-200	470-520
6.0-12.0		464		446	500	86-194	464-518
6.0-12.0		446-473	437-464	428-455	455-482	140-212	446-518
6.0-12.0		464		455	473	86-194	464-518
6.0-12.0		464-473	455-482	446-473	473-500	194	464-518
6.0-12.0		464-473	455-482	446-473	473-500	194	464-518
6.0-12.0		464-473	455-482	446-473	473-500	194	464-518
						86-248	405-477
6.0-12.0		464		455	473	86-194	464-518
6.0-12.0		464		455	473	86-194	464-518
6.0-12.0		482		473	491	86-194	464-500
6.0-12.0		482		473	491	86-194	464-482
6.0-12.0		464		455	473	86-194	464-482
5.0						150-250	470-510
4.0		450-525	450-525	450-525		150-180	450-475
4.0		450-525	450-525	450-525		150-180	450-475
4.0		450-525	450-525	450-525		150-180	450-475
4.0		450-525	450-525	450-525		150-180	450-475
4.0		450-525	450-525	450-525		150-180	450-475
4.0		450-525	450-525	450-525		150-180	450-475
4.0		450-525	450-525	450-525		150-180	450-475
4.0		450-525	450-525	450-525		150-180	450-475
4.0		450-525	450-525	450-525		150-180	450-475
4.0		450-525	450-525	450-525		150-180	450-475
3.0- 4.0		470	460	450	480	150-200	450-500
3.0- 4.0		470	460	450	480	150-200	450-500
3.0- 4.0		470	460	450	480	150-200	450-500
3.0- 4.0		470	460	450	480	150-200	450-500
3.0- 4.0		470	460	450	480	150-200	450-500
3.0- 4.0		470	460	450	480	150-200	450-500
3.0- 4.0		470	460	450	480	150-200	450-500

Grade	Filler	Sp Grav	Shrink, mils/in	Melt flow, g/10 min	Melt temp, °F	Back pres, psi	Drying temp, °F
2002		1.31	18.0-20.0			0- 50	250-280
2003		1.31	18.0-20.0			0- 50	250-280
2003K		1.31	18.0-20.0			0- 50	250-280
2006		1.31				0- 50	250-280
2008		1.31				0- 50	250-280
2012		1.43	18.0-20.0			0- 50	250-280
3112	GLF 13	1.52	5.0- 7.0			0- 50	250-280
3200	GLF 15	1.41	4.0- 6.0			0- 50	250-280
3210	GLF 18	1.62	4.0- 6.0			0- 50	250-280
3211	GLF 19	1.60	4.0- 6.0			0- 50	250-280
3310	GLF 30	1.66	3.0- 5.0			0- 50	250-280
3311	GLF 30	1.65	3.0- 5.0			0- 50	250-280
3400	GLF 40	1.61	3.0- 5.0			0- 50	250-280
4300	GLF 30	1.52	3.0- 5.0			0- 50	250-280
4330	GLF 30	1.53	2.5- 4.5			0- 50	250-280
5200	GLF 15	1.41	4.0- 6.0			0- 50	250-280
5300	GLF 30	1.54	4.0- 6.0			0- 50	250-280
6400	GMN 40	1.65	4.0- 6.0			0- 50	250-280
6406	GMN 40	1.58	4.0- 6.0			0- 50	250-280
6500	GMN 30	1.55	2.0- 4.0			0- 50	250-280
7304	GMN 35	1.74	5.0- 7.0			0- 50	250-280
7305	GMN 35	1.74	3.0- 5.0			0- 50	250-280
7700	GMN 35	1.74	5.0- 7.0			0- 50	250-280
J600	GMN 40	1.62	4.0- 6.0			0- 50	250-280
LW6443R	GMN 40	1.57	4.0- 6.0			0- 50	250-280
PBT		**Celstran**			**Hoechst**		
PBTG30-01-4	GLF 30	1.56					250
PBTG40-01-4	GLF 40	1.65					250
PBTG50-01-4	GLF 50	1.75					250
PBTG60-01-4	GLF 60	1.87					250
PBT		**Celstran S**			**Hoechst**		
PBTS10-01-4	STS 10	1.46	13.0-15.0				200
PBTS10-02-4	STS 10	1.42	3.0- 4.0				200
PBTS10-03-4	STS 10	1.77	3.0- 4.0				200
PBTS6-01-4	STS 6	1.41	15.0-17.0				200
PBTS6-02-4	STS 6	1.37	4.0- 5.0				200
PBTS6-03-4	STS 6	1.73	4.0- 5.0				200
PBT		**Comtuf**			**ComAlloy**		
415	GL 30	1.45	3.0- 6.0				250-280
PBT		**CTI General**			**CTI**		
CTXC-301	STS	1.64	10.0				220
PBT		**DSM PBT**			**DSM**		
J-1850/15	GLF 15	1.42	3.0				180-275
J-1850/30	GLF 30	1.51	1.5				180-275
PBT		**Electrafil**			**DSM**		
G-1854/SS/7	STS 7	1.37	13.0				180-275
J-1850/CF/30	CF 30	1.42	1.0				180-275
PBT		**Fiberfil VO**			**DSM**		
J-1850/30/VO	GLF 30	1.69	2.0- 3.0			50	250

Drying time, hr	Inj time, sec	Front temp, °F	Mid temp, °F	Rear temp, °F	Nozzle temp, °F	Mold temp, °F	Proc temp, °F
3.0- 4.0		470	460	450	480	150-200	450-500
3.0- 4.0		470	460	450	480	150-200	450-500
3.0- 4.0		470	460	450	480	150-200	450-500
3.0- 4.0		470	460	450	480	150-200	450-500
3.0- 4.0		470	460	450	480	150-200	450-500
3.0- 4.0		470	460	450	480	150-200	450-500
3.0- 4.0		470	460	450	480	150-200	450-500
3.0- 4.0		470	460	450	480	150-200	450-500
3.0- 4.0		470	460	450	480	150-200	450-500
3.0- 4.0		470	460	450	480	150-200	430-480
3.0- 4.0		470	460	450	480	150-200	450-500
3.0- 4.0		470	460	450	480	150-200	450-500
3.0- 4.0		470	460	450	480	150-200	450-500
3.0- 4.0		470	460	450	480	150-200	450-500
3.0- 4.0		470	460	450	480	150-200	450-500
3.0- 4.0		470	460	450	480	150-200	450-500
3.0- 4.0		470	460	450	480	150-200	450-500
3.0- 4.0		470	460	450	480	150-200	450-500
3.0- 4.0		470	460	450	480	150-200	450-500
3.0- 4.0		470	460	450	480	150-200	450-500
3.0- 4.0		470	460	450	480	150-200	450-500
3.0- 4.0		470	460	450	480	150-200	450-500
3.0- 4.0		470	460	450	480	150-200	450-500
3.0- 4.0		470	460	450	480	150-200	450-500
3.0- 4.0		470	460	450	480	150-200	450-500
4.0		490	490	480	500	180	
4.0		490	480	480	500	180	
4.0		500	490	480	510	180	
4.0		510	500	480	520	180	
4.0		500	490	490	500	180	
4.0		500	500	490	500	180	
4.0		500	490	480	500	180	
4.0		440	430	420	450	180	
4.0		500	500	490	500	180	
4.0		500	490	480	500	180	
3.0- 4.0		420-460	420-450	410-440	420-470	100-250	430-490
3.0- 4.0		480-500	460-480	450-470	470-480	150-200	
2.0-16.0		470-490	480-500	450-470	490-510	80-200	470-520
2.0-16.0		470-490	480-500	450-470	490-510	80-200	470-520
2.0-16.0		470-490	480-500	450-470	490-510	80-200	470-520
2.0-16.0		470-490	480-500	450-470	490-510	80-200	470-520
2.0- 4.0		430-470	440-480	420-450	440-460	140-180	550-480

Grade	Filler	Sp Grav	Shrink, mils/in	Melt flow, g/10 min	Melt temp, °F	Back pres, psi	Drying temp, °F
J-1875/30/VO	GLF 30	1.62	2.0- 6.0			50	250
PBT		**Hiloy**			**ComAlloy**		
411	MN 25	1.43	6.0-14.0				250-280
412	GL 15	1.42	7.0				250-280
413	GL 30	1.54	5.0				250-280
414	GL 45	1.67	4.0				250-280
416	GMN 40	1.65	6.0				250-280
417	GMN 50	1.76	5.0				250-280
418	MN	1.97	9.0-11.0				250-280
PBT		**Lubrilon**			**ComAlloy**		
410		1.42	13.0				250-280
PBT		**Lupox**			**Lucky**		
EE-4351F	GMN 35	1.69	3.0- 7.0			0- 568	248
EE-4400	GMN 40	1.65	3.0- 7.0			0- 568	248
EE-4401F	GMN 40	1.75	2.0- 5.0			0- 568	248
GP-1000		1.31	17.0-23.0				248
GP-1001F		1.44	13.0-20.0				248
GP-1006F		1.42	13.0-20.0				248
GP-1008F		1.42	13.0-20.0				248
GP-2150	GLF 15	1.41	6.0-14.0			0- 568	248
GP-2151F	GLF 15	1.53	4.0-14.0			0- 568	248
GP-2156F	GLF 15	1.52	4.0-14.0			0- 568	248
GP-2158F	GLF 15	1.52	4.0-14.0			0- 568	248
GP-2300	GLF 30	1.52	3.0-10.0			0- 568	248
GP-2301F	GLF 30	1.66	3.0-10.0			0- 568	248
GP-2306F	GLF 30	1.65	3.0-10.0			0- 568	248
GP-2308F	GLF 30	1.65	3.0-10.0			0- 568	248
HI-1002F		1.35	15.0-18.0				248
HI-2152	GLF 15	1.35	8.0-12.0			0- 568	248
HI-2302	GLF 30	1.44	4.0- 9.0			0- 568	248
HI-2302AF	GLF 30	1.57	4.0- 9.0			0- 568	248
HI-2302F	GLF 30	1.58	4.0- 9.0			0- 568	248
HI-2303	GLF 30	1.46	4.0- 8.0			0- 568	248
HI-5002BF		1.42	8.0-13.0			0- 568	248
HV-1010		1.31	17.0-23.0				248
LW-4302F	GMN 30	1.58	5.0- 7.0			0- 568	248
LW-5303	UNS 30	1.51	3.0- 7.0			0- 568	248
LW-5303F	UNS 30	1.55	3.0- 7.0			0- 568	248
SG-5150	UNS 15	1.43	7.0-11.0			0- 568	248
SG-5151F	UNS 15	1.52	7.0-11.0			0- 568	248
SG-5300	UNS 30	1.54	6.0- 9.0			0- 568	248
SG-5301F	UNS 30	1.66	6.0- 9.0			0- 568	248
TE-5001		1.22	7.0- 9.0			0- 568	248
TE-5010		1.22	7.0- 9.0			0- 568	248
TE-5011		1.22	7.0- 9.0			0- 568	248
TE-5020		1.22	7.0- 9.0			0- 568	248
PBT		**PermaStat**			**RTP**		
1000		1.28	18.0-26.0				
PBT		**Pocan**			**Albis**		
B 1505		1.30	17.0-20.0	10.40 T		30- 60	250
B 3225	GLF 20	1.46	5.0- 8.0	15.50 T		30- 60	250

Drying time, hr	Inj time, sec	Front temp, °F	Mid temp, °F	Rear temp, °F	Nozzle temp, °F	Mold temp, °F	Proc temp, °F
2.0- 4.0		430-470	440-480	420-450	440-460	140-180	550-480
3.0- 4.0		420-460	420-450	410-440	420-470	100-250	430-490
3.0- 4.0		420-460	420-450	410-440	420-470	100-250	430-490
3.0- 4.0		420-460	420-450	410-440	420-470	100-250	430-490
3.0- 4.0		420-460	420-450	410-440	420-470	100-250	430-490
3.0- 4.0		420-460	420-450	410-440	420-470	100-250	430-490
3.0- 4.0		420-460	420-450	410-440	420-470	100-250	430-490
3.0- 4.0		420-460	420-450	410-440	420-470	100-250	430-490
3.0- 4.0		420-460	420-450	410-440	420-470	100-250	430-490
5.0		464-482	455-482	446-473	464-500	122-212	464-500
5.0		464-482	455-482	446-473	464-500	122-212	464-500
5.0		464-482	455-482	446-473	464-500	122-212	464-500
5.0		455-473	446-464	437-455	455-473	104-176	455-473
5.0		455-473	446-464	437-455	455-473	104-176	455-473
5.0		455-473	446-464	437-455	455-473	104-176	455-473
5.0		455-473	446-464	437-455	455-473	104-176	455-473
5.0		464-482	455-482	446-473	464-500	122-212	464-500
5.0		464-482	455-482	446-473	464-500	122-212	464-500
5.0		464-482	455-482	446-473	464-500	122-212	464-500
5.0		464-482	455-482	446-473	464-500	122-212	464-500
5.0		464-482	455-482	446-473	464-500	122-212	464-500
5.0		464-482	455-482	446-473	464-500	122-212	464-500
5.0		464-482	455-482	446-473	464-500	122-212	464-500
5.0		464-482	455-482	446-473	464-500	122-212	464-500
5.0		455-473	446-464	437-455	455-473	104-176	455-473
5.0		464-482	455-482	446-473	464-500	122-212	464-500
5.0		464-482	455-482	446-473	464-500	122-212	464-500
5.0		464-482	455-482	446-473	464-500	122-212	464-500
5.0		464-482	455-482	446-473	464-500	122-212	464-500
5.0		464-482	455-482	446-473	464-500	122-212	464-500
5.0		464-482	455-482	446-473	464-500	122-212	464-500
5.0		455-473	446-464	437-455	455-473	104-176	455-473
5.0		464-482	455-482	446-473	464-500	122-212	464-500
5.0		464-482	455-482	446-473	464-500	122-212	464-500
5.0		464-482	455-482	446-473	464-500	122-212	464-500
5.0		464-482	455-482	446-473	464-500	122-212	464-500
5.0		464-482	455-482	446-473	464-500	122-212	464-500
5.0		464-482	455-482	446-473	464-500	122-212	464-500
5.0		464-482	455-482	446-473	464-500	122-212	464-500
5.0		464-482	455-482	446-473	464-500	122-212	464-500
5.0		464-482	455-482	446-473	464-500	122-212	464-500
5.0		464-482	455-482	446-473	464-500	122-212	464-500
		400-450	400-450	400-450		100-150	
2.0- 4.0		480-500	470-490	455-470	480-500	160-250	460-500
2.0- 4.0		480-500	470-490	455-470	480-500	160-250	460-500

Grade	Filler	Sp Grav	Shrink, mils/in	Melt flow, g/10 min	Melt temp, °F	Back pres, psi	Drying temp, °F
B 3235	GLF 30	1.46	4.0- 8.0	7.30 T		30- 60	250
B 4225	GLF 20	1.57	4.0- 8.0	17.00 T		30- 60	250
B 4235	GLF 30	1.65	4.0- 8.0	10.20 T		30- 60	250
S 1506		1.22	17.0-22.0	1.70 T		30- 60	250
PBT		**RTP Polymers**			**RTP**		
1000	GLF	1.31	18.0			25- 75	250
1000 FR A		1.42	23.0			25- 75	250
1000 GB 10	GLB	1.38	20.0			25- 75	250
1000 GB·20	GLB	1.45	20.0			25- 75	250
1000 GB 30	GLB	1.53	18.0			25- 75	250
1000 GB 40	GLB	1.62	16.0			25- 75	250
1001	GLF	1.38	9.0			25- 75	250
1001 FR A	GLF	1.50	7.0			25- 75	250
1001 GB 20	GLC	1.52	9.0			25- 75	250
1001 M 20	GLC	1.54	8.0			25- 75	250
1002 FR A	GLF	1.53	5.0			25- 75	250
1002 TFE 15	GLF	1.52	5.0			25- 75	250
1003	GLF	1.45	4.0			25- 75	250
1005	GLF	1.53	3.0			25- 75	250
1005 AG	GLF	1.51	3.0			25- 75	250
1005 FR A	GLF	1.63	2.0			25- 75	250
1005 TFE 15	GLF	1.65	2.0			25- 75	250
1005 TFE 15 FR	GLF	1.73	2.0			25- 75	250
1005 TFE 15 FR	GLF	1.73	2.0			25- 75	250
1007	GLF	1.63	2.0			25- 75	250
1007 GB 10	GLC	1.72	2.0			25- 75	250
1025	GLC	1.62	4.0			25- 75	250
1026	GLC	1.62	7.0			25- 75	250
1027	GLC	1.62	9.0			25- 75	250
1028	GLC	1.71	12.0			25- 75	250
1085	CF	1.42	1.0			50- 90	250
1085 TFE 15	CF	1.51	1.0			50- 90	250
1085 TFE 15	CF	1.51	1.0			50- 90	250
1085 TFE 15	CF	1.51	1.0				
1087	CF	1.48	1.0			50- 90	250
1099X52742H	GMN	1.68	4.0			25- 75	250
ESD-A-1000		1.28	15.0-18.0				
ESD-A-1080	CF	1.35	2.0				
ESD-C-1000		1.28	15.0-18.0				
ESD-C-1080	CF	1.36	2.0				
PBT		**Rynite**			**Du Pont**		
6125		1.31	16.0-20.0		425		225-275
6130		1.31	16.0-20.0		437		225-275
6131		1.31	16.0-20.0		437		225-275
6400		1.23	22.0-30.0		436		225-275
7015	GLF 15	1.42	6.0-10.0		435		225-275
7030	GLF 30	1.53	5.0- 8.0		440		225-275
FR6941		1.40	17.0-26.0		434		225-275
FR7915	GLF 15	1.61	6.0-10.0		437		225-275
FR7930	GLF 30	1.72	5.0- 8.0		435		225-275
FR7930F	GLF 30	1.69	5.0- 8.0		430		225-275
RE6944		1.45	15.0-18.0		435		225-275
RE7005	GLF 15	1.42	6.0-10.0		435		225-275

Drying time, hr	Inj time, sec	Front temp, °F	Mid temp, °F	Rear temp, °F	Nozzle temp, °F	Mold temp, °F	Proc temp, °F
2.0- 4.0		480-500	470-490	455-470	480-500	160-250	460-500
2.0- 4.0		480-500	470-490	455-470	480-500	160-250	460-500
2.0- 4.0		480-500	470-490	455-470	480-500	160-250	460-500
2.0- 4.0		480-500	470-490	455-470	480-500	160-250	460-500
4.0		480-550	460-530	440-500		120-250	
4.0		430-530	440-510	420-490		120-250	
4.0		480-550	460-530	440-500		120-250	
4.0		480-550	460-530	440-500		120-250	
4.0		480-550	460-530	440-500		120-250	
4.0		480-550	460-530	440-500		120-250	
4.0		480-550	460-530	440-500		120-250	
4.0		430-530	440-510	420-490		120-250	
4.0		480-550	460-530	440-500		120-250	
4.0		480-550	460-530	440-500		120-250	
4.0		430-530	440-510	420-490		120-250	
4.0		480-550	460-530	440-500		120-250	
4.0		480-550	460-530	440-500		120-250	
4.0		480-550	460-530	440-500		120-250	
4.0		480-550	460-530	440-500		120-250	
4.0		430-530	440-510	420-490		120-250	
4.0		480-550	460-530	440-500		120-250	
4.0		430-530	440-510	420-490		120-250	
4.0		430-530	440-510	420-490		120-250	
4.0		480-550	460-530	440-500		120-250	
4.0		480-550	460-530	440-500		120-250	
4.0		480-550	460-530	440-500		120-250	
4.0		480-550	460-530	440-500		120-250	
4.0		480-550	460-530	440-500		120-250	
4.0		480-550	460-530	440-500		120-250	
4.0		450-530	440-520	420-500		120-250	
4.0		450-530	440-520	420-500		120-250	
4.0		450-530	440-520	420-500		120-250	
		450-520	450-520	450-520		100-250	
4.0		450-530	440-520	420-500		120-250	
4.0		430-530	440-510	420-490		120-250	
		450-520	450-520	450-520		100-250	
		425-500	425-500	425-500		125-275	
		450-520	450-520	450-520		100-250	
		425-500	425-500	425-500		125-275	
2.0-16.0	1.0- 4.	460-500	460-500	450-480	460-500	90-150	480-500
2.0-16.0	1.0- 4.					90-150	480-500
2.0-16.0	1.0- 4.					90-150	480-500
2.0-16.0	1.0- 4.	460-500	460-500	450-480	460-500	90-150	480-500
2.0-16.0	1.0- 4.	460-500	460-500	450-480	460-500	90-150	480-500
2.0-16.0	1.0- 4.	460-500	460-500	450-480	460-500	90-150	480-500
2.0-16.0	1.0- 4.					90-150	480-510
2.0-16.0	1.0- 4.	460-500	460-500	450-480	460-500	90-150	480-500
2.0-16.0	1.0- 4.	460-500	460-500	450-480	460-500	90-150	480-500
2.0-16.0	1.0- 4.	460-500	460-500	450-480	460-500	90-150	480-500
2.0-16.0	1.0- 4.					90-150	480-500
2.0-16.0	1.0- 4.					90-150	480-510

Grade	Filler	Sp Grav	Shrink, mils/in	Melt flow, g/10 min	Melt temp, °F	Back pres, psi	Drying temp, °F
PBT		**Shinite**			**Shinkong**		
D201		1.31	20.0-22.0		439		250
D201G30	GLF 30	1.52	3.0-14.0		435		250
D202		1.42	20.0-22.0		435		250
D202G30	GLF 30	1.59	3.0-13.0		432		250
PBT		**Topex**			**Tong Yang**		
4010		1.31	17.0-23.0		432	142	248
4322SWN		1.42	10.0-17.0		432	142	248
4410GF	GLF	1.52	3.0- 6.0		432	142	248
4413GWN	GLF	1.65	3.0- 6.0		432	142	248
PBT		**Toray PBT**			**Toray**		
1101G30	GL 30	1.55	3.0-10.0			853-2132	248-266
1104G30	GL 30	1.68	3.0-10.0			853-2132	248-266
1401X06		1.31	17.0-23.0			853-2132	248-266
1404X04		1.48	17.0-23.0			853-2132	248-266
PBT		**Ultradur**			**BASF**		
B2550		1.30	11.0-22.0		428-437	50- 100	210-285
B4300 G10	GLF 50	1.71	2.0- 3.0		428-437	50- 100	210-285
B4300 G2	GLF 10	1.37	8.0-10.0		428-437	50- 100	210-285
B4300 G4	GLF 20	1.45	6.0- 8.0		428-437	50- 100	210-285
B4300 G6	GLF 30	1.53	3.0- 5.0		428-437	50- 100	210-285
B4300 K4	GLB 20	1.45	9.0-22.0		428-437	50- 100	210-285
B4300 K6	GLB 30	1.53	9.0-20.0		428-437	50- 100	210-285
B4306		1.45	11.0-12.0		428-437	50- 100	210-285
B4306 G2	GLF 10	1.50	8.0-10.0		428-437	50- 100	210-285
B4306 G4	GLF 20	1.55	6.0- 8.0		428-437	50- 100	210-285
B4306 G6	GLF 30	1.65	3.0- 5.0		428-437	50- 100	210-285
B4406		1.45	11.0-12.0		428-437	50- 100	210-285
B4406 G2	GLF 10	1.50	7.0- 9.0		428-437	50- 100	210-285
B4406 G4	GLF 20	1.55	6.0- 8.0		428-437	50- 100	210-285
B4406 G6	GLF 30	1.65	3.0- 5.0		428-437	50- 100	210-285
B4500		1.30	9.0-22.0		428-437	50- 100	210-285
B4520		1.30	11.0-22.0		428-437	50- 100	210-285
KR4001	MN 25	1.51	11.0-13.0		428-437	50- 100	210-285
KR4011	GLF 20	1.55	4.0- 5.0		428-437	50- 100	210-285
KR4036		1.30	9.0-22.0		428-437	50- 100	210-285
KR4070		1.20	17.0-21.0		428-437	50- 100	210-285
KR4071		1.20	17.0-21.0		428-437	50- 100	210-285
KR4075	MN 10	1.36	14.0-15.0		428-437	50- 100	210-285
PBT		**Valox**			**GE**		
215HP		1.31	6.0-16.0			0- 50	250
215HPR		1.31	6.0-16.0			0- 50	250
295		1.31	6.0-16.0			0- 50	250
310SEO		1.39	6.0-16.0			25- 50	250-260
310SEOU		1.39	6.0-16.0			25- 50	250-260
311		1.31	6.0-16.0			25- 50	250-260
312		1.31	6.0-16.0			25- 50	250-260
315		1.31	6.0-16.0			0- 50	250
321		1.31	6.0-16.0			25- 50	250-260
325		1.31	6.0-16.0			25- 50	250-260
325E		1.31	6.0-16.0			25- 50	250-260

Drying time, hr	Inj time, sec	Front temp, °F	Mid temp, °F	Rear temp, °F	Nozzle temp, °F	Mold temp, °F	Proc temp, °F
2.0- 4.0		450-500	450-500	450-500		140-250	465-500
2.0- 4.0		450-500	450-500	450-500		140-250	465-500
2.0- 4.0		450-500	450-500	450-500		140-250	465-500
2.0- 4.0		450-500	450-500	450-500		140-250	465-500
6.0	5.0	446	446	410	446	140	
6.0	5.0	455	455	410	455	140-176	
6.0	5.0	491	482	464	482	176	
6.0	5.0	482	473	455	478	176	
3.0- 5.0		464-482	464-482	464-482		176-194	
3.0- 5.0		464-482	464-482	464-482		176-194	
3.0- 5.0		446-464	446-464	446-464		104-176	
3.0- 5.0		446-464	446-464	446-464		104-176	
3.0- 4.0						176-212	473-518
3.0- 4.0						176-212	473-518
3.0- 4.0						176-212	473-518
3.0- 4.0						176-212	473-518
3.0- 4.0						176-212	473-518
3.0- 4.0						104-176	473-509
3.0- 4.0						104-176	473-509
3.0- 4.0						104-140	473-518
3.0- 4.0						176-212	473-518
3.0- 4.0						176-212	473-518
3.0- 4.0						176-212	473-518
3.0- 4.0						104-140	473-518
3.0- 4.0						176-212	473-518
3.0- 4.0						176-212	473-518
3.0- 4.0						176-212	473-518
3.0- 4.0						104-140	473-518
3.0- 4.0						104-140	473-518
3.0- 4.0						104-140	473-509
3.0- 4.0						140-194	473-518
3.0- 4.0						104-140	482-518
3.0- 4.0						104-140	482-518
3.0- 4.0						104-140	482-518
3.0- 4.0						104-194	473-518
3.0- 4.0		455-480	450-470	440-460	460-480	60-150	460-480
3.0- 4.0		455-480	450-470	440-460	460-480	60-150	460-480
3.0- 4.0		460-490	450-480	440-470	460-490	110-140	455-490
3.0- 4.0		460-490	450-480	440-470	460-490	110-140	455-490
3.0- 4.0		460-490	450-480	440-470	460-490	110-140	455-490
3.0- 4.0		460-490	450-480	440-470	460-490	110-140	455-490
3.0- 4.0		460-490	450-480	440-470	460-490	110-140	455-490
3.0- 4.0		460-490	450-480	440-470	460-490	110-140	455-490
3.0- 4.0		460-490	450-480	440-470	460-490	110-140	455-490
3.0- 4.0		460-490	450-480	440-470	460-490	110-140	455-490
3.0- 4.0		460-490	450-480	440-470	460-490	110-140	455-490

Grade	Filler	Sp Grav	Shrink, mils/in	Melt flow, g/10 min	Melt temp, °F	Back pres, psi	Drying temp, °F
325M		1.31	6.0-16.0			25- 50	250-260
326		1.29				25- 50	250
327		1.29	17.0-21.0			25- 50	250-260
327E		1.29	17.0-21.0			25- 50	250-260
357		1.34	3.0-12.0			25- 50	250-260
357M		1.34	3.0-12.0			25- 50	250-260
357R		1.34	3.0-12.0			25- 50	250-260
357U		1.34	3.0-12.0			25- 50	250-260
360		1.31	3.0-12.0			25- 50	250-260
360E		1.31	3.0-12.0			25- 50	250-260
365		1.33	3.0-12.0			25- 75	250-275
412	GL 20	1.45	4.0- 8.0			0- 50	250
414	GL 40	1.63	3.0- 5.0			25- 50	250
414U	GL 40	1.63				25- 50	250
417	GL 17	1.43	4.0- 9.0			25- 50	250
420	GL 30	1.53	3.0- 8.0			25- 50	250-260
420D	GL 30	1.53	3.0- 8.0			25- 50	250-260
420E	GL 30	1.53	3.0- 8.0			25- 50	250-260
420K	GL 30	1.53	3.0- 8.0			25- 50	250-260
420M	GL 30	1.53	3.0- 8.0			25- 50	250-260
420P	GL 30	1.53	3.0- 8.0			25- 50	250-260
420R	GL 30	1.53	3.0- 8.0			25- 50	250-260
420SEO	GL 30	1.60	3.0- 8.0			25- 50	250-260
420SEOM	GL 30	1.60	3.0- 8.0			25- 50	250-260
420SEOU	GL 30	1.60	3.0- 8.0			25- 50	250-260
420SEOX	GL 30	1.62	3.0- 8.0			25- 50	250-260
420U	GL 30	1.53	3.0- 8.0			25- 50	250-260
430	GL 33	1.52	3.0- 8.0			25- 50	250-260
430E	GL 33	1.52	3.0- 8.0			25- 50	250-260
451	GL 20	1.53	4.0- 8.0			25- 50	250-260
451E	GL 20	1.53	4.0- 9.0			25- 50	250-260
457	GL 8	1.43	6.0-10.0			25- 50	250-260
457X	GL 8	1.43	6.0-10.0			25- 50	250-260
466	GL 30	1.69	3.0- 8.0			25- 50	250-260
469	GL 30	1.65	3.0- 8.0			0- 50	250
508	GL 30	1.50	3.0- 8.0			25- 50	250-260
508R	GL 30	1.50	3.0- 8.0			25- 50	250-260
508U	GL 30	1.50	3.0- 8.0			25- 50	250-260
553	GL 30	1.58	3.0- 8.0			25- 50	250-260
553E	GL 30	1.58	3.0- 8.0			25- 50	250-260
553M	GL 30	1.58	3.0- 8.0			25- 50	250-260
553U	GL 30	1.58	3.0- 8.0			25- 50	250-260
701	GMN 35	1.59	3.0- 6.0			25- 50	250
711	GL 30	1.55	3.0- 8.0			25- 50	250
711M	GL 30	1.55	3.0- 8.0			25- 50	250
730	GMN 35	1.55	3.0- 6.0			25- 50	250
732	GMN 30	1.51	3.0- 6.0			25- 50	250
732E	GMN 30	1.51	3.0- 6.0			25- 50	250
735	GMN 40	1.62	3.0- 6.0			25- 50	250
735X	GMN 40	1.62	3.0- 6.0			25- 50	250
736	GMN 45	1.69	3.0- 5.0			25- 50	250
744	MNF 10	1.36	6.0-16.0			25- 50	250-260
745	MNF 25	1.46	6.0-10.0			25- 50	250
745U		1.46	8.0-10.0			25- 50	250
750	GMN 45	1.80	3.0- 7.0			25- 50	250
751	GMN 45	1.83	3.0- 7.0			25- 50	250

Drying time, hr	Inj time, sec	Front temp, °F	Mid temp, °F	Rear temp, °F	Nozzle temp, °F	Mold temp, °F	Proc temp, °F
3.0- 4.0		460-490	450-480	440-470	460-490	110-140	455-490
3.0- 4.0		455-480	450-470	440-460	460-480	60-150	460-480
3.0- 4.0		460-490	450-480	440-470	460-490	110-140	455-490
3.0- 4.0		460-490	450-480	440-470	460-490	110-140	455-490
3.0- 4.0		460-490	450-480	440-470	460-490	110-140	455-490
3.0- 4.0		460-490	450-480	440-470	460-490	110-140	455-490
3.0- 4.0		460-490	450-480	440-470	460-490	110-140	455-490
3.0- 4.0		460-490	450-480	440-470	460-490	110-140	455-490
3.0- 4.0		460-490	450-480	440-470	460-490	110-140	455-490
3.0- 4.0		460-490	450-480	440-470	460-490	110-140	455-490
4.0- 8.0		470-545	465-540	455-530	475-550	150-170	475-550
3.0- 4.0		470-490	460-480	450-470	470-500	150-190	460-500
3.0- 4.0		470-490	460-480	450-470	470-500	150-190	460-500
3.0- 4.0		470-490	460-480	450-470	470-500	150-190	460-500
3.0- 4.0		460-500	450-480	440-470	460-490	160-190	470-500
2.0- 4.0		460-500	450-480	440-470	460-490	160-190	470-500
2.0- 4.0		460-500	450-480	440-470	460-490	160-190	470-500
2.0- 4.0		460-500	450-480	440-470	460-490	160-190	470-500
2.0- 4.0		460-500	450-480	440-470	460-490	160-190	470-500
2.0- 4.0		460-500	450-480	440-470	460-490	160-190	470-500
2.0- 4.0		460-500	450-480	440-470	460-490	160-190	470-500
2.0- 4.0		460-500	450-480	440-470	460-490	160-190	470-500
2.0- 4.0		460-500	450-480	440-470	460-490	160-190	470-500
2.0- 4.0		460-500	450-480	440-470	460-490	160-190	470-500
2.0- 4.0		460-500	450-480	440-470	460-490	160-190	470-500
2.0- 4.0		460-500	450-480	440-470	460-490	160-190	470-500
2.0- 4.0		460-500	450-480	440-470	460-490	160-190	470-500
2.0- 4.0		460-500	450-480	440-470	460-490	160-190	470-500
2.0- 4.0		460-500	450-480	440-470	460-490	160-190	470-500
2.0- 4.0		460-500	450-480	440-470	460-490	160-190	470-500
2.0- 4.0		460-500	450-480	440-470	460-490	160-190	470-500
2.0- 4.0		460-500	450-480	440-470	460-490	160-190	470-500
2.0- 4.0		460-500	450-480	440-470	460-490	160-190	470-500
2.0- 4.0		460-500	450-480	440-470	460-490	160-190	470-500
2.0- 4.0		460-490	460-480	450-470	470-500	150-190	460-500
2.0- 4.0		460-500	450-480	440-470	460-490	160-190	470-500
2.0- 4.0		460-500	450-480	440-470	460-490	160-190	470-500
2.0- 4.0		460-500	450-480	440-470	460-490	160-190	470-500
2.0- 4.0		460-500	450-480	440-470	460-490	160-190	470-500
2.0- 4.0		460-500	450-480	440-470	460-490	160-190	470-500
2.0- 4.0		460-500	450-480	440-470	460-490	160-190	470-500
2.0- 4.0		460-500	450-480	440-470	460-490	160-190	470-500
4.0- 6.0		490-510	480-500	470-490	490-520	160-200	480-530
4.0- 6.0		490-510	480-500	470-490	490-520	160-200	480-530
4.0- 6.0		490-510	480-500	470-490	490-520	160-200	480-530
4.0- 6.0		490-510	480-500	470-490	490-520	160-200	480-530
4.0- 6.0		490-510	480-500	470-490	490-520	160-200	480-530
4.0- 6.0		490-510	480-500	470-490	490-520	160-200	480-530
4.0- 6.0		490-510	480-500	470-490	490-520	160-200	480-530
4.0- 6.0		490-510	480-500	470-490	490-520	160-200	480-530
4.0- 6.0		490-510	480-500	470-490	490-520	160-200	480-530
3.0- 4.0		460-490	450-480	440-470	460-490	110-140	455-490
4.0- 6.0		490-510	480-500	470-490	490-520	160-200	480-530
4.0- 6.0		490-510	480-500	470-490	490-520	160-200	480-530
4.0- 6.0		490-510	480-500	470-490	490-520	160-200	480-530
4.0- 6.0		490-510	480-500	470-490	490-520	160-200	480-530

Grade	Filler	Sp Grav	Shrink, mils/in	Melt flow, g/10 min	Melt temp, °F	Back pres, psi	Drying temp, °F
760	MN 25	1.55	4.0- 9.0			25- 50	250
761	MN 25	1.55	4.0- 9.0			25- 50	250
780	GMN 40	1.77	3.0- 6.0			25- 50	250
815	GL 15	1.43	4.0- 9.0			25- 50	250
815U	GL 15	1.43	4.0- 9.0			25- 50	250
820	GL 20	1.41				25- 75	230
830	GL 30	1.54	3.0- 8.0			25- 50	250
830U	GL 30	1.54	3.0- 8.0			25- 50	250
850	GL 15	1.51	4.0- 9.0			25- 50	250
855	GL 15	1.54	4.0- 9.0			25- 50	250
865	GL 15	1.66	3.0- 8.0			25- 50	250
DR48	GL 15	1.53	4.0- 9.0			25- 50	250-260
DR51	GL 15	1.41	4.0- 9.0			25- 50	250-260
DR51M	GL 15	1.41	4.0- 9.0			25- 50	250-260
DR51R	GL 15	1.41	4.0- 9.0			25- 50	250-260
DR51U	GL 15	1.41	4.0- 9.0			25- 50	250-260
HS433	GL 37	1.59	3.0- 8.0			25- 50	250
HV7075	MNF 68	2.40				50- 100	265
PBT		**Voloy**			**ComAlloy**		
411	MN 25	1.55	6.0-14.0				250-280
412	GLF 15	1.55	7.0				250-280
413	GLF 30	1.62	5.0				250-280
414	GL 30	1.57	3.0- 6.0				250-280
416	GL 45	1.77	4.0				250-280
417	GMN	1.62	8.0				250-280
PBT Alloy		**Comtuf**			**ComAlloy**		
431	GMN 40	1.65	3.0- 6.0				250-280
432	GL 30	1.53	3.0- 6.0				250-280
461	UNS 30	1.55	3.0- 7.0				250-280
462	GLF	1.57	2.0- 6.0				250-280
PBT Alloy		**Hiloy**			**ComAlloy**		
431	GL 15	1.43	3.0- 7.0				250-280
432	GL 30	1.55	3.0- 6.0				250-280
433	GLF 45	1.68	3.0- 7.0				250-280
434	GMN 40	1.67	3.0- 6.0				250-280
435	UNS	1.72	2.0- 6.0				250-280
461	UNS 40	1.55	3.0- 6.0				250-280
463	UNS	1.62	2.0- 5.0				250-280
PBT Alloy		**Lubrilon**			**ComAlloy**		
431	UNS	1.37	7.0				250-280
432	UNS 30	1.46	5.0				250-280
PBT Alloy		**Lumax**			**Lucky**		
GP-5000H		1.21	10.0-12.0			284- 711	176-230
GP-5006F		1.25	5.0- 6.0			284- 711	176-230
GP-5100	GLF 10	1.20	4.0- 6.0			0- 568	176-230
GP-5106F	GLF 10	1.30	4.0- 6.0			0- 568	176-230
GP-5200	GLF 20	1.32	3.0- 5.0			0- 568	176-230
GP-5206F	GLF 20	1.42	3.0- 5.0			0- 568	176-230
GP-5300	GLF 30	1.39	2.0- 4.0			0- 568	176-230
GP-5306F	GLF 30	1.50	2.0- 4.0			0- 568	176-230
HF-5008		1.10	5.0- 6.0			284- 711	176-230

Drying time, hr	Inj time, sec	Front temp, °F	Mid temp, °F	Rear temp, °F	Nozzle temp, °F	Mold temp, °F	Proc temp, °F
4.0- 6.0		490-510	480-500	470-490	490-520	160-200	480-530
4.0- 6.0		490-510	480-500	470-490	490-520	160-200	480-530
4.0- 6.0		490-510	480-500	470-490	490-520	160-200	480-530
3.0- 4.0		470-500	460-490	450-470	470-510	150-190	470-510
3.0- 4.0		470-500	460-490	450-470	470-510	150-190	470-510
4.0- 6.0						170-190	480-520
4.0- 6.0		470-500	460-490	450-470	470-510	150-190	470-510
4.0- 6.0		470-500	460-490	450-470	470-510	150-190	470-510
4.0- 6.0		470-500	460-490	450-470	470-510	150-190	470-510
4.0- 6.0		470-500	460-490	450-470	470-510	150-190	470-510
4.0- 6.0		470-500	460-490	450-470	470-510	150-190	470-510
2.0- 4.0		460-500	450-480	440-470	460-490	160-190	470-500
2.0- 4.0		460-500	450-480	440-470	460-490	160-190	470-500
2.0- 4.0		460-500	450-480	440-470	460-490	160-190	470-500
2.0- 4.0		460-500	450-480	440-470	460-490	160-190	470-500
2.0- 4.0		460-500	450-480	440-470	460-490	160-190	470-500
2.0- 4.0		490-510	480-500	460-480	490-510	180-200	460-520
3.0- 4.0		480-500	470-490	470-490	500-520	150-200	500-530
3.0- 4.0		420-460	420-450	410-440	420-470	100-250	430-490
3.0- 4.0		420-460	420-450	410-440	420-470	100-250	430-490
3.0- 4.0		420-460	420-450	410-440	420-470	100-250	430-490
3.0- 4.0		420-460	420-450	410-440	420-470	100-250	430-490
3.0- 4.0		420-460	420-450	410-440	420-470	100-250	430-490
3.0- 4.0		420-460	420-450	410-440	420-470	100-250	430-490
3.0- 4.0		500-560	500-560	490-530	510-560	150-250	500-560
3.0- 4.0		500-560	500-560	490-530	510-560	150-250	500-560
3.0- 4.0		500-560	500-560	490-530	510-560	150-250	500-560
3.0- 4.0		500-560	500-560	490-530	510-560	150-250	500-560
3.0- 4.0		500-560	500-560	490-530	510-560	150-250	500-560
3.0- 4.0		500-560	500-560	490-530	510-560	150-250	500-560
3.0- 4.0		500-560	500-560	490-530	510-560	150-250	500-560
3.0- 4.0		500-560	500-560	490-530	510-560	150-250	500-560
3.0- 4.0		500-560	500-560	490-530	510-560	150-250	500-560
3.0- 4.0		500-560	500-560	490-530	510-560	150-250	500-560
3.0- 4.0		500-560	500-560	490-530	510-560	150-250	500-560
3.0- 4.0		500-560	500-560	490-530	510-560	150-250	500-600
3.0- 4.0		500-560	500-560	490-530	510-560	150-250	500-600
3.0- 4.0		446-482	428-464	374-428	446-482	140-176	446-482
3.0- 4.0		446-482	428-464	374-428	446-482	140-176	446-482
3.0- 4.0		464-482	446-473	437-455	464-482	140-194	464-491
3.0- 4.0		464-482	446-473	437-455	464-482	140-194	464-491
3.0- 4.0		464-482	446-473	437-455	464-482	140-194	464-491
3.0- 4.0		464-482	446-473	437-455	464-482	140-194	464-491
3.0- 4.0		464-482	446-473	437-455	464-482	140-194	464-491
3.0- 4.0		464-482	446-473	437-455	464-482	140-194	464-491
3.0- 4.0		446-482	428-464	374-428	446-482	140-176	446-482

Grade	Filler	Sp Grav	Shrink, mils/in	Melt flow, g/10 min	Melt temp, °F	Back pres, psi	Drying temp, °F
HF-5300	GLF 30	1.29	1.0- 2.0			0- 568	176-230
HF-5306F	GLF 30	1.40	1.0- 3.0			0- 568	176-230
HR-5007		1.08	5.0- 6.0			284- 711	176-230
HR-5300	GLF 30	1.42	3.0- 5.0			0- 568	176-230
HR-5306F	GLF 30	1.52	3.0- 5.0			0- 568	176-230
PBT Alloy		**Voloy**			**ComAlloy**		
415	GMN 40	1.72	5.0				250-280
430	GL 30	1.62	3.0- 7.0				250-280
431	UNS	1.74	3.0- 7.0				250-280
460	UNS	1.62	2.0- 5.0				250-280
PBT+PET Alloy		**Valox**			**GE**		
HV7062	UNS 38	1.80				50- 100	265
HV7062HP	UNS 38	1.80				50- 100	265
HV7064	UNS 53	2.10				50- 100	265
HV7064HP	UNS 53	2.10				50- 100	265
HV7065	MNF 63	2.40				50- 100	265
HV7065HP	MNF 63	2.40				50- 100	265
HV7085	GMN 68	2.40				50- 100	265
PC		**Apec**			**Miles**		
HT DP9-9308		1.20	7.0- 9.0			50- 100	265
HT DP9-9330		1.18	7.0- 8.0	8.50		50- 100	265
HT DP9-9331		1.18	7.0- 8.0	8.50		50- 100	265
HT DP9-9333		1.18	7.0- 8.0	8.00		50- 100	265
HT DP9-9340		1.17	7.0- 8.0	6.50		50- 100	265
HT DP9-9341		1.17	7.0- 8.0	6.50		50- 100	265
HT DP9-9343		1.17	7.0- 8.0	7.00		50- 100	265
HT DP9-9350		1.15	8.0- 9.0	3.50		50- 100	265
HT DP9-9351		1.15	8.0- 9.0	3.50		50- 100	265
HT DP9-9352		1.15	8.0- 9.0	3.50		50- 100	265
HT DP9-9353		1.15	8.0- 9.0	3.50		50- 100	265
HT DP9-9354T		1.15	8.0- 9.0	3.50		50- 100	265
HT DP9-9360		1.15	8.0- 9.0	2.00		50- 100	265
HT DP9-9361		1.15	8.0- 9.0	2.00		50- 100	265
HT DP9-9363		1.15	8.0- 9.0	2.00		50- 100	265
HT DP9-9370		1.14	8.0- 9.0	1.50		50- 100	265
HT DP9-9371		1.14	8.0- 9.0	1.25		50- 100	265
HT DP9-9373		1.14	8.0- 9.0	1.00		50- 100	265
PC		**AVP**			**Polymerland**		
KLL10	GL 10	1.28	3.0- 5.0	9.00 O		50- 200	250
RLL10	GL 10	1.27	3.0- 5.0	10.00 O		50- 200	250
RLL20	GL 20	1.34	2.0- 3.0	7.00 O		50- 200	250
RLL30	GL 30	1.43	1.0- 2.0	4.00 O		50- 200	250
TLL03		1.21	5.0- 7.0	25.00 O		50	250
TLL11		1.21	5.0- 7.0	3.50 O		50	250
TLL12		1.21	5.0- 7.0	12.00 O		50	250
TLL80		1.21	5.0- 7.0	20.00 O		50	250
ZLL01		1.21	5.0- 7.0	15.00 O		50	250
ZLL08		1.21	5.0- 7.0	8.00 O		50	250
ZLL12		1.21	5.0- 7.0	9.00 O		50	250
PC		**Bay Resins Polycarbonate**				**Bay Resins**	
PC-1100		1.20	6.0			50- 100	250

Drying time, hr	Inj time, sec	Front temp, °F	Mid temp, °F	Rear temp, °F	Nozzle temp, °F	Mold temp, °F	Proc temp, °F
3.0- 4.0		464-482	446-473	437-455	464-482	140-194	464-491
3.0- 4.0		464-482	446-473	437-455	464-482	140-194	464-491
3.0- 4.0		446-482	428-464	374-428	446-482	140-176	446-482
3.0- 4.0		464-482	446-473	437-455	464-482	140-194	464-491
3.0- 4.0		464-482	446-473	437-455	464-482	140-194	464-491
3.0- 4.0		420-460	420-450	410-440	420-470	100-250	430-490
3.0- 4.0		500-560	500-560	490-530	510-560	150-250	500-560
3.0- 4.0		500-560	500-560	490-530	510-560	150-250	500-560
3.0- 4.0		500-560	500-560	490-530	510-560	150-250	500-560
3.0- 4.0		480-500	470-490	470-490	500-520	150-200	500-530
3.0- 4.0		480-500	470-490	470-490	500-520	150-200	500-530
3.0- 4.0		480-500	470-490	470-490	500-520	150-200	500-530
3.0- 4.0		480-500	470-490	470-490	500-520	150-200	500-530
3.0- 4.0		480-500	470-490	470-490	500-520	150-200	500-530
3.0- 4.0		480-500	470-490	470-490	500-520	150-200	500-530
3.0- 4.0		480-500	470-490	470-490	500-520	150-200	500-530
4.0		590	580	570	600	175-250	600
4.0		590	580	570	600	175-250	600
4.0		590	580	570	600	175-250	600
4.0		590	580	570	600	175-250	600
4.0		610	605	600	615	190-275	610
4.0		610	605	600	615	190-275	610
4.0		610	605	600	615	190-275	610
4.0		625	615	610	615	200	630
4.0		625	615	610	635	190-275	620
4.0		625	615	610	635	190-275	620
4.0		625	615	610	635	190-275	620
4.0		625	615	610	635	190-275	620
4.0		645	635	625	655	200-300	635
4.0		645	635	625	655	200-300	635
4.0		645	635	625	655	200-300	635
4.0		660	655	645	670	200-300	645
4.0		660	655	645	670	200-300	645
4.0		660	655	645	670	200-300	645
6.0						180-240	580-630
6.0						180-240	580-630
6.0						180-240	600-650
6.0						180-240	600-650
1.0						160-200	500-540
4.0						180-240	600-660
4.0						170-210	550-600
4.0						160-200	500-540
5.0						160-200	530-580
5.0						180-230	575-625
5.0						180-210	550-600
4.0		580-650	580-650	580-650		180-250	580-620

Grade	Filler	Sp Grav	Shrink, mils/in	Melt flow, g/10 min	Melt temp, °F	Back pres, psi	Drying temp, °F
PC-1100G10	GLF 10	1.25	3.0			50- 100	250
PC-1100G20	GLF 20	1.34	1.5			50- 100	250
PC-1100G30	GLF 30	1.43	1.0			50- 100	250
PC-1100G30TF15	GLF 30	1.55	1.0			50- 100	250
PC-1100G40	GLF 40	1.52	1.0			50- 100	250
PC-1100H30	CF 30	1.33	1.5			50- 100	250
PC-1100TF15		1.28	6.0			50- 100	250
PC-1700G10FR	GLF 10	1.25	3.0			50- 100	250
PC		**Calibre**			**Dow**		
200-10		1.20	5.0- 7.0	10.00 O			250
200-15		1.20	5.0- 7.0	15.00 O			250
200-22		1.20	6.0	22.00 O			250
200-3		1.20	5.0- 7.0	3.00 O			250
200-4		1.20	6.0	4.00 O			250
200-6		1.20	5.0- 7.0	6.00 O			250
201-10		1.20	6.0	10.00 O			250
201-15		1.20	6.0	15.00 O			250
201-22		1.20	6.0	22.00 O			250
201-4		1.20	6.0	4.00 O			250
201-6		1.20	6.0	6.00 O			250
202-10		1.20	6.0	10.00 O			250
202-15		1.20	6.0	15.00 O			250
202-22		1.20	6.0	22.00 O			250
202-4		1.20	6.0	4.00 O			250
202-6		1.20	6.0	6.00 O			250
203-10		1.20	6.0	10.00 O			250
203-15		1.20	6.0	15.00 O			250
203-22		1.20	6.0	22.00 O			250
203-4		1.20	6.0	4.00 O			250
203-6		1.20	6.0	6.00 O			250
2060-10		1.20	6.0	10.00 O			250
2060-15		1.20	6.0	15.00 O			250
2060-22		1.20	6.0	22.00 O			250
2060-4		1.20	6.0	4.00 O			250
2060-6		1.20	6.0	6.00 O			250
2061-10		1.20	6.0	10.00 O			250
2061-15		1.20	6.0	15.00 O			250
2061-22		1.20	6.0	22.00 O			250
2061-4		1.20	6.0	4.00 O			250
2061-6		1.20	6.0	6.00 O			250
2062-10		1.20	6.0	10.00 O			250
2062-15		1.20	6.0	15.00 O			250
2062-22		1.20	6.0	22.00 O			250
2062-4		1.20	6.0	4.00 O			250
2062-6		1.20	6.0	6.00 O			250
2063-10		1.20	6.0	10.00 O			250
2063-15		1.20	6.0	15.00 O			250
2063-22		1.20	6.0	22.00 O			250
2063-4		1.20	6.0	4.00 O			250
2063-6		1.20	6.0	6.00 O			250
2080-10		1.20	6.0	10.00 O			250
2080-15		1.20	6.0	15.00 O			250
2080-4		1.20	6.0	4.00 O			250
2080-6		1.20	6.0	6.00 O			250
2081-10		1.20	6.0	10.00 O			250

Drying time, hr	Inj time, sec	Front temp, °F	Mid temp, °F	Rear temp, °F	Nozzle temp, °F	Mold temp, °F	Proc temp, °F
4.0		580-650	580-650	580-650		180-250	580-620
4.0		580-650	580-650	580-650		180-250	580-620
4.0		580-650	580-650	580-650		180-250	580-620
4.0		580-650	580-650	580-650		180-250	580-620
4.0		580-650	580-650	580-650		180-250	580-620
4.0		580-650	580-650	580-650		180-250	580-620
4.0		580-650	580-650	580-650		180-250	580-620
4.0		580-650	580-650	580-650		180-250	580-620
3.0		550-600	510-550	500-520	550-600	160-220	550-600
3.0		530-560	500-540	470-530	530-560	160-220	530-560
3.0		500-530	490-520	470-500	500-530	160-200	500-530
3.0		600-640	560-590	550-580	600-640	180-240	600-640
3.0		600-640	560-590	550-580	600-640	180-240	600-640
3.0		570-620	530-570	520-550	570-620	170-230	570-620
3.0		550-600	510-550	500-520	550-600	160-220	550-600
3.0		530-560	500-540	470-530	530-560	160-220	530-560
3.0		500-530	490-520	470-500	500-530	160-200	500-530
3.0		600-640	560-590	550-580	600-640	180-240	600-640
3.0		570-620	530-570	520-550	570-620	170-230	570-620
3.0		550-600	510-550	500-520	550-600	160-220	550-600
3.0		530-560	500-540	470-530	530-560	160-220	530-560
3.0		500-530	490-520	470-500	500-530	160-200	500-530
3.0		600-640	560-590	550-580	600-640	180-240	600-640
3.0		570-620	530-570	520-550	570-620	170-230	570-620
3.0		550-600	510-550	500-520	550-600	160-220	550-600
3.0		530-560	500-540	470-530	530-560	160-220	530-560
3.0		500-530	490-520	470-500	500-530	160-200	500-530
3.0		600-640	560-590	550-580	600-640	180-240	600-640
3.0		570-620	530-570	520-550	570-620	170-230	570-620
3.0		550-600	510-550	500-520	550-600	160-220	550-600
3.0		530-560	500-540	470-530	530-560	160-220	530-560
3.0		500-530	490-520	470-500	500-530	160-200	500-530
3.0		600-640	560-590	550-580	600-640	180-240	600-640
3.0		570-620	530-570	520-550	570-620	170-230	570-620
3.0		550-600	510-550	500-520	550-600	160-220	550-600
3.0		530-560	500-540	470-530	530-560	160-220	530-560
3.0		500-530	490-520	470-500	500-530	160-200	500-530
3.0		600-640	560-590	550-580	600-640	180-240	600-640
3.0		570-620	530-570	520-550	570-620	170-230	570-620
3.0		550-600	510-550	500-520	550-600	160-220	550-600
3.0		530-560	500-540	470-530	530-560	160-220	530-560
3.0		500-530	490-520	470-500	500-530	160-200	500-530
3.0		600-640	560-590	550-580	600-640	180-240	600-640
3.0		570-620	530-570	520-550	570-620	170-230	570-620
3.0		550-600	510-550	500-520	550-600	160-220	550-600
3.0		530-560	500-540	470-530	530-560	160-220	530-560
3.0		600-640	560-590	550-580	600-640	180-240	600-640
3.0		570-620	530-570	520-550	570-620	170-230	570-620
3.0		550-600	510-550	500-520	550-600	160-220	550-600

Grade	Filler	Sp Grav	Shrink, mils/in	Melt flow, g/10 min	Melt temp, °F	Back pres, psi	Drying temp, °F
2081-15		1.20	6.0	15.00 O			250
2081-6		1.20	6.0	6.00 O			250
300-10		1.20	6.0	10.00 O			250
300-15		1.20	5.0- 7.0	15.00 O			250
300-22		1.20	6.0	22.00 O			250
300-3		1.20	5.0- 7.0	3.00 O			250
300-4		1.20	6.0	4.00 O			250
300-6		1.20	5.0- 7.0	6.00 O			250
301-10		1.20	6.0	10.00 O			250
301-15		1.20	6.0	15.00 O			250
301-22		1.20	6.0	22.00 O			250
301-4		1.20	6.0	4.00 O			250
301-6		1.20	6.0	6.00 O			250
302-10		1.20	6.0	10.00 O			250
302-15		1.20	6.0	15.00 O			250
302-22		1.20	6.0	22.00 O			250
302-4		1.20	6.0	4.00 O			250
302E		1.20	5.0- 7.0	6.00 O			250
303-10		1.20	6.0	10.00 O			250
303-15		1.20	6.0	15.00 O			250
303-22		1.20	6.0	22.00 O			250
303-4		1.20	6.0	4.00 O			250
303-6		1.20	6.0	6.00 O			250
400-10		1.18	7.0	10.00 O			250
400-6		1.20		6.00 O			250
510	GLF	1.27	2.0- 5.0				250
5200	GLF	1.33		3.00 O			250
550	GLF	1.27	2.0- 5.0				250
700-10		1.20	5.0- 7.0	10.00 O			250
700-15		1.20	5.0- 7.0	15.00 O			250
700-3		1.20	5.0- 7.0	3.00 O			250
700-4		1.20	6.0	4.00 O			250
700-6		1.20	5.0- 7.0	6.00 O			250
701-10		1.20	6.0	10.00 O			250
701-15		1.20	6.0	15.00 O			250
701-4		1.20	6.0	4.00 O			250
701-6		1.20	6.0	6.00 O			250
702-10		1.20	6.0	10.00 O			250
702-15		1.20	6.0	15.00 O			250
702-4		1.20	6.0	4.00 O			250
702-6		1.20	6.0	6.00 O			250
703-10		1.20	6.0	10.00 O			250
703-15		1.20	6.0	15.00 O			250
703-4		1.20	6.0	4.00 O			250
703-6		1.20	6.0	6.00 O			250
800-10		1.21	5.0- 7.0	10.00 O			250
800-15		1.21	5.0- 7.0	15.00 O			250
800-3		1.21	5.0- 7.0	3.00 O			250
800-4		1.21	6.0	4.00 O			250
800-6		1.21	5.0- 7.0	6.00 O			250
801-10		1.21	6.0	10.00 O			250
801-15		1.21	6.0	15.00 O			250
801-4		1.21	6.0	4.00 O			250
801-6		1.21	6.0	6.00 O			250
802-10		1.21	6.0	10.00 O			250
802-15		1.21	6.0	15.00 O			250

Drying time, hr	Inj time, sec	Front temp, °F	Mid temp, °F	Rear temp, °F	Nozzle temp, °F	Mold temp, °F	Proc temp, °F
3.0		530-560	500-540	470-530	530-560	160-220	530-560
3.0		570-620	530-570	520-550	570-620	170-230	570-620
3.0		550-600	510-550	500-520	550-600	160-220	550-600
3.0		530-560	500-540	470-530	530-560	160-220	530-560
3.0		500-530	490-520	470-500	500-530	160-200	500-530
3.0		600-640	560-590	550-580	600-640	180-240	600-640
3.0		600-640	560-590	550-580	600-640	180-240	600-640
3.0		570-620	530-570	520-550	570-620	170-230	570-620
3.0		550-600	510-550	500-520	550-600	160-220	550-600
3.0		530-560	500-540	470-530	530-560	160-220	530-560
3.0		500-530	490-520	470-500	500-530	160-200	500-530
3.0		600-640	560-590	550-580	600-640	180-240	600-640
3.0		570-620	530-570	520-550	570-620	170-230	570-620
3.0		550-600	510-550	500-520	550-600	160-220	550-600
3.0		530-560	500-540	470-530	530-560	160-220	530-560
3.0		500-530	490-520	470-500	500-530	160-200	500-530
3.0		600-640	560-590	550-580	600-640	180-240	600-640
3.0		570-620	530-570	520-550	570-620	170-230	570-620
3.0		550-600	510-550	500-520	550-600	160-220	550-600
3.0		530-560	500-540	470-530	530-560	160-220	530-560
3.0		500-530	490-520	470-500	500-530	160-200	500-530
3.0		600-640	560-590	550-580	600-640	180-240	600-640
3.0		570-620	530-570	520-550	570-620	170-230	570-620
3.0		550-600	510-550	500-520	550-600	160-220	550-600
3.0		570-620	530-570	520-550	570-620	170-230	570-620
3.0		550-600	540-580	540-560	550-600	180-240	550-600
3.0		550-600	540-580	540-560	550-600	180-240	550-600
3.0		550-600	540-580	540-560	550-600	180-240	550-600
3.0		550-600	510-550	500-520	550-600	160-220	550-600
3.0		530-560	500-540	470-530	530-560	160-220	530-560
3.0		600-640	560-590	550-580	600-640	180-240	600-640
3.0		600-640	560-590	550-580	600-640	180-240	600-640
3.0		570-620	530-570	520-550	570-620	170-230	570-620
3.0		550-600	510-550	500-520	550-600	160-220	550-600
3.0		530-560	500-540	470-530	530-560	160-220	530-560
3.0		600-640	560-590	550-580	600-640	180-240	600-640
3.0		570-620	530-570	520-550	570-620	170-230	570-620
3.0		550-600	510-550	500-520	550-600	160-220	550-600
3.0		530-560	500-540	470-530	530-560	160-220	530-560
3.0		600-640	560-590	550-580	600-640	180-240	600-640
3.0		570-620	530-570	520-550	570-620	170-230	570-620
3.0		550-600	510-550	500-520	550-600	160-220	550-600
3.0		530-560	500-540	470-530	530-560	160-220	530-560
3.0		600-640	560-590	550-580	600-640	180-240	600-640
3.0		570-620	530-570	520-550	570-620	170-230	570-620
3.0		550-600	510-550	500-520	550-600	160-220	550-600
3.0		530-560	500-540	470-530	530-560	160-220	530-560
3.0		600-640	560-590	550-580	600-640	180-240	600-640
3.0		600-640	560-590	550-580	600-640	180-240	600-640
3.0		570-620	530-570	520-550	570-620	170-230	570-620
3.0		550-600	510-550	500-520	550-600	160-220	550-600
3.0		530-560	500-540	470-530	530-560	160-220	530-560
3.0		600-640	560-590	550-580	600-640	180-240	600-640
3.0		570-620	530-570	520-550	570-620	170-230	570-620
3.0		550-600	510-550	500-520	550-600	160-220	550-600
3.0		530-560	500-540	470-530	530-560	160-220	530-560

Grade	Filler	Sp Grav	Shrink, mils/in	Melt flow, g/10 min	Melt temp, °F	Back pres, psi	Drying temp, °F
802-4		1.21	6.0	4.00 O			250
802-6		1.21	6.0	6.00 O			250
803-10		1.21	6.0	10.00 O			250
803-15		1.21	6.0	15.00 O			250
803-4		1.21	6.0	4.00 O			250
803-6		1.21	6.0	6.00 O			250
LG2010		1.20		6.00 O			250
LG2020		1.20		10.00 O			250
MegaRad 2080-10		1.20	5.0- 7.0	10.00 O			250
MegaRad 2080-15		1.20	5.0- 7.0	15.00 O			250
MegaRad 2080-4		1.20	5.0- 7.0	4.00 O			250
PC		**Celstran**			**Hoechst**		
PCG30-01-4	GLF 30	1.43					250
PCG40-01-4	GLF 40	1.52					250
PCG50-01-4	GLF 50	1.63					250
PC		**Celstran S**			**Hoechst**		
PCS10-01-4	STS 10	1.35	4.0- 6.0				250
PCS10-02-4	STS 10	1.35	4.0- 6.0				250
PCS6-01-4	STS 6	1.28	4.0- 6.0				250
PCS6-02-4	STS 6	1.28	4.0- 6.0				250
PC		**CTI Polycarbonate**			**CTI**		
PC		1.20	5.0- 7.0				250
PC-000/10T	PTF 10	1.26	5.0- 7.0				250
PC-000/15T	PTF 15	1.26	5.0- 7.0				250
PC-000/20T	PTF 20	1.32	5.0- 7.0				250
PC-000/5T	PTF 5	1.23	5.0- 7.0				250
PC-10CF/000	CF 10	1.23	2.0				250
PC-10CF/0002	CF 10	1.23	2.0				250
PC-10CF/15T FCR	CF 10	1.35	2.0- 5.0				250
PC-10GB/000	GLB 10	1.26	6.0				250
PC-10GF/000 FR	GLF 10	1.27	2.0- 4.0				250
PC-15CF/15T FCR	CF 15	1.35	1.5- 2.0				250
PC-15MCF/000 FCR	CF 15	1.23	2.0				250
PC-20CF	CF 20		1.5				250
PC-20CF/15T	CF 20	1.37	1.0				250
PC-20CF/20T	CF 20	1.41	0.9- 1.0				250
PC-20GB/15T	GLB 20	1.45	3.5- 4.0				250
PC-20GF	GLF 20	1.34	1.5				250
PC-20GF/15T	GLF 20	1.45	1.5				250
PC-20NCF/000 FCR	CFN 20	1.36	1.5				250
PC-25GB-5GF/000	GLF 5	1.43	3.5- 4.5				250
PC-30CF	CF 30	1.32	1.5				250
PC-30CF/20T	CF 30	1.46	0.7				250
PC-30CF/20T2	CF 30	1.46	1.0- 1.5				250
PC-30GF	GLF 30	1.43	1.0				250
PC-30GF/15-2S	GLF 30	1.55	1.0				250
PC-30GF/15T	GLF 30	1.55	2.5				250
PC-30GF/15T2	GLF 30	1.54	1.0				250
PC-30GF/OOO FR	GLF 30	1.45	2.0				250
PC-40CF	CF 40						250
PC-40GF	GLF 30	1.51	1.0				250
PC-40GF/15T	GLF 40	1.64	1.0				250
PC-40GM/000	GL 40	1.52	4.0- 4.5				250

Drying time, hr	Inj time, sec	Front temp, °F	Mid temp, °F	Rear temp, °F	Nozzle temp, °F	Mold temp, °F	Proc temp, °F
3.0		600-640	560-590	550-580	600-640	180-240	600-640
3.0		570-620	530-570	520-550	570-620	170-230	570-620
3.0		550-600	510-550	500-520	550-600	160-220	550-600
3.0		530-560	500-540	470-530	530-560	160-220	530-560
3.0		600-640	560-590	550-580	600-640	180-240	600-640
3.0		570-620	530-570	520-550	570-620	170-230	570-620
3.0		570-620	530-570	520-550	570-620	170-230	570-620
3.0		550-600	510-550	500-520	550-600	160-220	550-600
3.0		550-600	510-550	500-520	550-600	160-220	550-600
3.0		530-560	500-540	470-530	530-560	160-220	530-560
3.0		600-640	560-590	550-580	600-640	180-240	600-640
4.0		590	580	580	600	250	
4.0		600	590	580	610	250	
4.0		610	600	590	620	250	
4.0		540	510	500	540	180	
4.0		540	510	500	540	180	
4.0		510	500	490	510	180	
4.0		510	500	490	510	180	
4.0		580	570	550	580	175	
4.0		580	570	550	580	175	
4.0		580	570	550	580	175	
4.0		580	570	550	580	175	
4.0		580	570	550	580	175	
4.0		580-630	570-600	550-590	580-630	175-240	
4.0		580-630	570-600	550-590	580-630	175-240	
4.0		580-630	570-600	550-590	580-630	175-240	
4.0		580-630	570-600	550-590	580-630	175-240	
4.0		580-630	570-600	550-590	580-630	175-240	
4.0		580-630	570-600	550-590	580-630	175-240	
4.0		580-630	570-600	550-590	580-630	175-240	
4.0		580-630	570-600	550-590	580-630	175-240	
4.0		580-630	570-600	550-590	580-630	175-240	
4.0		580-630	570-600	550-590	580-630	175-240	
4.0		580-630	570-600	550-590	580-630	175-240	
4.0		580-630	570-600	550-590	580-630	175-240	
4.0		580-630	570-600	550-590	580-630	175-240	
4.0		580-630	570-600	550-590	580-630	175-240	
4.0		580-630	570-600	550-590	580-630	175-240	
4.0		580-630	570-600	550-590	580-630	175-240	
4.0		580-630	570-600	550-590	580-630	175-240	
4.0		580-630	570-600	550-590	580-630	175-240	
4.0		580-630	570-600	550-590	580-630	175-240	
4.0		580-630	570-600	550-590	580-630	175-240	
4.0		580-630	570-600	550-590	580-630	175-240	
4.0		580-630	570-600	550-590	580-630	175-240	
4.0		580-630	570-600	550-590	580-630	175-240	
4.0		580-630	570-600	550-590	580-630	175-240	
4.0		580-630	570-600	550-590	580-630	175-240	
4.0		580-630	570-600	550-590	580-630	175-240	
4.0		580-630	570-600	550-590	580-630	175-240	

Grade	Filler	Sp Grav	Shrink, mils/in	Melt flow, g/10 min	Melt temp, °F	Back pres, psi	Drying temp, °F
PC-40GM/05T	GL 40	1.57	4.0- 4.5				250
PC-40GM/10T	GL 40	1.61	4.0- 4.5				250
PC-40GM/15T	GL 40	1.66	4.0- 4.5				250
PC		**DSM Polycarbonate**				**DSM**	
F-50/10	GLF 10		6.0				250
F-50/20	GLF 20		5.0				250
F-50/30	GLF 30		4.0				250
F-50/40	GLF 40		3.0				250
FG-50/20	GLF 20		3.0				250
J-50/20	GLF 20	1.34	2.0				250
J-50/30	GLF 30	1.43	2.0				250
J-50/40	GLF 40	1.05	1.0				250
PC		**Electrafil**			**DSM**		
G-50/CF/20	CF 20	1.28	0.5				250
G-50/CF/40	CF 40	1.44	0.5				250
G-50/CF/5/SS/5	CF 5	1.28	1.5				250
G-50/SS/10	STS 10	1.33	3.0				250
G-50/SS/5	STS 5	1.26	5.0				250
G-50/SS/5/FR	STS 5	1.27	4.0				250
G-50/SS/5/K	STS 5	1.26	5.0				250
J-50/20/CF/10	CF 10	1.39	1.0				250
J-50/5/CF/10/HF	CF 10	1.28	2.0				250
J-50/CF/10	CF 10	1.24	1.5				250
J-50/CF/10/TF	CF 10	1.34	1.5				250
J-50/CF/20	CF 20	1.28	1.0				250
J-50/CF/20/FR	CF 20	1.30	1.0				250
J-50/CF/30	CF 30	1.33	0.5				250
J-50/CF/30/TF	CF 30	1.42	1.0				250
J-50/CF/40	CF 40	1.38	0.5				250
PC-50/EC	CBL	1.21	5.0				250
PC-50/EC/VO	CBL	1.30	5.0				250
PC-50/EC/VO/HI	CBL	1.30	4.0			25- 50	250
PC		**Fiberfil**			**DSM**		
F-50/10	GLF 10		6.0				250
F-50/20	GLF 20		5.0				250
F-50/30	GLF 30		4.0				250
F-50/40	GLF 40		3.0				250
J-50/10	GLF 10	1.28	3.0				250
J-50/20	GLF 20	1.35	3.0				250
J-50/30	GLF 30	1.45	1.0				250
J-50/40	GLF 40	1.54	1.0				250
PC		**Fiberfil VO**			**DSM**		
G-50/20/FR	GLF 20	1.36	3.0- 4.0				250
J-50/20/FR	GLF 20	1.36	3.0- 4.0				250
J-50/30/FR	GLF 30	1.45	2.0- 3.0				250
J-54/10	GLF 10	1.26	3.0				250
PC		**Fiberstran**			**DSM**		
G-50/20	GLF 20	1.34	2.0			10- 50	200-250
G-50/20/FR	GLF 20	1.36	3.0				250
G-50/20/TF/10	GLF 20	1.43	2.0				250
G-50/20/TF/15	GLF 20	1.50	2.0				250

Drying time, hr	Inj time, sec	Front temp, °F	Mid temp, °F	Rear temp, °F	Nozzle temp, °F	Mold temp, °F	Proc temp, °F
4.0		580-630	570-600	550-590	580-630	175-240	
4.0		580-630	570-600	550-590	580-630	175-240	
4.0		580-630	570-600	550-590	580-630	175-240	
16.0						160-180	580
16.0						160-180	580
16.0						160-180	580
16.0						160-180	580
16.0						160-180	580
16.0		600-630	590-650	575-600	590-630	160-190	580-625
16.0		600-630	590-650	575-600	590-630	160-190	580-625
16.0		600-630	590-650	575-600	590-630	160-190	580-625
16.0		600-630	590-650	575-600	590-630	160-190	580-625
16.0		600-630	590-650	575-600	590-630	160-190	580-625
16.0		600-630	590-650	575-600	590-630	160-190	580-625
16.0		600-630	590-650	575-600	590-630	160-190	580-625
16.0		600-630	590-650	575-600	590-630	160-190	580-625
16.0		540-560	550-590	530-550	530-560	180-200	540-570
16.0		600-630	590-650	575-600	590-630	160-190	580-625
16.0		600-630	590-650	575-600	590-630	160-190	580-625
16.0		600-630	590-650	575-600	590-630	160-190	580-625
16.0		600-630	590-650	575-600	590-630	160-190	580-625
16.0		540-560	550-590	530-550	530-560	180-200	540-570
16.0		600-630	590-650	575-600	590-630	160-190	580-625
16.0		540-560	550-590	530-550	530-560	180-200	540-570
16.0		600-630	590-650	575-600	590-630	160-190	580-625
16.0		600-630	590-650	575-600	590-630	160-190	580-625
16.0		600-630	590-650	575-600	590-630	160-190	580-625
16.0		600-630	585-650	575-590	590-620	160-190	580-625
16.0		600-630	585-650	575-590	590-620	160-190	580-625
4.0		550-580	560-590	540-570	550-580	150-180	540-590
16.0		600-630	590-650	575-600	590-630	160-190	580-625
16.0		600-630	590-650	575-600	590-630	160-190	580-625
16.0		600-630	590-650	575-600	590-630	160-190	580-625
16.0		600-630	590-650	575-600	590-630	160-190	580-625
16.0		600-630	590-650	575-600	590-630	160-190	580-625
16.0		600-630	590-650	575-600	590-630	160-190	580-625
16.0		600-630	590-650	575-600	590-630	160-190	580-625
16.0		600-630	590-650	575-600	590-630	160-190	580-625
16.0		540-560	550-590	530-550	530-560	180-200	540-570
16.0		540-560	550-590	530-550	530-560	180-200	540-570
16.0		540-560	550-590	530-550	530-560	180-200	540-570
16.0		540-560	550-590	530-550	530-560	180-200	540-570
4.0-16.0		580-610	590-620	570-590	580-610	175-225	590-630
16.0		600-630	590-650	575-600	590-630	160-190	580-625
16.0		600-630	590-650	575-600	590-630	160-190	580-625
16.0		600-630	590-650	575-600	590-630	160-190	580-625

Grade	Filler	Sp Grav	Shrink, mils/in	Melt flow, g/10 min	Melt temp, °F	Back pres, psi	Drying temp, °F
G-50/30	GLF 30	1.43	1.0			10- 50	200-250
G-50/30/CF/10	CF 10	1.48	0.5				250
G-50/CF/20	CF 20	1.28	0.5				250
G-50/CF/40	CF 40	1.44	0.5				250
G-50/CF/5/SS/5	CF 5	1.28	1.5				250
G-50/SS/10	STS 10	1.33	3.0				250
G-50/SS/5	STS 5	1.26	5.0				250
G-50/SS/5/FR	STS 5	1.27	4.0				250

PC FR-PC Lucky

Grade	Filler	Sp Grav	Shrink, mils/in	Melt flow, g/10 min	Melt temp, °F	Back pres, psi	Drying temp, °F
GP-2100	GLF 10	1.25	2.0- 4.0				248
GP-2200	GLF 20	1.35	1.0- 3.0				248
GP-2300	GLF 30	1.43	1.0- 2.5				248
GP-2400	GLF 40	1.52	1.0- 2.0				248

PC Iupilon Mitsubishi Gas

Grade	Filler	Sp Grav	Shrink, mils/in	Melt flow, g/10 min	Melt temp, °F	Back pres, psi	Drying temp, °F
GS-2010M	GLF 10	1.27	3.0				248
GS-2020M	GLF 20	1.33	1.0				248
GS-2020MN-1	GLF 20	1.33	1.0				248
GS-2030M	GLF 30	1.42	0.5				248
H-3000		1.20	5.0- 8.0				248
H-3000R		1.20	5.0- 8.0				248
LS-2010	PTF 5	1.22	5.0- 8.0				248
LS-2020	PTF 10	1.25	5.0- 8.0				248
LS-2030	PTF 15	1.28	5.0- 8.0				248
ML-100		1.20	5.0- 8.0				248
ML-200		1.20	5.0- 8.0				248
ML-300		1.20	5.0- 8.0				248
N-3		1.22	5.0- 8.0				248
S-1000		1.20	5.0- 8.0				248
S-1000R		1.20	5.0- 8.0				248
S-1000U		1.20	5.0- 8.0				248
S-1001		1.20	5.0- 8.0				248
S-1003		1.20	5.0- 8.0				248
S-2000		1.20	5.0- 8.0				248
S-2000R		1.20	5.0- 8.0				248
S-2000U		1.20	5.0- 8.0				248
S-2001		1.20	5.0- 8.0				248
S-2003		1.20	5.0- 8.0				248
S-3000		1.20	5.0- 8.0				248
S-3000R		1.20	5.0- 8.0				248
S-3000U		1.20	5.0- 8.0				248
S-3001		1.20	5.0- 8.0				248
S-3003		1.20	5.0- 8.0				248

PC Lexan GE

Grade	Filler	Sp Grav	Shrink, mils/in	Melt flow, g/10 min	Melt temp, °F	Back pres, psi	Drying temp, °F
101		1.20	5.0- 7.0	7.00 O		50- 100	250
101R		1.20	5.0- 7.0	7.00 O		50- 100	250
103		1.20	5.0- 7.0	7.00 O		50- 100	250
103R		1.20	5.0- 7.0	7.00 O		50- 100	250
104		1.20	5.0- 7.0	7.00 O		50- 100	250
104R		1.20	5.0- 7.0	7.00 O		50- 100	250
121		1.20	5.0- 7.0	17.50 O		50- 100	250
121R		1.20	5.0- 7.0	17.50 O		50- 100	250
123		1.20	5.0- 7.0	17.50 O		50- 100	250
123R		1.20	5.0- 7.0	17.50 O		50- 100	250

Drying time, hr	Inj time, sec	Front temp, °F	Mid temp, °F	Rear temp, °F	Nozzle temp, °F	Mold temp, °F	Proc temp, °F
4.0-16.0		580-610	590-620	570-590	580-610	175-225	590-630
16.0		600-630	590-650	575-600	590-630	160-190	580-625
16.0		600-630	590-650	575-600	590-630	160-190	580-625
16.0		600-630	590-650	575-600	590-630	160-190	580-625
16.0		600-630	590-650	575-600	590-630	160-190	580-625
16.0		600-630	590-650	575-600	590-630	160-190	580-625
16.0		600-630	590-650	575-600	590-630	160-190	580-625
16.0		600-630	590-650	575-600	590-630	160-190	580-625
5.0		545	536	518	554	212-248	
5.0		554	545	536	554	212-248	
5.0		554	545	536	554	212-248	
5.0		572	554	554	572	212-248	
4.0						176-248	536-608
4.0						176-248	536-608
4.0						176-248	536-608
4.0						176-248	536-608
4.0						158-248	500-608
4.0						158-248	500-608
4.0						176-248	536-608
4.0						176-248	536-608
4.0						176-248	536-608
4.0						158-248	500-608
4.0						158-248	500-608
4.0						158-248	500-608
4.0						158-248	500-608
4.0						158-248	500-608
4.0						158-248	500-608
4.0						158-248	500-608
4.0						158-248	500-608
4.0						158-248	500-608
4.0						158-248	500-608
4.0						158-248	500-608
4.0						158-248	500-608
4.0						158-248	500-608
4.0						158-248	500-608
4.0						158-248	500-608
4.0						158-248	500-608
4.0						158-248	500-608
4.0						158-248	500-608
4.0						158-248	500-608
3.0- 5.0		600-640	560-600	550-580	580-630	180-240	600-650
3.0- 5.0		600-640	560-600	550-580	580-630	180-240	600-650
3.0- 5.0		600-640	560-600	550-580	580-630	180-240	600-650
3.0- 5.0		600-640	560-600	550-580	580-630	180-240	600-650
3.0- 5.0		600-640	560-600	550-580	580-630	180-240	600-650
3.0- 5.0		600-640	560-600	550-580	580-630	180-240	600-650
3.0- 5.0		530-560	510-530	490-525	520-560	160-200	540-580
3.0- 5.0		530-560	510-530	490-525	520-560	160-200	540-580
3.0- 5.0		530-560	510-530	490-525	520-560	160-200	540-580
3.0- 5.0		530-560	510-530	490-525	520-560	160-200	540-580

Grade	Filler	Sp Grav	Shrink, mils/in	Melt flow, g/10 min	Melt temp, °F	Back pres, psi	Drying temp, °F
124		1.20	5.0- 7.0	17.50 O		50- 100	250
124R		1.20	5.0- 7.0	17.50 O		50- 100	250
131		1.20	5.0- 7.0	3.50 O		50- 100	250
131R		1.20	5.0- 7.0	3.50 O		50- 100	250
133		1.20	5.0- 7.0	3.50 O		50- 100	250
133R		1.20	5.0- 7.0	3.50 O		50- 100	250
134		1.20	5.0- 7.0	3.50 O		50- 100	250
134R		1.20	5.0- 7.0	3.50 O		50- 100	250
141		1.20	5.0- 7.0	10.50 O		50- 100	250
141L		1.20	5.0- 7.0	12.50 O		50- 100	250
141LR		1.20	5.0- 7.0	12.50 O		50- 100	250
141R		1.20	5.0- 7.0	10.50 O		50- 100	250
141S		1.20	5.0- 7.0	10.50 O		50- 100	250
143		1.20	5.0- 7.0	10.50 O		50- 100	250
143L		1.20	5.0- 7.0	12.50 O		50- 100	250
143LR		1.20	5.0- 7.0	12.50 O		50- 100	250
143R		1.20	5.0- 7.0	10.50 O		50- 100	250
143S		1.20	5.0- 7.0	10.50 O		50- 100	250
144		1.20	5.0- 7.0	10.50 O		50- 100	250
144L		1.20	5.0- 7.0	12.50 O		50- 100	250
144LR		1.20	5.0- 7.0	12.50 O		50- 100	250
144R		1.20	5.0- 7.0	10.50 O		50- 100	250
144S		1.20	5.0- 7.0	10.50 O		50- 100	250
1500		1.20	5.0- 7.0	5.50 O		50- 200	250
1500H		1.20	5.0- 7.0	5.50 O		50- 200	250
1500R		1.20	5.0- 7.0	5.50 O		50- 200	250
161		1.20	5.0- 7.0	8.50 O		50- 100	250
161R		1.20	5.0- 7.0	8.50 O		50- 100	250
163		1.20	5.0- 7.0	8.50 O		50- 100	250
163R		1.20	5.0- 7.0	8.50 O		50- 100	250
164		1.20	5.0- 7.0	8.50 O		50- 100	250
164R		1.20	5.0- 7.0	8.50 O		50- 100	250
181		1.20	5.0- 7.0	5.50 O		50- 100	250
181R		1.20	5.0- 7.0	5.50 O		50- 100	250
183		1.20	5.0- 7.0	5.50 O		50- 100	250
183R		1.20	5.0- 7.0	5.50 O		50- 100	250
184		1.20	5.0- 7.0	5.50 O		50- 100	250
184R		1.20	5.0- 7.0	5.50 O		50- 100	250
191		1.19	5.0- 7.0	8.50 O		50	250
191R		1.19	5.0- 7.0	8.50 O		50- 100	250
193		1.19	5.0- 7.0	8.50 O		50- 100	250
193R		1.19	5.0- 7.0	8.50 O		50- 100	250
194		1.19	5.0- 7.0	8.50 O		50- 100	250
194R		1.19	5.0- 7.0	8.50 O		50- 100	250
201		1.20	5.0- 7.0	7.00 O		50- 100	250
2014		1.24	5.0- 7.0	10.00 O		50- 100	250
2014R		1.24	5.0- 7.0	10.00 O		50- 100	250
201R		1.20	5.0- 7.0	7.00 O		50- 100	250
203		1.20	5.0- 7.0	7.00 O		50- 100	250
2034		1.20	5.0- 7.0	10.00 O		50- 100	250
203R		1.20	5.0- 7.0	7.00 O		50- 100	250
204		1.20	5.0- 7.0	7.00 O		50- 100	250
204R		1.20	5.0- 7.0	7.00 O		50- 100	250
221		1.20	5.0- 7.0	17.50 O		50- 100	250
221R		1.20	5.0- 7.0	17.50 O		50- 100	250
223		1.20	5.0- 7.0	17.50 O		50- 100	250

Drying time, hr	Inj time, sec	Front temp, °F	Mid temp, °F	Rear temp, °F	Nozzle temp, °F	Mold temp, °F	Proc temp, °F
3.0- 5.0		530-560	510-530	490-525	520-560	160-200	540-580
3.0- 5.0		530-560	510-530	490-525	520-560	160-200	540-580
3.0- 4.0		610-650	590-630	570-610	600-640	180-210	600-650
3.0- 4.0		610-650	590-630	570-610	600-640	180-210	600-650
3.0- 4.0		610-650	590-630	570-610	600-640	180-210	600-650
3.0- 4.0		610-650	590-630	570-610	600-640	180-210	600-650
3.0- 4.0		610-650	590-630	570-610	600-640	180-210	600-650
3.0- 4.0		610-650	590-630	570-610	600-640	180-210	600-650
3.0- 5.0		540-570	520-540	510-530	530-580	160-200	550-600
3.0- 5.0		530-560	510-530	490-525	520-560	160-200	540-580
3.0- 5.0		530-560	510-530	490-525	520-560	160-200	540-580
3.0- 5.0		540-570	520-540	510-530	530-580	160-200	550-600
3.0- 5.0		540-570	520-540	510-530	530-580	160-200	550-600
3.0- 5.0		540-570	520-540	510-530	530-580	160-200	550-600
3.0- 5.0		530-560	510-530	490-525	520-560	160-200	540-580
3.0- 5.0		530-560	510-530	490-525	520-560	160-200	540-580
3.0- 5.0		540-570	520-540	510-530	530-580	160-200	550-600
3.0- 5.0		540-570	520-540	510-530	530-580	160-200	550-600
3.0- 5.0		540-570	520-540	510-530	530-580	160-200	550-600
3.0- 5.0		530-560	510-530	490-525	520-560	160-200	540-580
3.0- 5.0		530-560	510-530	490-525	520-560	160-200	540-580
3.0- 5.0		540-570	520-540	510-530	530-580	160-200	550-600
3.0- 5.0		540-570	520-540	510-530	530-580	160-200	550-600
3.0- 4.0		610-650	570-610	560-590	590-640	180-240	610-660
3.0- 4.0		610-650	570-610	560-590	590-640	180-240	610-660
3.0- 4.0		610-650	570-610	560-590	590-640	180-240	610-660
3.0- 5.0		570-605	540-560	530-555	555-605	160-220	575-625
3.0- 5.0		570-605	540-560	530-555	555-605	160-220	575-625
3.0- 5.0		570-605	540-560	530-555	555-605	160-220	575-625
3.0- 5.0		570-605	540-560	530-555	555-605	160-220	575-625
3.0- 5.0		570-605	540-560	530-555	555-605	160-220	575-625
3.0- 5.0		570-605	540-560	530-555	555-605	160-220	575-625
3.0- 4.0		600-640	580-620	560-600	590-630	180-240	590-640
3.0- 4.0		600-640	580-620	560-600	590-630	180-240	590-640
3.0- 4.0		600-640	580-620	560-600	590-630	180-240	590-640
3.0- 4.0		600-640	580-620	560-600	590-630	180-240	590-640
3.0- 4.0		600-640	580-620	560-600	590-630	180-240	590-640
3.0- 4.0		600-640	580-620	560-600	590-630	180-240	590-640
3.0- 4.0		540-570	520-540	510-530	530-580	160-200	550-600
3.0- 5.0		540-570	520-540	510-530	530-580	160-200	550-560
3.0- 5.0		540-570	520-540	510-530	530-580	160-200	550-560
3.0- 5.0		540-570	520-540	510-530	530-580	160-200	550-560
3.0- 5.0		540-570	520-540	510-530	530-580	160-200	550-560
3.0- 5.0		540-570	520-540	510-530	530-580	160-200	550-560
3.0- 5.0		600-640	560-600	550-580	580-630	180-240	600-650
3.0- 5.0		530-560	510-530	490-525	520-560	160-200	540-580
3.0- 5.0		530-560	510-530	490-525	520-560	160-200	540-580
3.0- 5.0		600-640	560-600	550-580	580-630	180-240	600-650
3.0- 5.0		600-640	560-600	550-580	580-630	180-240	600-650
3.0- 5.0		530-560	510-530	490-525	520-560	160-200	540-580
3.0- 5.0		600-640	560-600	550-580	580-630	180-240	600-650
3.0- 5.0		600-640	560-600	550-580	580-630	180-240	600-650
3.0- 5.0		600-640	560-600	550-580	580-630	180-240	600-650
3.0- 5.0		530-560	510-530	490-525	520-560	160-200	540-580
3.0- 5.0		530-560	510-530	490-525	520-560	160-200	540-580
3.0- 5.0		530-560	510-530	490-525	520-560	160-200	540-580

Grade	Filler	Sp Grav	Shrink, mils/in	Melt flow, g/10 min	Melt temp, °F	Back pres, psi	Drying temp, °F
223R		1.20	5.0- 7.0	17.50 O		50- 100	250
224		1.20	5.0- 7.0	17.50 O		50- 100	250
224R		1.20	5.0- 7.0	17.50 O		50- 100	250
241		1.20	5.0- 7.0	10.50 O		50- 100	250
241L		1.20	5.0- 7.0	12.50 O		50- 100	250
241LR		1.20	5.0- 7.0	12.50 O		50- 100	250
241R		1.20	5.0- 7.0	10.50 O		50- 100	250
243		1.20	5.0- 7.0	10.50 O		50- 100	250
243L		1.20	5.0- 7.0	12.50 O		50- 100	250
243LR		1.20	5.0- 7.0	12.50 O		50- 100	250
243R		1.20	5.0- 7.0	10.50 O		50- 100	250
244		1.20	5.0- 7.0	10.50 O		50- 100	250
244L		1.20	5.0- 7.0	12.50 O		50- 100	250
244LR		1.20	5.0- 7.0	12.50 O		50- 100	250
244R		1.20	5.0- 7.0	10.50 O		50- 100	250
261		1.20	5.0- 7.0	8.50 O		50- 100	250
261R		1.20	5.0- 7.0	8.50 O		50- 100	250
263		1.20	5.0- 7.0	8.50 O		50- 100	250
263R		1.20	5.0- 7.0	8.50 O		50- 100	250
264		1.20	5.0- 7.0	8.50 O		50- 100	250
264R		1.20	5.0- 7.0	8.50 O		50- 100	250
281		1.20	5.0- 7.0	5.50 O		50- 100	250
2814		1.28	2.0- 4.0			50- 200	250
281R		1.20	5.0- 7.0	5.50 O		50- 100	250
283		1.20	5.0- 7.0	5.50 O		50- 100	250
283R		1.20	5.0- 7.0	5.50 O		50- 100	250
284		1.20	5.0- 7.0	5.50 O		50- 100	250
284R		1.20	5.0- 7.0	5.50 O		50- 100	250
303		1.20	5.0- 7.0	5.50 O		50- 100	250
303R		1.20	5.0- 7.0	5.50 O		50- 100	250
3412	GLF 20	1.35	1.0- 3.0			50- 200	250
3412R	GLF 20	1.35	1.0- 3.0			50- 200	250
3413	GLF 30	1.43	1.0- 3.0			50- 200	250
3413R	GLF 30	1.43	1.0- 3.0			50- 200	250
3414	GLF 40	1.52	1.0- 2.0			50- 200	250
3414R	GLF 40	1.52	1.0- 2.0			50- 200	250
3432	GLF 20	1.35	1.0- 3.0			50- 200	250
3433	GLF 30	1.43	1.0- 3.0			50- 200	250
500	GLF 10	1.25	2.0- 4.0			50- 200	250
500R	GLF 10	1.25	2.0- 4.0			50- 200	250
503	GLF 10	1.25	2.0- 4.0			50- 200	250
920		1.21	5.0- 7.0	14.50		50- 200	250
920A		1.21	5.0- 7.0	14.50 O		50- 200	250
920AR		1.21	5.0- 7.0	14.50 O		50- 200	250
920R		1.21	5.0- 7.0	14.50 O		50- 200	250
923		1.21	5.0- 7.0	14.50 O		50- 200	250
923A		1.21	5.0- 7.0	14.50 O		50- 200	250
923AR		1.21	5.0- 7.0	14.50 O		50- 200	250
923R		1.21	5.0- 7.0	14.50 O		50- 200	250
940		1.21	5.0- 7.0	10.00 O		50- 100	250
940A		1.21	5.0- 7.0	10.00 O		50- 100	250
940AR		1.21	5.0- 7.0	10.00 O		50- 100	250
940R		1.21	5.0- 7.0	10.00 O		50- 100	250
943		1.21	5.0- 7.0	10.00 O		50- 100	250
943A		1.21	5.0- 7.0	10.00 O		50- 100	250
943AR		1.21	5.0- 7.0	10.00 O		50- 100	250

Drying time, hr	Inj time, sec	Front temp, °F	Mid temp, °F	Rear temp, °F	Nozzle temp, °F	Mold temp, °F	Proc temp, °F
3.0- 5.0		530-560	510-530	490-525	520-560	160-200	540-580
3.0- 5.0		530-560	510-530	490-525	520-560	160-200	540-580
3.0- 5.0		530-560	510-530	490-525	520-560	160-200	540-580
3.0- 5.0		540-570	520-540	510-530	530-580	160-200	550-600
3.0- 5.0		540-570	520-540	510-530	530-580	160-200	550-600
3.0- 5.0		530-560	510-530	490-525	520-560	160-200	540-580
3.0- 5.0		540-570	520-540	510-530	530-580	160-200	550-600
3.0- 5.0		540-570	520-540	510-530	530-580	160-200	550-600
3.0- 5.0		530-560	510-530	490-525	520-560	160-200	540-580
3.0- 5.0		530-560	510-530	490-525	520-560	160-200	540-580
3.0- 5.0		540-570	520-540	510-530	530-580	160-200	550-600
3.0- 5.0		540-570	520-540	510-530	530-580	160-200	550-600
3.0- 5.0		530-560	510-530	490-525	520-560	160-200	540-580
3.0- 5.0		530-560	510-530	490-525	520-560	160-200	540-580
3.0- 5.0		540-570	520-540	510-530	530-580	160-200	550-600
3.0- 5.0		570-605	540-560	530-555	555-605	160-220	575-625
3.0- 5.0		570-605	540-560	530-555	555-605	160-220	575-625
3.0- 5.0		570-605	540-560	530-555	555-605	160-220	575-625
3.0- 5.0		570-605	540-560	530-555	555-605	160-220	575-625
3.0- 5.0		570-605	540-560	530-555	555-605	160-220	575-625
3.0- 5.0		570-605	540-560	530-555	555-605	160-220	575-625
3.0- 4.0		600-640	580-620	560-600	590-630	180-240	590-640
4.0- 6.0		550-600	540-560		550-600	180-240	570-620
3.0- 4.0		600-640	580-620	560-600	590-630	180-240	590-640
3.0- 4.0		600-640	580-620	560-600	590-630	180-240	590-640
3.0- 4.0		600-640	580-620	560-600	590-630	180-240	590-640
3.0- 4.0		600-640	580-620	560-600	590-630	180-240	590-640
3.0- 4.0		600-640	580-620	560-600	590-630	180-240	590-640
3.0- 5.0		600-640	560-600	550-580	580-630	180-240	600-650
3.0- 5.0		600-640	560-600	550-580	580-630	180-240	600-650
4.0- 6.0		580-620	560-600	540-580	580-620	180-240	600-640
4.0- 6.0		580-620	560-600	540-580	580-620	180-240	600-640
4.0- 6.0		580-640	560-600	540-580	580-640	180-240	600-660
4.0- 6.0		580-640	560-600	540-580	580-640	180-240	600-660
4.0- 6.0		580-640	570-600	550-580	580-640	180-240	600-660
4.0- 6.0		580-640	570-600	550-580	580-640	180-240	600-660
4.0- 6.0		580-620	560-600	540-580	580-620	180-240	600-640
4.0- 6.0		580-640	560-600	540-580	580-640	180-240	600-660
4.0- 6.0		550-600	540-560	540-560	550-600	180-240	570-620
4.0- 6.0		550-600	540-560		550-600	180-240	570-620
4.0- 6.0		550-600	540-560		550-600	180-240	570-620
3.0- 4.0		520-550	500-520	480-520	510-540	160-200	530-560
3.0- 4.0		520-550	500-520	480-520	510-540	160-200	530-560
3.0- 4.0		520-550	500-520	480-520	510-540	160-200	530-560
3.0- 4.0		520-550	500-520	480-520	510-540	160-200	530-560
3.0- 4.0		520-550	500-520	480-520	510-540	160-200	530-560
3.0- 4.0		520-550	500-520	480-520	510-540	160-200	530-560
3.0- 4.0		520-550	500-520	480-520	510-540	160-200	530-560
3.0- 4.0		520-550	500-520	480-520	510-540	160-200	530-560
3.0- 4.0		540-570	520-540	510-530	530-580	160-200	550-560
3.0- 4.0		540-570	520-540	510-530	530-580	160-200	550-560
3.0- 4.0		540-570	520-540	510-530	530-580	160-200	550-560
3.0- 4.0		540-570	520-540	510-530	530-580	160-200	550-560
3.0- 4.0		540-570	520-540	510-530	530-580	160-200	550-560
3.0- 4.0		540-570	520-540	510-530	530-580	160-200	550-560
3.0- 4.0		540-570	520-540	510-530	530-580	160-200	550-560

Grade	Filler	Sp Grav	Shrink, mils/in	Melt flow, g/10 min	Melt temp, °F	Back pres, psi	Drying temp, °F
943R		1.21	5.0- 7.0	10.00 O		50- 100	250
950		1.21	5.0- 7.0	7.00 O		50- 200	250
950A		1.21	5.0- 7.0	7.00 O		50- 200	250
950AR		1.21	5.0- 7.0	7.00 O		50- 200	250
950R		1.21	5.0- 7.0	7.00 O		50- 200	250
953		1.21	5.0- 7.0	7.00 O		50- 200	250
953A		1.21	5.0- 7.0	7.00 O		50- 200	250
953AR		1.21	5.0- 7.0	7.00 O		50- 200	250
953R		1.21	5.0- 7.0	7.00 O		50- 200	250
BE1130		1.21	6.0- 8.0	19.50 O		50- 100	250
BE1130A		1.21	6.0- 8.0	19.50 O		50- 100	250
BE2130		1.21	6.0- 8.0	25.00 O		50- 100	250
BE2130R		1.21	6.0- 8.0	19.50 O		50- 100	250
C4600		1.20	5.0- 7.0	14.00 O		50- 200	250
EM1210		1.19	5.0- 7.0	13.00 O		50	250
EM2212	GLF 10	1.24	3.0- 5.0			50- 150	250
EM2214	GLF 20	1.24	2.0- 3.0			50- 200	250
EM3110		1.19	5.0- 7.0	20.00 O		50	250
GR1110		1.20	5.0- 7.0	22.00 O		0- 100	250
GR1210		1.20	5.0- 7.0	15.00 O		0- 100	250
GR1310		1.20	5.0- 7.0	10.00 O		0- 100	250
HF1110		1.20	5.0- 7.0	25.00 O		50- 100	250
HF1110R		1.20	5.0- 7.0	25.00 O		50- 100	250
HF1130		1.20	5.0- 7.0	25.00 O		50- 100	250
HF1130R		1.20	5.0- 7.0	25.00 O		50- 100	250
HF1140		1.20	5.0- 7.0	25.00 O		50- 100	250
HF1140R		1.20	5.0- 7.0	25.00 O		50- 100	250
HP1		1.20	5.0- 7.0	25.00 O		50- 100	250
HP1R		1.20	5.0- 7.0	25.00 O		50- 200	250
HP1X		1.20	5.0- 7.0	25.00 O		50- 100	250
HP2		1.20	5.0- 7.0	17.50 O		50- 200	250
HP2R		1.20	5.0- 7.0	17.50 O		50- 200	250
HP2X		1.20	5.0- 7.0	17.50 O		50- 200	250
HP3		1.20	5.0- 7.0	12.50 O		50- 200	250
HP3X		1.20	5.0- 7.0	12.50 O		50- 200	250
HP4		1.20	5.0- 7.0	10.00 O		50- 100	250
HP4X		1.20	5.0- 7.0	10.50 O		50- 100	250
HP5		1.20	5.0- 7.0	8.50 O		50- 100	250
HP5X		1.20	5.0- 7.0	8.50 O		50- 100	250
HPS1		1.20	5.0- 7.0	25.00 O		50- 100	250
HPS1R		1.20	5.0- 7.0	25.00 O		50- 200	250
HPS2		1.20	5.0- 7.0	17.50 O		50- 200	250
HPS2R		1.20	5.0- 7.0	17.50 O		50- 200	250
HPS3		1.20	5.0- 7.0	12.50 O		50- 200	250
HPS3R		1.20	5.0- 7.0	12.50 O		50- 200	250
HPS6		1.20	5.0- 7.0	7.00 O		50- 100	250
HW1210		1.20	5.0- 7.0	16.00 O			250
HW1240		1.20	5.0- 7.0	16.00 O			250
HW2210	GLM	1.20	5.0- 7.0	16.00		50- 200	250
HW2240		1.20	5.0- 7.0	16.00		50- 200	250
LS1		1.20	5.0- 7.0	17.50 O		50- 100	250
LS2		1.20	5.0- 7.0	11.00 O		50- 100	250
LS3		1.20	5.0- 7.0	7.00 O		50- 200	250
OQ1020		1.20	5.0- 7.0	60.00 O		50- 60	250
OQ1020L		1.20		70.00 O		50- 60	250
OQ2220		1.20	5.0- 7.0	17.50 O			250

Drying time, hr	Inj time, sec	Front temp, °F	Mid temp, °F	Rear temp, °F	Nozzle temp, °F	Mold temp, °F	Proc temp, °F
3.0- 4.0		540-570	520-540	510-530	530-580	160-200	550-560
3.0- 5.0		600-640	560-600	550-580	580-630	180-240	600-650
3.0- 4.0		600-640	560-600	550-580	580-630	180-240	600-650
3.0- 4.0		600-640	560-600	550-580	580-630	180-240	600-650
3.0- 4.0		600-640	560-600	550-580	580-630	180-240	600-650
3.0- 4.0		600-640	560-600	550-580	580-630	180-240	600-650
3.0- 4.0		600-640	560-600	550-580	580-630	180-240	600-650
3.0- 4.0		600-640	560-600	550-580	580-630	180-240	600-650
3.0- 4.0		600-640	560-600	550-580	580-630	180-240	600-650
3.0- 6.0		540-580	530-560	500-530	530-570	150-220	520-600
3.0- 6.0		540-580	530-560	500-530	530-570	150-220	520-600
3.0- 6.0		530-575	520-550	480-520	520-565	150-220	510-575
3.0- 6.0		540-580	530-560	500-530	530-570	150-220	520-600
3.0- 4.0		520-550	500-520	480-520	510-540	160-200	530-560
3.0- 4.0		520-550	500-520	480-520	510-540	160-200	520-600
3.0- 4.0		530-580	520-540	520-540	530-580	180-240	520-580
3.0- 4.0		560-600	540-580	520-560	560-600	180-240	540-600
3.0- 4.0		500-530	480-520	470-510	490-520	160-200	500-580
3.0- 4.0		570-625	550-605	530-585	570-625	150-225	550-625
3.0- 4.0		570-625	550-605	530-585	570-625	150-225	550-625
3.0- 4.0		570-625	550-605	530-585	570-625	150-225	550-625
3.0- 4.0		500-520	480-500	470-500	490-520	140-180	500-540
3.0- 4.0		500-520	480-500	470-500	490-500	140-180	500-540
3.0- 4.0		500-520	480-500	470-500	490-520	140-180	500-540
3.0- 4.0		500-520	480-500	470-500	490-500	140-180	500-540
3.0- 4.0		500-520	480-500	470-500	490-520	140-180	500-540
3.0- 4.0		500-520	480-500	470-500	490-520	140-180	500-540
3.0- 4.0		500-520	480-500	470-500	490-520	140-180	500-540
3.0- 4.0		480-500	460-480	450-480	470-500	160-200	480-520
3.0- 4.0		500-520	480-500	470-500	490-520	140-180	500-540
3.0- 4.0		520-550	500-520	480-520	510-540	160-200	530-560
3.0- 4.0		520-550	500-520	480-520	510-540	160-200	530-560
3.0- 4.0		520-550	500-520	480-520	510-540	160-200	530-560
3.0- 4.0		530-560	510-530	490-525	520-560	160-200	540-580
3.0- 4.0		530-560	510-530	490-525	520-560	160-200	540-580
3.0- 4.0		540-570	520-540	510-530	530-580	160-200	550-600
3.0- 4.0		540-570	520-540	510-530	530-580	160-200	550-600
3.0- 4.0		570-605	540-560	530-555	555-605	170-220	575-625
3.0- 4.0		570-605	540-560	530-555	555-605	170-220	575-625
3.0- 4.0		500-520	480-500	470-500	490-520	140-180	
3.0- 4.0		480-500	460-480	450-480	470-500	160-200	480-520
3.0- 4.0		520-550	500-520	480-520	510-540	160-200	530-560
3.0- 4.0		520-550	500-520	480-520	510-540	160-200	530-560
3.0- 4.0		530-560	510-530	490-525	520-560	160-200	540-580
3.0- 4.0		530-560	510-530	490-525	520-560	160-200	540-580
3.0- 5.0		600-640	560-600	550-580	580-630	180-240	600-650
3.0- 4.0		500-520	480-500	470-500	490-520	140-180	500-540
3.0- 4.0		500-520	480-500	470-500	490-520	140-180	500-540
3.0- 4.0		520-550	500-520	480-520	510-540	160-200	530-560
3.0- 4.0		520-550	500-520	480-520	510-540	160-200	530-560
3.0- 4.0		530-560	510-530	490-525	520-560	160-200	540-580
3.0- 4.0		540-570	520-540	510-530	530-580	160-200	550-560
3.0- 4.0		600-640	560-600	550-580	580-630	180-240	600-650
3.0- 4.0		540-560	540-560	480-500		150-200	520-560
3.0- 4.0		540-560	540-560	480-500		150-200	520-560
4.0- 6.0		580	580	580		180-210	

Grade	Filler	Sp Grav	Shrink, mils/in	Melt flow, g/10 min	Melt temp, °F	Back pres, psi	Drying temp, °F
OQ2320		1.20	5.0- 7.0	12.50 O		50	250
OQ2720		1.20	5.0- 7.0	7.50 O		50	250
OQ3120		1.20	5.0- 7.0	25.00 O		50- 200	250
OQ3220		1.20	5.0- 7.0	18.00 O		50- 100	250
OQ3420		1.20	5.0- 7.0	10.50 O		50- 200	250
OQ3820		1.20	5.0- 7.0	7.00 O		50- 200	250
OQ4120		1.20	5.0- 7.0	25.00 O		50- 100	250
OQ4220		1.20	5.0- 7.0	17.50 O		50- 100	250
OQ4320		1.20	5.0- 7.0	12.50 O		50- 100	250
PK2040		1.20	5.0- 7.0	40.00 O		50- 100	250
PK2140		1.20		22.00 O		50- 200	250
PK2640		1.20	5.0- 7.0	7.00 O		50- 200	250
WD1110		1.20	5.0- 7.0	27.00 O		100	250
WR1210	PTF	1.22	5.0- 7.0	13.00 O			250
WR2210		1.19	5.0- 7.0	16.00 O		50	250
WR2310		1.19	5.0- 7.0	13.00 O		50	250
WR4210		1.23	5.0- 7.0	20.00 O		50- 100	250
WR4210R		1.23	5.0- 7.0	20.00 O		50- 100	250

PC Lexan SP GE

Grade	Filler	Sp Grav	Shrink, mils/in	Melt flow, g/10 min	Melt temp, °F	Back pres, psi	Drying temp, °F
SP1010		1.18	5.0- 7.0			50- 100	220-230
SP1010R		1.18	5.0- 7.0			50- 100	220-230
SP1030		1.18	5.0- 7.0			50- 100	220-230
SP1030R		1.18	5.0- 7.0			50- 100	220-230
SP1110		1.18	5.0- 7.0	22.00 O		50- 100	220-230
SP1210		1.18	5.0- 7.0	16.00 O		50- 100	220-230
SP1210R		1.18	5.0- 7.0	16.00 O		50- 100	220-230
SP1212	GLF 10	1.24				50- 100	220-230
SP1230		1.18	5.0- 7.0	16.00 O		50- 100	220-230
SP1230R		1.18	5.0- 7.0	16.00 O		50- 100	220-230
SP1252	GLF 10	1.27	2.5- 4.5			50- 100	220-230
SP1310		1.18	5.0- 7.0	10.00 O		50- 100	220-230
SP1310R		1.18	5.0- 7.0	10.00 O		50- 100	220-230
SP1352	GLF 10	1.27	2.5- 4.5			50- 100	220-230

PC Lubricomp LNP

Grade	Filler	Sp Grav	Shrink, mils/in	Melt flow, g/10 min	Melt temp, °F	Back pres, psi	Drying temp, °F
DFL-4034 EM MG	GL 10	1.46	2.0- 3.0			50- 100	250

PC Makrolon Miles

Grade	Filler	Sp Grav	Shrink, mils/in	Melt flow, g/10 min	Melt temp, °F	Back pres, psi	Drying temp, °F
2500		1.20	6.0- 8.0	15.00 O		50- 100	250
2503		1.20	6.0- 8.0	15.00 O		50- 100	250
2505		1.20	6.0- 8.0	15.00 O		50- 100	250
2507		1.20	6.0- 8.0	15.00 O		50- 100	250
2508		1.20	6.0- 8.0	15.00 O		50- 100	250
2558		1.20	6.0- 8.0	15.00 O		50- 100	250
2600		1.20	6.0- 8.0	12.00 O		50- 100	250
2603		1.20	6.0- 8.0	12.00 O		50- 100	250
2605		1.20	6.0- 8.0	12.00 O		50- 100	250
2607		1.20	6.0- 8.0	12.00 O		50- 100	250
2608		1.20	6.0- 8.0	12.00 O		50- 100	250
2658		1.20	6.0- 8.0	12.00 O		50- 100	250
2800		1.20	6.0- 8.0	10.00 O		50- 100	250
2803		1.20	6.0- 8.0	10.00 O		50- 100	250
2805		1.20	6.0- 8.0	10.00 O		50- 100	250
2807		1.20	6.0- 8.0	10.00 O		50- 100	250
2808		1.20	6.0- 8.0	10.00 O		50- 100	250

Drying time, hr	Inj time, sec	Front temp, °F	Mid temp, °F	Rear temp, °F	Nozzle temp, °F	Mold temp, °F	Proc temp, °F
4.0- 6.0		540-570	520-540	510-530	530-580	180-210	550-560
4.0- 6.0		600-640	560-600	550-580	580-630	180-210	600-650
3.0- 4.0		500-520	480-500	470-500	490-520	160-200	500-540
3.0- 5.0		520-550	500-520	490-520	520-550		520-550
4.0- 6.0		540-570	520-540	510-530	530-580	160-200	550-600
3.0- 4.0		600-640	560-600	550-580	580-630	180-240	600-650
3.0- 4.0		500-520	480-500	470-500	490-520	140-180	500-540
3.0- 5.0		530-560	510-530	490-525	520-560	160-200	540-580
3.0- 5.0		530-560	510-530	490-525	520-560	160-200	540-580
3.0- 4.0		500-520	480-500	470-500	490-520	140-180	500-540
3.0- 4.0		480-500	460-480	450-480	470-500	160-200	480-520
3.0- 5.0		600-640	560-600	550-580	580-630	180-240	600-650
4.0						160-200	550
3.0- 4.0		520-550	500-520	480-510	510-540	160-200	510-550
3.0- 4.0		540-580	520-560	500-540	540-580	160-200	540-580
3.0- 4.0		540-570	520-540	480-510	530-580	160-200	510-550
3.0		540-580	520-560	500-540		160-200	540-580
3.0		540-580	520-560	500-540		160-200	540-580
3.0- 4.0		450-540	430-525	410-500	440-535	120-180	450-555
3.0- 4.0						120-180	450-555
3.0- 4.0						120-180	450-555
3.0- 4.0						120-180	450-555
3.0- 4.0		480-550	460-530	440-510	470-540	120-180	480-550
3.0- 4.0		500-580	480-550	470-520	490-570	120-180	500-580
3.0- 4.0						120-180	500-580
3.0- 4.0						120-180	520-580
3.0- 4.0						120-180	500-580
3.0- 4.0						120-180	500-580
3.0- 4.0		520-570	520-550	500-530	520-570	120-200	520-590
3.0- 4.0		550-600	490-570	480-540	540-590	120-180	550-600
3.0- 4.0						120-180	550-600
3.0- 4.0		520-570	520-550	500-530	520-570	120-200	520-590
4.0		560-630	560-630	560-630		175-225	580-620
4.0		535-575	515-550	465-510	515-585	150-220	540-575
4.0		535-575	515-550	465-510	515-585	150-220	540-575
4.0		535-575	515-550	465-510	515-585	150-220	540-575
4.0		535-575	515-550	465-510	515-585	150-220	540-575
4.0		535-575	515-550	465-510	515-585	150-220	540-575
4.0		535-575	515-550	465-510	515-585	150-220	540-575
4.0		545-585	520-560	480-520	515-585	150-220	550-580
4.0		545-585	520-560	480-520	515-585	150-220	550-580
4.0		545-585	520-560	480-520	515-585	150-220	550-580
4.0		545-585	520-560	480-520	515-585	150-220	550-580
4.0		545-585	520-560	480-520	515-585	150-220	550-580
4.0		545-585	520-560	480-520	515-585	150-220	550-580
4.0		555-595	510-550	500-540	535-595	150-220	560-590
4.0		555-595	510-550	500-540	535-595	150-220	560-590
4.0		555-595	510-550	500-540	535-595	150-220	560-590
4.0		555-595	510-550	500-540	535-595	150-220	560-590
4.0		555-595	510-550	500-540	535-595	150-220	560-590

Grade	Filler	Sp Grav	Shrink, mils/in	Melt flow, g/10 min	Melt temp, °F	Back pres, psi	Drying temp, °F
2858		1.20	6.0- 8.0	10.00 O		50- 100	250
3100		1.20	6.0- 8.0	6.50 O		50- 100	250
3103		1.20	6.0- 8.0	6.50 O		50- 100	250
3105		1.20	6.0- 8.0	6.50 O		50- 100	250
3107		1.20	6.0- 8.0	6.50 O		50- 100	250
3108		1.20	6.0- 8.0	6.50 O		50- 100	250
3158		1.20	6.0- 8.0	6.50 O		50- 100	250
3200		1.20	6.0- 8.0	4.50 O		50- 150	250
3203		1.20	6.0- 8.0	4.50 O		50- 150	250
3205		1.20	6.0- 8.0	4.50 O		50- 150	250
3207		1.20	6.0- 8.0	4.50 O		50- 150	250
3208		1.20	6.0- 8.0	4.50 O		50- 150	250
3258		1.20	6.0- 8.0	4.50 O		50- 150	250
5303		1.20	5.0- 7.0	11.00 O			250
5308		1.20	5.0- 7.0	11.00 O			250
5V-6485		1.20	5.0- 7.0	12.00 O		50- 100	250
6265		1.19	5.0- 7.0	19.00 O			250
6355		1.20	6.0- 8.0	16.00 O		50- 100	250
6455		1.20	6.0- 8.0	12.00 O		50- 100	250
6465		1.21	5.0- 7.0	11.00 O		50- 100	250
8325	GLF 20	1.35	3.0- 5.0	5.00 O			250
9415	GLF 10	1.27	3.0- 5.0	7.00 O			250
AL-2447		1.20	5.0- 7.0	19.00 O		50- 100	250
AL-2647		1.20	6.0- 8.0	12.00 O		50- 100	250
CD-2005		1.11	5.5			150- 575	250
DP1-1096		1.20	5.0- 7.0	20.00 O		50- 100	250
DP1-1096		1.20	5.0- 7.0	20.00 O		50- 100	250
DP1-1413		1.20	6.0- 8.0	12.00 O		50- 100	250
DP1-1413		1.20	6.0- 8.0	12.00 O		50- 100	250
FCR-2405		1.20	5.0- 7.0	20.50 O		50- 100	250
FCR-2407		1.20	5.0- 7.0	20.50 O		50- 100	250
FCR-2458		1.20	5.0- 7.0	20.50 O		50- 100	250
FCR-6255		1.20	5.0- 7.0	20.50 O		50- 100	250
LQ-2847		1.20	6.0- 8.0	12.00 O		50- 100	250
LQ-3147		1.20	6.0- 8.0	6.50 O		50- 100	250
LQ-3187		1.20	6.0- 8.0	6.50 O		50- 100	250
LTG-2623		1.20	5.0- 7.0	11.00 O		50- 100	250
LTG-3123		1.20	6.0- 8.0	6.50 O		50- 100	250
Rx-2530		1.20	6.0- 8.0	15.00 O			
T-7855	GLF 10	1.19	6.0- 8.0	12.00 O		0- 100	250
T-7955		1.19	5.0- 7.0	7.50 O		0- 100	250

PC Nyloy Nytex Compos.

Grade	Filler	Sp Grav	Shrink, mils/in	Melt flow, g/10 min	Melt temp, °F	Back pres, psi	Drying temp, °F
C-0010N-VO		1.21	6.0				
C-0200		1.19	6.0				
CG-0010	GLF 10	1.27	3.0				
CG-0020	GLF 20	1.34	2.0				

PC Panlite Teijin Chemical

Grade	Filler	Sp Grav	Shrink, mils/in	Melt flow, g/10 min	Melt temp, °F	Back pres, psi	Drying temp, °F
G-3110	GLF 10	1.27	3.0- 5.0				
G-3115	GLF 15	1.30	2.0- 4.0				
G-3115E	GLF 15	1.43	2.0- 4.0				
G-3120	GLF 20	1.34	1.0- 3.0				
G-3130	GLF 30	1.43	2.0				
G-3130E	GLF 30	1.41	2.0				
GN-3110	GLF 10	1.28	3.0- 5.0				

Drying time, hr	Inj time, sec	Front temp, °F	Mid temp, °F	Rear temp, °F	Nozzle temp, °F	Mold temp, °F	Proc temp, °F
4.0		555-595	510-550	500-540	535-595	150-220	560-590
4.0		565-605	540-580	520-560	540-560	150-220	570-600
4.0		565-605	540-580	520-560	540-560	150-220	570-600
4.0		565-605	540-580	520-560	540-560	150-220	570-600
4.0		565-605	540-580	520-560	540-560	150-220	570-600
4.0		565-605	540-580	520-560	540-560	150-220	570-600
4.0		565-605	540-580	520-560	540-560	150-220	570-600
4.0		600	580	565	570	150-220	600
4.0		600	580	565	570	150-220	600
4.0		600	580	565	570	150-220	600
4.0		600	580	565	570	150-220	600
4.0		600	580	565	570	150-220	600
4.0		600	580	565	570	150-220	600
1.0							
1.0							
4.0		545-585	520-560	480-520	515-585	150-220	550-580
						150-220	540-620
4.0		545-585	520-560	480-520	515-585	150-220	550-580
4.0		545-585	520-560	480-520	515-585	150-220	550-580
4.0		545-585	520-560	480-520	515-585	150-220	550-580
						150-220	540-620
						150-220	540-620
5.0		530-570	510-550	445-495	510-530	150-220	535-565
5.0		545-585	520-560	480-520	515-585	150-220	550-580
4.0		580-620	555-595	515-555	570-590	175-230	580-630
4.0		530-570	510-550	445-495	510-530	150-220	535-565
4.0		530-570	510-550	445-495	510-530	150-220	535-565
4.0		545-585	520-560	480-520	515-585	150-220	550-580
4.0		545-585	520-560	480-520	515-585	150-220	550-580
4.0		530-570	510-550	445-495	510-530	150-220	535-565
4.0		530-570	510-550	445-495	510-530	150-220	535-565
4.0		530-570	510-550	445-495	510-530	150-220	535-565
4.0		530-570	510-550	445-495	510-530	150-220	535-565
5.0		555-595	510-550	500-540	535-595	150-220	560-590
5.0		565-605	540-580	520-560	540-560	150-220	570-600
5.0		565-605	540-580	520-560	540-560	150-220	570-600
5.0		545-585	520-560	480-520	515-585	150-220	550-580
5.0		565-605	540-580	520-560	540-560	150-220	570-600
						150-220	540-620
3.0- 4.0		550-575	540-570	530-560	520-560	120-200	550-590
3.0- 4.0		550-575	540-570	530-560	520-560	120-200	550-590
							545
							545
							554
							554
						176-248	518-608
						176-248	518-608
						176-248	518-608
						176-248	518-608
						176-248	518-608
						176-248	518-608
						176-248	518-608

Grade	Filler	Sp Grav	Shrink, mils/in	Melt flow, g/10 min	Melt temp, °F	Back pres, psi	Drying temp, °F
GN-3130	GLF 30	1.44	2.0				
GS-3130	GLF 30	1.50	2.0- 5.0				
K-1285		1.20	5.0- 7.0				
K-1300		1.20	5.0- 7.0				
K-1300Z		1.20	5.0- 7.0				
KE-1300		1.19	6.0- 9.0				
L-1225		1.20	5.0- 7.0				
L-1225T		1.20	5.0- 7.0				
L-1225Y		1.20	5.0- 7.0				
L-1225Z		1.20	5.0- 7.0				
L-1250		1.20	5.0- 7.0				
L-1250J		1.20	5.0- 7.0				
L-1250Y		1.20	5.0- 7.0				
L-1250Z		1.20	5.0- 7.0				
LE-1250		1.18	6.0- 9.0				
LN-1250		1.22	5.0- 7.0				
LS-1250	GLF	1.25	5.0- 7.0				

PC Plaslube DSM

Grade	Filler	Sp Grav	Shrink, mils/in	Melt flow, g/10 min	Melt temp, °F	Back pres, psi	Drying temp, °F
G-50/20/TF/10	GLF 20	1.43	2.0				250
G-50/20/TF/15	GLF 20	1.50	2.0				250
J-50/20/TF/10	GLF 20	1.43	2.0				250
J-50/20/TF/15	GLF 20	1.50	2.0				250
J-50/30/S1/2	GLF 30	1.46	2.0				250
J-50/30/TF/10	GLF 30	1.53	1.0				250
J-50/30/TF/15	GLF 30	1.55	1.0				250
J-50/CF/10/TF	CF 10	1.34	1.5				250
J-50/CF/20/TF/15	CF 20	1.37	1.0				250

PC PMC Polycarbonate PMC

Grade	Filler	Sp Grav	Shrink, mils/in	Melt flow, g/10 min	Melt temp, °F	Back pres, psi	Drying temp, °F
PC-000		1.20	7.0		485	50- 150	250
PC-010	GLF 10	1.27	5.0		485	50- 200	250
PC-020	GLF 20	1.34	3.0		495	50- 200	250
PC-030	GLF 30	1.43	2.0		495	50- 200	250
PC-040	GLF 40	1.51	2.0		495	50- 200	250

PC RTP Polymers RTP

Grade	Filler	Sp Grav	Shrink, mils/in	Melt flow, g/10 min	Melt temp, °F	Back pres, psi	Drying temp, °F
300		1.20	7.0			25- 50	250
300 FR		1.24	6.0			25- 50	250
300 GB 20 SE	GLB	1.35	6.0			25- 50	250
300 TFE 10	GLF	1.26	7.0			25- 50	250
300 TFE 15 SE	PTF	1.29	6.0			25- 50	250
300 TFE 15 SI	GLF	1.28	6.0			25- 50	250
300 TFE 20 SE	PTF	1.33	6.0			25- 50	250
301	GLF 10	1.27	15.0			25- 50	250
301 FR	GLB	1.28	3.0			25- 50	250
301 TFE 5	GLF	1.30	4.0			25- 50	250
301-500	GLF	1.27	3.0			25- 50	250
302 TFE 10	GLF	1.38	2.0			25- 50	250
302 TFE 15	GLF	1.42	2.0			25- 50	250
303	GLF 20	1.34	3.0			25- 50	250
303 FR	GLB	1.36	2.0			25- 50	250
303 TFE 10	GLF	1.41	2.0			25- 50	250
303 TFE 15 SE	GLF	1.46	2.0			25- 50	250
303 TFE 20 SE	GLF	1.50	2.0			25- 50	250
303 TFE 20 SI2	GLF	1.50	2.0			25- 50	250

Drying time, hr	Inj time, sec	Front temp, °F	Mid temp, °F	Rear temp, °F	Nozzle temp, °F	Mold temp, °F	Proc temp, °F
						176-248	518-608
						176-248	518-608
						176-248	518-608
						176-248	518-608
						176-248	518-608
						176-248	518-608
						176-248	518-608
						176-248	518-608
						176-248	518-608
						176-248	518-608
						176-248	518-608
						176-248	518-608
						176-248	518-608
						176-248	518-608
						176-248	518-608
						176-248	518-608
						176-248	518-608
16.0		600-630	590-650	575-600	590-630	160-190	580-625
16.0		600-630	590-650	575-600	590-630	160-190	580-625
16.0		600-630	585-650	575-590	590-620	180-200	580-625
16.0		600-630	585-650	575-590	590-620	180-200	580-625
16.0		600-630	585-650	575-590	590-620	180-200	580-625
16.0		600-630	585-650	575-590	590-620	180-200	580-625
16.0		600-630	585-650	575-590	590-620	180-200	580-625
16.0		600-630	585-650	575-590	590-620	180-200	580-625
16.0		600-630	585-650	575-590	590-620	180-200	580-625
3.0- 4.0		540-570	520-540	510-530	530-580	160-200	
3.0- 4.0		580-640	570-600	550-580	580-640	180-240	
3.0- 4.0		580-640	570-600	550-580	580-640	180-240	
3.0- 4.0		580-640	570-600	550-580	580-640	180-240	
3.0- 4.0		580-640	570-600	550-580	580-640	180-240	
4.0		570-650	560-640	550-630		150-250	
4.0		525-565	515-545	500-535		150-250	
4.0		560-600	530-570	510-550		150-250	
4.0		560-600	530-570	510-550		150-250	
4.0		530-560	520-540	500-530		150-200	
4.0		560-600	530-570	510-550		150-250	
4.0		530-560	520-540	500-530		150-200	
4.0		560-600	530-570	510-550		150-250	
4.0		525-565	515-545	500-535		150-250	
4.0		560-600	530-570	510-550		150-250	
4.0		525-565	515-545	500-535		150-250	
4.0		560-600	530-570	510-550		150-250	
4.0		560-600	530-570	510-550		150-250	
4.0		560-600	530-570	510-550		150-250	
4.0		525-565	515-545	500-535		150-250	
4.0		560-600	530-570	510-550		150-250	
4.0		560-600	530-570	510-550		150-250	
4.0		560-600	530-570	510-550		150-250	
4.0		560-600	530-570	510-550		150-250	

Grade	Filler	Sp Grav	Shrink, mils/in	Melt flow, g/10 min	Melt temp, °F	Back pres, psi	Drying temp, °F
304 FR	GLF	1.38	2.0			25- 50	250
304 SE	GLF	1.37	2.0			25- 50	250
305	GLF 30	1.43	2.0			25- 50	250
305 FR	GLB	1.45	1.0			25- 50	250
305 FR L	GLF	1.45	1.0			25- 50	250
305 TFE 13 SI2	GLF	1.54	1.0			25- 50	250
305 TFE 15 SE	GLF	1.55	1.0			25- 50	250
305Z	GLF 30	1.43	2.0			25- 50	250
305Z	GLF 30	1.43	1.0			25- 50	250
306 TFE 20	GLF	1.61	1.0			25- 50	250
307	GLF 40	1.52	2.0			25- 50	250
307 FR	GLB	1.54	1.0			25- 50	250
307Z	GLF 40	1.52	2.0			25- 50	250
307Z	GLF 40	1.52	1.0			25- 50	250
383	CF	1.28	1.0			25- 50	250
383 FR	CF	1.31	1.0			25- 50	250
385	CF	1.32	1.0			25- 50	250
385 TFE 13 SI2	CF	1.42	1.0			25- 50	250
387	CF	1.36	1.0			25- 50	250
399X52604F	CF	1.36	1.0			25- 50	250
399X52604F	CF	1.36	1.0			25- 50	250
399X52604G	CF	1.32	1.0			25- 50	250
399X52604G	CF	1.36	1.0			25- 50	250
399X55715	STS	1.35	5.0			100- 200	250
399X55715	STS	1.35	5.0			100- 200	250
EMI-360.5	STS	1.25	5.0				
EMI-360.5 FR	STS	1.30	5.0				
EMI-361	STS	1.32	5.0				
EMI-361 FR	STS	1.35	5.0				
EMI-380.5	GRN	1.30	3.0				
EMI-381	GRN	1.36	2.0				
ESD 300 EM FR	GLF	1.30	5.0				
ESD 301 EM FR	GLF	1.37	3.0				
ESD 303 EM FR	GLF	1.45	2.0				
ESD-A-380	CF	1.24	1.0				
ESD-C-360	STS	1.25	6.0				
ESD-C-380	CF	1.26	1.0				

PC Shuman Polycarbonate Shuman

Grade	Filler	Sp Grav	Shrink, mils/in	Melt flow, g/10 min	Melt temp, °F	Back pres, psi	Drying temp, °F
910		1.20		10.00			250-260
920		1.20		10.00			250-260
930		1.20		15.00			250-260

PC Texapol PC Texapol

Grade	Filler	Sp Grav	Shrink, mils/in	Melt flow, g/10 min	Melt temp, °F	Back pres, psi	Drying temp, °F
8610 A		1.20	6.0	9.50 T	267		250
8620		1.20	6.0	16.50 T	267		250

PC Thermocomp LNP

Grade	Filler	Sp Grav	Shrink, mils/in	Melt flow, g/10 min	Melt temp, °F	Back pres, psi	Drying temp, °F
DF-1006 MR SM	GLF 30	1.42	1.0- 2.5			50- 100	250

PC Thermofil Polycarbonate Thermofil

Grade	Filler	Sp Grav	Shrink, mils/in	Melt flow, g/10 min	Melt temp, °F	Back pres, psi	Drying temp, °F
R-10FG-5100	GLC 10	1.26	2.0	10.00 O		100	260
R-20FG-0103	GLC 20	1.35	2.0	5.00 O		100	260
R-20FG-5103	GLC 20	1.35	2.0	9.00 O		100	260
R-30FG-0100	GLC 30	1.44	1.0	4.50 O		100	260
R-9900-8138		1.19	6.0	12.00 O		100	260

Drying time, hr	Inj time, sec	Front temp, °F	Mid temp, °F	Rear temp, °F	Nozzle temp, °F	Mold temp, °F	Proc temp, °F
4.0		525-565	515-545	500-535		150-250	
4.0		560-600	530-570	510-550		150-250	
4.0		560-600	530-570	510-550		150-250	
4.0		525-565	515-545	500-535		150-250	
4.0		525-565	515-545	500-535		150-250	
4.0		560-600	530-570	510-550		150-250	
4.0		560-600	530-570	510-550		150-250	
4.0		560-600	530-570	510-550		150-250	
4.0		560-600	530-570	510-550		150-250	
4.0		560-600	530-570	510-550		150-250	
4.0		560-600	530-570	510-550		150-250	
4.0		525-565	515-545	500-535		150-250	
4.0		560-600	530-570	510-550		150-250	
4.0		560-600	530-570	510-550		150-250	
4.0		560-590	530-560	510-540		150-250	
4.0		525-565	515-545	500-535		150-250	
4.0		560-590	530-560	510-540		150-250	
4.0		560-590	530-560	510-540		150-250	
4.0		560-590	530-560	510-540		150-250	
4.0		525-565	515-545	500-535		150-250	
4.0		525-565	515-545	500-535		150-250	
4.0		525-565	515-545	500-535		150-250	
4.0		525-565	515-545	500-535		150-250	
4.0		530-550	540-560	550-570		160-200	
4.0		530-550	540-560	550-570		160-200	
		550-650	550-650	550-650		150-250	
		480-550	480-550	480-550		150-250	
		550-650	550-650	550-650		150-250	
		480-550	480-550	480-550		150-250	
		525-625	525-625	525-625		175-275	
		525-625	525-625	525-625		175-275	
		525-600	525-600	525-600		150-250	
		525-600	525-600	525-600		150-250	
		525-600	525-600	525-600		150-250	
		525-625	525-625	525-625		175-275	
		550-650	550-650	550-650		150-250	
		525-625	525-625	525-625		175-275	
2.0-24.0							510-550
2.0-24.0							510-550
2.0-24.0							510-550
3.0- 4.0		540	520	510	530		
3.0- 4.0		540	520	510	530		
4.0		560-630	560-630	560-630		175-225	580-620
4.0		560-590	540-570	520-550	580-600	200-250	
4.0		560-590	540-570	520-550	580-600	200-250	
4.0		560-590	540-570	520-550	580-600	200-250	
4.0		560-590	540-570	520-550	580-600	200-250	
4.0		560-590	540-570	520-550	580-600	200-250	

Grade	Filler	Sp Grav	Shrink, mils/in	Melt flow, g/10 min	Melt temp, °F	Back pres, psi	Drying temp, °F
R2-20FG-0100	GLC 20	1.40	3.0	4.00 O		50	265
R2-20FG-5100	GLC 20	1.40	3.0	7.50 O		50	265
R2-20FG-8100	GLC 20	1.40	3.0	5.50 O		50	265
R2-30FG-0843	GLC 30	1.48	2.0	5.00 O		50	265
R2-9900-0100		1.25	7.0	10.00 O		50	265
R2-9900-5100		1.24	8.0	15.00 O		50	265
R2-9900-8100		1.25	5.0	16.00 O		50	265
PC		**Trirex**			**Sam Yang**		
3022I		1.20	4.0- 7.0				248
3025I		1.20	4.0- 7.0				248
3027I		1.20	4.0- 7.0	4.50 O			248
3500G20	GLF 20	1.34	1.0- 6.0				248
3500G30	GLF 30	1.43	1.0- 6.0				248
PC+PBT Alloy		**AVP**			**Polymerland**		
GLV80		1.20	8.0-10.0	12.00 I		25- 50	230
GLV8U		1.20	8.0-10.0	12.00 I		25- 50	230
PC+PBT Alloy		**Xenoy**			**GE**		
1104		1.17	6.0- 9.0			25- 50	230
1760	GL	1.30	4.0- 6.0				250
2230		1.22	6.0- 9.0			25- 50	250
5220		1.21	8.0-10.0			25- 50	230
5220U		1.21	8.0-10.0			25- 50	230
5230		1.22	6.0- 9.0			25- 50	230
5230U		1.22	6.0- 9.0			25- 50	230
5720		1.17	6.0- 9.0			0- 50	230
5720U		1.17	6.0- 9.0			0- 50	230
6120		1.25	16.0-21.0			25- 50	250
6120M		1.25	16.0-21.0			25- 50	250
6120U		1.25	16.0-21.0			25- 50	250
6121		1.26				25- 50	250
6123		1.24	12.0-15.0			25- 50	250
6123M		1.24	12.0-15.0			25- 50	250
6230	GL 10	1.30	8.0- 9.0			25- 50	250
6240	GL 10	1.30	7.0- 9.0			25- 50	250
6240U	GL 10	1.30	7.0- 9.0			25- 50	250
6302	CF 20	1.32	1.0- 3.0			0- 100	250
6370	GL 30	1.44	3.0- 5.0			25- 50	250
6380	GL 30	1.51	4.0- 6.0			25- 50	250
6620		1.20	16.0-18.0			25- 50	250
6620E		1.20	16.0-18.0			25- 50	250
6620U		1.20	16.0-18.0			25- 50	250
DX1500E		1.22	6.0- 9.0			25- 50	230
DX1500M		1.22	9.0-11.0			25- 50	230
DX5230E		1.22	6.0- 9.0			25- 50	230
DX5230EU		1.22	6.0- 9.0			25- 50	230
DX5230M		1.22	6.0- 9.0			25- 50	230
DX6123U		1.24	12.0-15.0			25- 50	250
DX6230U	GL 10	1.30	8.0- 9.0			25- 50	250
DX6380R		1.51	4.0- 6.0			25- 50	250
PC+PET Alloy		**Makroblend**			**Miles**		
DP4-1368		1.30	5.0- 7.0	19.00		50- 100	210
DP4-1370		1.22	5.0- 7.0	17.00		50- 100	210

Drying time, hr	Inj time, sec	Front temp, °F	Mid temp, °F	Rear temp, °F	Nozzle temp, °F	Mold temp, °F	Proc temp, °F
4.0		500-550	490-530	470-510	510-560	100-180	
4.0		500-550	490-530	470-510	510-560	100-180	
4.0		500-550	490-530	470-510	510-560	100-180	
4.0		500-550	490-530	470-510	510-560	100-180	
4.0		500-550	490-530	470-510	510-560	100-180	
4.0		500-550	490-530	470-510	510-560	100-180	
4.0		500-550	490-530	470-510	510-560	100-180	
6.0		518-590	518-590	464-536	518-572	158-230	518-590
6.0		518-590	518-590	464-536	518-572	158-230	518-590
6.0		518-590	518-590	464-536	518-572	158-530	518-590
6.0		518-590	518-590	464-536	518-572	158-230	518-590
6.0		518-590	518-590	464-536	518-572	158-230	518-590
4.0						100-180	480-510
4.0						100-180	480-510
2.0- 4.0						150-180	470-535
3.0- 4.0		490	490	490		150	
2.0- 4.0		470-525	465-520	465-510	475-530	150-190	485-530
2.0- 4.0		480-510	465-490	450-475	480-520	150-190	480-520
2.0- 4.0		480-510	465-490	450-475	480-520	150-190	480-520
2.0- 4.0		480-510	465-490	450-475	480-520	100-180	480-520
2.0- 4.0		480-510	465-490	450-475	480-520	100-180	480-520
2.0- 4.0		480-520	470-500	460-490	480-525	150-180	470-535
2.0- 4.0		480-520	470-500	460-490	480-525	150-180	470-535
2.0- 4.0		450-480	450-475	440-475	460-495	150-180	460-500
2.0- 4.0		450-480	450-475	440-475	460-495	150-180	460-500
2.0- 4.0		450-480	450-475	440-475	460-495	150-180	460-500
2.0- 4.0		450-480	450-475	440-475	460-495	150-180	460-500
2.0- 4.0		450-480	450-475	440-475	460-495	150-190	470-510
2.0- 4.0		450-480	450-475	440-475	460-495	150-190	470-510
2.0- 4.0		450-480	450-475	440-475	460-495	160-190	470-500
2.0- 4.0		450-480	450-475	440-475	460-495	150-190	470-500
2.0- 4.0		450-480	450-475	440-475	460-495	150-190	470-500
2.0- 4.0						150-190	490-540
2.0- 4.0		450-480	450-475	440-475	460-495	150-190	460-500
2.0- 4.0		480-510	465-490	450-475	480-520	150-190	480-525
2.0- 4.0		470-525	465-520	465-510	475-530	130-170	485-550
2.0- 4.0		470-525	465-520	465-510	475-530	130-170	485-550
2.0- 4.0		470-525	465-520	465-510	475-530	130-170	485-550
2.0- 4.0		480-510	465-490	450-475	480-520	150-190	480-525
2.0- 4.0		480-510	465-490	450-475	480-520	150-190	480-525
2.0- 4.0						100-180	480-520
2.0- 4.0						100-180	480-520
2.0- 4.0						100-180	480-520
2.0- 4.0						150-190	470-510
2.0- 4.0						160-190	470-500
2.0- 4.0						150-190	480-520
4.0		500-545	500-545	500-545	490-535	130-160	530-555
4.0		500-545	500-545	500-545	490-535	130-160	530-555

Grade	Filler	Sp Grav	Shrink, mils/in	Melt flow, g/10 min	Melt temp, °F	Back pres, psi	Drying temp, °F
UT 1018		1.22	6.0- 9.0	10.00		50- 100	210-215
UT 400		1.22	6.0- 8.0	30.00		50- 100	190-230
UT 403		1.22	6.0- 8.0	30.00		50- 100	190-230
UT 620 G	GLF 10	1.31	1.0- 6.0			50- 100	190-230
UT 640 G	GLF 20	1.37	1.0- 6.0			50- 100	190-230
PC+PET Alloy		**Xenoy**			**GE**		
1103U		1.20	8.0-10.0			25- 50	230
2230M		1.22	6.0- 9.0			25- 50	250
2730U		1.21				25- 50	250
2735		1.21	5.0- 8.0			25- 50	250
6125		1.12	15.0-17.0			0- 50	250
6125U		1.12	15.0-17.0			0- 50	250
DX1600		1.22	8.0-10.0			0- 50	250
DX1600E		1.22	8.0-10.0			0- 50	250
DX1605		1.22	8.0-10.0			0- 50	250
PC+Polyester		**Ektar MB**			**Eastman**		
DA003		1.20	5.0- 7.0			50- 200	190-200
DA007		1.21	4.0- 6.0				190-200
DA009		1.20	5.0- 7.0				190-200
DA105		1.20	4.0- 7.0				200-210
DA107-9220K		1.21	4.0- 7.0				160-180
DA111		1.18	5.0- 8.0				230
EA001		1.21	5.0- 7.0				225-250
OS500		1.18	4.0- 7.0				230
OS550		1.19	4.0- 7.0			50- 100	280
PC+Polyester		**Sabre**			**Dow**		
1628		1.20	6.0- 9.0	7.00			200-220
1647		1.22	6.0- 9.0	11.00			200-240
1664		1.20	6.0- 9.0	8.00			200-240
PC+Polyester		**Xenoy**			**GE**		
1101		1.21	8.0-10.0			25- 50	230
1102		1.20	8.0-10.0			25- 50	230
1102U		1.20	8.0-10.0			25- 50	230
1200		1.20	16.0-18.0			25- 50	230
1200E		1.20	16.0-18.0			25- 50	230
5770	GMN 20	1.39				50- 100	250
DX1500		1.22	9.0-11.0			25- 50	230
PC+PUR Alloy		**Texin**			**Miles**		
3203		1.22	8.0			0- 200	180-220
3215		1.21	6.0			0- 200	180-220
4203		1.21	8.0			0- 200	180-220
4206		1.21	8.0			0- 200	180-220
4210		1.21	8.0			0- 200	180-220
4210-M		1.21	8.0			0- 200	180-220
PCT		**Ektar FB**			**Eastman**		
AG215	GLF 15	1.32	6.0				300-330
AG215-8901A			5.0				330
AG220	GLF 20	1.37	6.0				300-330
AG230	GLF 30	1.43	2.0				300-330
CG002	GL 30						150-160

Drying time, hr	Inj time, sec	Front temp, °F	Mid temp, °F	Rear temp, °F	Nozzle temp, °F	Mold temp, °F	Proc temp, °F
4.0- 6.0		510	500	490	515	65-165	500-530
6.0		490	480	470	495	65-180	500-550
4.0- 6.0		490	480	470	495	65-180	500-550
4.0- 6.0		510	500	490	515	165-200	500-530
4.0- 6.0		510	500	490	515	165-200	500-530
2.0- 4.0		480-510	465-490	450-475	480-520	150-190	480-525
2.0- 4.0		470-525	465-520	465-510	475-530	150-190	485-530
2.0- 4.0		470-525	465-520	465-510	475-530	150-190	485-530
2.0- 4.0		470-525	465-520	465-510	475-530	150-190	480-520
2.0- 4.0		470-525	465-520	465-510	475-530	150-190	485-530
2.0- 4.0		470-525	465-520	465-510	475-530	150-190	485-530
2.0- 4.0						150-200	540-560
2.0- 4.0						150-200	540-560
2.0- 4.0						150-200	540-560
4.0- 6.0		520	520	520	520	90-150	520-550
5.0- 6.0						80-100	520-560
4.0- 6.0						60-120	530-560
4.0- 6.0						70-130	540-580
4.0- 6.0						80-100	540-560
4.0- 6.0						80-150	565-630
4.0- 6.0						70-100	540-580
2.0- 7.0						90-130	565-620
6.0		565	565	550		110-150	565-580
4.0						150-170	480-530
4.0						150-170	480-530
4.0						150-190	500-560
2.0- 4.0						100-180	480-520
2.0- 4.0						100-180	480-520
2.0- 4.0		480-510	465-490	450-475	480-520	150-190	480-525
2.0- 4.0						130-180	460-475
2.0- 4.0						130-180	460-475
2.0- 4.0		485-525	480-510	475-500	485-530	150-190	495-530
2.0- 4.0		480-510	465-490	450-475	480-520	150-190	480-525
1.0- 3.0		420-460	415-460	410-455	425-465	60-110	430
1.0- 3.0		440-460	440-460	430-450	450-475	80-110	465
1.0- 3.0		420-460	415-460	410-455	425-465	60-110	430
1.0- 3.0		420-460	415-460	410-455	425-465	60-110	440
1.0- 3.0		440-460	440-460	430-450	450-475	80-110	455
1.0- 3.0		440-460	440-460	430-450	450-475	80-110	455
4.0						300-350	560-590
4.0						300-350	560-590
4.0						300-350	560-590
4.0						300-350	560-590
6.0	2.0- 4.					225-275	565-595

Grade	Filler	Sp Grav	Shrink, mils/in	Melt flow, g/10 min	Melt temp, °F	Back pres, psi	Drying temp, °F
CG004	GLF 35	1.51	2.0- 5.0				150-160
CG007	GLF 30	1.46	1.0- 4.0				150-160
CG011	GLF 15	1.33	1.0- 4.0			100	150-160
CG011-9308C	GLF 15	1.33	1.0- 4.0			100	150-160
CG041	GLF 40	1.53	0.5- 3.0			100	150-160
CG041-9307C	GLF 40	1.53	0.5- 3.0			100	150-160
CG053	GLF 20	1.43	2.0- 5.0			100	150-160
CG053-9312C	GLF 20	1.43	2.0- 5.0			100	150-160
CG220-9322C	GLF 20	1.39					310
CG220-9323C	GLF 20	1.40					310
CG905	GLF 30	1.60	2.0				150-160
CG906	GLF 40	1.75					150-160
CG907	GLF 30	1.62	1.0- 4.0			100	150-160
CG920	GLF 20						150-160
CG921	GLF 20	1.54	1.0- 4.0				150-160
CG941	GLF 40	1.70	1.0- 3.0				150-160
CG941-9303C	GLF 40	1.70	1.0- 3.0				150-160
DG901	GLF 10	1.39	1.0- 2.0				150-170
DG902	GLF 20	1.47	1.0- 3.0				150-170
DG903	GLF 30	1.56	1.0- 3.0				150-170
K0860-309A	GLF 40	1.71	2.0- 3.0				
K0860-310A	GLF 40	1.72	2.0- 4.0				150-160
PCT		**Valox**			**GE**		
9715M	GL 15	1.48	4.0- 9.0			50- 100	250-275
9730M	GL 30	1.61	3.0- 8.0			50- 100	250-275
PCTG		**Ektar**			**Eastman**		
DN001		1.23	2.0- 5.0				160-170
DN003		1.23	2.0- 5.0				160-170
DN004		1.23	2.0- 5.0			44- 363	158-167
DN007		1.22	2.0- 5.0				160-170
PCTG		**Ektar FB**			**Eastman**		
DG001	GLF 10	1.29	1.0- 2.0				150-160
DG002	GLF 20	1.35	1.0- 2.0				150-170
DG003	GLF 30	1.45	1.0- 2.0				150-170
PCTG		**Ektar MB**			**Eastman**		
OS150		1.23	2.0- 5.0				220-230
PE		**Ferrene**			**Ferro**		
RPE10HW	GLF 10	1.02	6.0	6.00 L		20- 50	200
RPE20HW	GLF 20	1.10		4.50 L		20- 50	200
PE		**Microthene**			**Quantum**		
MR SP001	CBL 3			5.50			
MR SP002	CBL 3			5.50			
PE		**Retain**			**Dow**		
PE-1001		0.96		4.90			
PE		**RTP Polymers**			**RTP**		
700 FR		1.43	15.0			50	175

Drying time, hr	Inj time, sec	Front temp, °F	Mid temp, °F	Rear temp, °F	Nozzle temp, °F	Mold temp, °F	Proc temp, °F
6.0	2.0- 4.					225-275	565-595
4.0- 6.0	2.0- 4.					225-275	565-595
4.0- 6.0	1.0- 4.	535-570	535-570	520-555	565-590	200-250	565-590
4.0- 6.0	1.0- 4.	535-570	535-570	520-555	565-590	200-250	565-590
4.0- 6.0		535-570	535-570	520-555	565-590	200-250	565-590
4.0- 6.0		535-570	535-570	520-555	565-590	200-250	565-590
4.0- 6.0		535-570	535-570	520-555	565-590	200-250	565-590
4.0- 6.0		535-570	535-570	520-555	565-590	200-250	565-590
6.0						250	560-590
6.0						250	560-590
	1.0- 4.					200-250	555-580
6.0	2.0- 4.					200-250	555-580
4.0	1.0- 4.	535-570	535-570	520-555	565-590	200-250	565-590
6.0	2.0- 4.					200-250	555-580
4.0		535-570	535-570	520-555	565-590		565-590
4.0		535-570	535-570	520-555	565-590		565-590
4.0		535-570	535-570	520-555	565-590		565-590
4.0						70-100	540-560
4.0						70-100	540-560
4.0						70-100	540-560
4.0- 6.0		535-570	535-570	520-555	565-590		565-590
4.0- 6.0		535-570	535-570	520-555	565-590		565-590
4.0- 6.0		540-560	530-550	520-540	540-560	180-220	530-560
4.0- 6.0		540-560	530-550	520-540	540-560	180-220	530-560
4.0- 6.0						60-100	520-560
4.0- 6.0						60-100	530-560
6.0		520	520	520	520	59- 95	518-554
6.0						60-100	530-560
4.0- 6.0		535-570	535-570	520-555	565-590	60-100	530-560
4.0						70-100	540-560
4.0						70-100	540-560
2.0- 7.0						60-100	520-560
2.0		420-480	400-480	390-430	420-480	80-150	
2.0		420-480	400-480	390-430	420-480	80-150	
							450-550
							450-550
						77	392
2.0		380-410	360-380	340-360		75-150	

Grade	Filler	Sp Grav	Shrink, mils/in	Melt flow, g/10 min	Melt temp, °F	Back pres, psi	Drying temp, °F
PEBA		**Pebax**			**Atochem**		
2533 SA 00		1.01		7.00 Q	212-302	50- 100	158
2533 SN 00		1.01		7.00 Q	212-302	50- 100	158
3533 SA 00		1.01		8.00 Q	228-309	50- 100	158
3533 SN 00		1.01		8.00 Q	228-309	50- 100	158
4033 SA 00		1.01		3.00 Q	246-338	50- 100	158
4033 SN 00		1.01		3.00 Q	246-338	50- 100	158
5512 MA 00		1.10		2.00 Q	313-405	50- 100	176
5512 MN 00		1.10		2.00 Q	313-405	50- 100	176
5533 SA 00		1.01		3.00 Q	262-338	50- 100	176
5533 SN 00		1.01		3.00 Q	262-338	50- 100	176
5562 MA 00		1.06		8.00 Q	167-266	50- 100	176
6312 MA 00		1.10		1.00 Q	340-426	50- 100	176
6312 MN 00		1.10		1.00 Q	340-426	50- 100	176
6333 SA 00		1.01		2.00 Q	284-347	50- 100	176
6333 SN 00		1.01		2.00 Q	284-347	50- 100	176
PECTFE		**Halar**			**Ausimont**		
500		1.68	20.0-25.0		464		
520		1.68	20.0-25.0		464		
PEEK		**CTI General**			**CTI**		
PK-10GF/000	GLF 10	1.39	4.0				302
PK-20CF/000	CF 20	1.38	1.0- 8.0				302
PK-20GF/000	GLF 20	1.44	2.0				302
PK-30GF/000	GLF 30	1.54	2.0				302
PK-40CF/000	CF 40	1.44	0.5- 0.9				302
PX-10CF-4CP/15T	CF 10	1.26	1.2				302
PX-10CF-5CC/15T	CF 10	1.22	1.2				302
PEEK		**Electrafil**			**DSM**		
J-1105/CF/30	CF 30	1.43	0.5				220-300
PEEK		**RTP Polymers**			**RTP**		
2200 HF	GLF	1.32	10.0			50- 100	300
2200 LF	GLF	1.32	10.0			50- 100	300
2201 HF	GLF	1.37	4.0			50- 100	300
2201 LF	GLF	1.39	4.0			50- 100	300
2203 HF	GLF	1.44	3.0			50- 100	300
2203 LF	GLF	1.45	3.0			50- 100	300
2205 HF	GLF	1.54	2.0			50- 100	300
2205 LF	GLF	1.51	2.0			50- 100	300
2207 HF	GLF	1.61	2.0			50- 100	300
2207 LF	GLF	1.61	2.0			50- 100	300
2283	CF	1.38	1.0			50- 100	300
2285	CF	1.42	0.5			50- 100	300
2287	CF	1.46	0.5			50- 100	300
PEI		**CTI General**			**CTI**		
BB-10GF/15T	GLF 10	1.28	2.0- 3.0			0- 100	275
PI-20GF/000	GLF 20	1.41	3.0			0- 100	275
PI-30CF/000	CF 30	1.39	2.0			0- 100	275
PI-30GF/000	GLF 30	1.49	2.0- 3.0			0- 100	275
PI-40CF/000	CF 40	1.44	1.0			0- 100	275
PI-40GF/000	GLF 40	1.59	2.0			0- 100	275

Drying time, hr	Inj time, sec	Front temp, °F	Mid temp, °F	Rear temp, °F	Nozzle temp, °F	Mold temp, °F	Proc temp, °F
6.0						60-140	315-430
6.0						60-140	315-430
6.0						60-140	330-460
6.0						60-140	330-460
6.0						60-140	365-460
6.0						60-140	365-460
4.0						60-140	415-460
4.0						60-140	415-460
4.0						60-140	365-480
4.0						60-140	365-480
4.0						60-140	280-360
4.0						60-140	425-500
4.0						60-140	425-500
4.0						60-140	375-480
4.0						60-140	375-480
		500	470	450	490	200-300	510
		500	470	450	490	200-300	510
3.0		720-750	700-730	660-700	720-750	300	
3.0		720-750	700-730	660-700	720-750	300	
3.0		720-750	700-730	660-700	720-750	300	
3.0		720-750	700-730	660-700	720-750	300	
3.0		720-750	700-730	660-700	720-750	300	
3.0		720-750	700-730	660-700	720-750	300	
3.0		720-750	700-730	660-700	720-750	300	
2.0-16.0		690-710	680-700	660-680	700-720	270-350	700-750
4.0		725-800	690-785	675-750		250-400	
4.0		725-800	690-785	675-750		250-400	
4.0		725-800	690-785	675-750		250-400	
4.0		725-800	690-785	675-750		250-400	
4.0		725-800	690-785	675-750		250-400	
4.0		725-800	690-785	675-750		250-400	
4.0		725-800	690-785	675-750		250-400	
4.0		725-800	690-785	675-750		250-400	
4.0		725-800	690-785	675-750		250-400	
4.0		725-800	690-785	675-750		250-400	
4.0		725-800	690-785	675-750		250-400	
4.0		725-800	690-785	675-750		250-400	
4.0		725-800	690-785	675-750		250-400	
4.0		690-750	680-720	660-690	710-730	150-350	725
4.0		690-750	680-720	660-690	710-730	150-350	725
4.0		690-750	680-720	660-690	710-730	150-350	725
4.0		690-750	680-720	660-690	710-730	150-350	725
4.0		690-750	680-720	660-690	710-730	150-350	725
4.0		690-750	680-720	660-690	710-730	150-350	725

Grade	Filler	Sp Grav	Shrink, mils/in	Melt flow, g/10 min	Melt temp, °F	Back pres, psi	Drying temp, °F
PEI		**DSM Polyetherimide**				**DSM**	
J-1106/20	GLF 20	1.44					220-300
J-1106/30	GLF 30	1.53					220-300
J-1106/40	GLF 40	1.61					220-300
PEI		**Electrafil**			**DSM**		
J-1106/CF/30	CF 30	1.39	0.5				220-300
PEI		**Fiberfil**			**DSM**		
J-1106/20	GLF 20	1.44					220-300
J-1106/30	GLF 30	1.53					220-300
J-1106/40	GLF 40	1.61					220-300
PEI		**RTP Polymers**			**RTP**		
2100	GLF	1.27	5.0			50- 75	300
2101	GLF	1.34	4.0			50- 75	300
2103	GLF	1.41	2.0			50- 75	300
2105	GLF	1.49	1.0			50- 75	300
2107	GLF	1.59	1.0			50- 75	300
2183	CF	1.35	1.5			50- 75	300
2185	CF	1.39	1.0			50- 75	300
2187	CF	1.44	0.5			50- 75	300
ESD-A-2180	CF	1.32	2.0			50- 75	300
ESD-C-2180	CF	1.34	1.0			50- 75	300
PEI		**Ultem**			**GE**		
1000		1.27	5.0- 7.0	8.90		50- 200	300
1000F		1.27	5.0- 7.0			50- 200	300
1000FP		1.27	5.0- 7.0			50- 200	300
1000P		1.27	5.0- 7.0			50- 200	300
1000R		1.27	5.0- 7.0			50- 200	300
1000S		1.27	5.0- 7.0			50- 200	300
1010		1.27	5.0- 7.0	17.80		50- 200	300
1010F		1.27	5.0- 7.0			50- 200	300
1010P		1.27	5.0- 7.0			50- 200	300
1010R		1.27	5.0- 7.0			50- 200	300
1010S		1.27	5.0- 7.0			50- 200	300
1100		1.36	5.0- 7.0	7.90		50- 200	300
1100F		1.36	5.0- 7.0			50- 200	300
1100R		1.36	5.0- 7.0			50- 200	300
1110		1.36	5.0- 7.0	16.50		50- 200	300
1110F		1.36	5.0- 7.0			50- 200	300
2100	GL 10	1.34	5.0- 6.0	6.30		50- 100	300
2100R	GL 10	1.34	5.0- 6.0	7.10		50- 100	300
2110	GL 10	1.34	5.0- 6.0	12.00		50- 100	300
2110R	GL 10	1.34		12.20		50- 100	300
2112	SGL 10	1.35		15.60		50- 100	300
2200	GL 20	1.42	3.0- 5.0	5.30		50- 100	300
2200R	GL 20	1.42	3.0- 5.0	5.60		50- 100	300
2210	GL 20	1.42	3.0- 5.0	8.70		50- 100	300
2210R	GL 20	1.42		9.30		50- 100	300
2212	GL 20	1.43		12.70		50- 100	300
2212R	GL 20	1.43		13.80		50- 100	300
2240	GL 20	1.44		34.80		50- 100	300
2242R	GL 20	1.41		53.60		50- 100	300

Drying time, hr	Inj time, sec	Front temp, °F	Mid temp, °F	Rear temp, °F	Nozzle temp, °F	Mold temp, °F	Proc temp, °F
4.0-16.0		680-710	670-720	640-660	700-720	200-350	680-720
4.0-16.0		680-710	670-720	640-660	700-720	200-350	680-720
4.0-16.0		680-710	670-720	640-660	700-720	200-350	680-720
2.0-16.0		680-710	670-720	640-660	700-720	200-350	680-720
2.0-16.0		680-710	670-720	640-660	700-720	200-350	680-720
2.0-16.0		680-710	670-720	640-660	700-720	200-350	680-720
2.0-16.0		680-710	670-720	640-660	700-720	200-350	680-720
4.0		675-750	650-730	625-725		200-350	
4.0		675-750	650-730	625-725		200-350	
4.0		675-750	650-730	625-725		200-350	
4.0		675-750	650-730	625-725		200-350	
4.0		675-750	650-730	625-725		200-350	
4.0		675-750	650-730	625-725		200-350	
4.0		675-750	650-730	625-725		200-350	
4.0		675-750	650-730	625-725		200-350	
4.0		675-750	650-730	625-725		200-350	
4.0		675-750	650-730	625-725		200-350	
4.0		650-800	620-750	590-700	650-775	250-350	640-800
4.0		650-800	620-750	590-700	650-775	250-350	640-800
4.0		650-800	620-750	590-700	650-775	250-350	640-800
4.0		650-800	620-750	590-700	650-775	250-350	640-800
4.0		650-800	620-750	590-700	650-775	250-350	640-800
4.0		650-800	620-750	590-700	650-775	250-350	640-800
4.0		650-800	620-750	590-700	650-775	250-350	640-800
4.0		650-800	620-750	590-700	650-775	250-350	640-800
4.0		650-800	620-750	590-700	650-775	250-350	640-800
4.0		650-800	620-750	590-700	650-775	250-350	640-800
4.0		650-800	620-750	590-700	650-775	250-350	640-800
4.0		650-800	620-750	590-700	650-775	250-350	640-800
4.0		650-800	620-750	590-700	650-775	250-350	640-800
4.0		650-800	620-750	590-700	650-775	250-350	640-800
4.0		650-800	620-750	590-700	650-775	250-350	640-800
4.0		650-800	620-750	590-700	650-775	250-350	640-800
6.0		650-775	650-750	650-700	650-775	225-350	660-775
6.0		650-775	650-750	650-700	650-775	225-350	660-775
6.0		650-775	650-750	650-700	650-775	225-350	660-775
6.0		650-775	650-750	650-700	650-775	225-350	660-775
4.0		650-800	620-750	590-700	650-775	150-350	600-800
6.0		650-775	650-750	650-700	650-775	200-350	660-800
6.0		650-775	650-750	650-700	650-775	200-350	660-800
6.0		650-775	650-750	650-700	650-775	200-350	660-800
6.0		650-775	650-750	650-700	650-775	200-350	660-800
4.0		650-800	620-750	590-700	650-775	150-350	600-800
4.0		650-800	620-750	590-700	650-775	150-350	600-800
6.0		680-760	680-740	680-720	680-760	150-350	680-760
4.0		650-800	620-750	590-700	650-775	250-350	600-800

Grade	Filler	Sp Grav	Shrink, mils/in	Melt flow, g/10 min	Melt temp, °F	Back pres, psi	Drying temp, °F
2300	GL 30	1.51	2.0- 4.0	4.20		50- 100	300
2300R	GL 30	1.51	2.0- 4.0	4.50		50- 100	300
2310	GL 30	1.51	2.0- 4.0	7.10		50- 100	300
2310R	GL 30	1.51		7.40		50- 100	300
2312	GL 30	1.51	3.0- 4.0	10.10		50- 100	300
2313	GL 30	1.52	3.0- 4.0			50- 100	300
2340	GL 30	1.50		27.10		50- 100	300
2342	GL 30	1.50		46.70		50- 100	300
2400	GL 40	1.61	1.0- 3.0	3.10		50- 100	300
2400R	GL 40	1.61	1.0- 3.0	3.30		50- 100	300
2410	GL 40	1.61	1.0- 3.0	5.20		50- 100	300
2410R	GL 40			5.30		50- 100	300
2440	GL 40	1.60		16.20		50- 100	300
3254	MI	1.42	3.0- 4.0	11.90		50- 200	300
3452	GL	1.68				50- 200	300
3453	GSA 45	1.67				50- 100	300
3455	GMI 45	1.67				50- 100	300
4000	GL	1.70	2.0- 3.0	2.80		50- 200	300
4001		1.33	5.0- 7.0	9.90		50- 200	300
4211	GL 20	1.45				50- 100	300
6000		1.29	5.0- 7.0			50- 200	300
6100	GLF 10	1.35	5.0- 6.0			50- 200	300
6200	GLF 20	1.43	3.0- 5.0			50- 200	300
6202	MN 20	1.42	5.0- 7.0			50- 200	300
7201	CF 20	1.34	1.5- 3.0			50- 200	300
7700	CF 15	1.33	3.0- 5.0			50- 100	300
7801	CF 25	1.37	1.5- 3.0			50- 200	300
8015		1.29	4.0- 5.0	3.10		50- 100	300
9075		1.27	3.0- 5.0	2.40		50- 100	300
9076		1.27	3.0- 5.0	1.40		50- 100	300
AR9100	GLF 10	1.32	5.0- 6.0	6.90		50- 100	300
AR9200	GLF 20	1.40	3.0- 5.0	5.20		50- 100	300
AR9300	GLF 30	1.49	2.0- 4.0	4.10		50- 100	300
AR9400	GLF 40	1.59	1.0- 3.0	3.20		50- 100	300
CRS5001		1.28	4.0- 7.0	4.20		50- 150	300
CRS5001R		1.28	4.0- 7.0	6.40		50- 150	300
CRS5011		1.28	4.0- 7.0	10.80		50- 150	300
CRS5101	GL 10	1.35		3.00		50- 100	300
CRS5111	GL 10	1.36		7.30		50- 100	300
CRS5201	GL 20	1.42		2.50		50- 150	300
CRS5211	GL 20	1.42		5.10		50- 150	300
CRS5212	GL 20	1.42		7.20		50- 150	300
CRS5301	GL 30	1.51	2.0- 4.0	1.80		50- 150	300
CRS5311	GL 30	1.52		3.90		50- 150	300
CRS5312	GL 30	1.51		6.50		50- 100	300

PEI+PC Alloy Ultem GE

Grade	Filler	Sp Grav	Shrink, mils/in	Melt flow, g/10 min	Melt temp, °F	Back pres, psi	Drying temp, °F
HP700		1.27	5.0- 7.0	8.70		50- 200	300
HP750		1.32	5.0- 7.0	8.60		50- 200	300
LTX 100A		1.31	4.0- 5.0	3.10		50- 100	300
LTX 100B		1.31		2.20		50- 100	300
LTX 200B		1.26				50- 100	300
LTX 233B	GL 33	1.50		23.90		50- 100	300
LTX 300A		1.27	3.0- 5.0	2.60		50- 100	300
LTX 300B		1.27	3.0- 5.0	2.00		50- 100	300

Drying time, hr	Inj time, sec	Front temp, °F	Mid temp, °F	Rear temp, °F	Nozzle temp, °F	Mold temp, °F	Proc temp, °F
6.0		650-775	650-750	650-700	650-775	225-350	660-800
6.0		650-775	650-750	650-700	650-775	225-350	660-800
6.0		650-775	650-750	650-700	650-775	200-350	660-800
6.0		650-775	650-750	650-700	650-775	200-350	660-800
6.0		650-775	650-750	650-700	650-775	200-350	660-800
6.0		650-775	650-750	650-700	650-775	225-350	660-775
4.0- 6.0		680-760	680-740	680-720	680-760	150-350	680-760
4.0- 6.0		680-760	680-740	680-720	680-760	150-350	680-760
6.0		650-775	650-750	650-700	650-775	200-350	660-800
6.0		650-775	650-750	650-700	650-775	200-350	660-800
6.0		650-775	650-750	650-700	650-775	200-350	660-800
6.0		650-775	650-750	650-700	650-775	200-350	660-800
4.0- 6.0		680-760	680-740	680-720	680-760	150-350	680-760
4.0		650-800	620-750	590-700	650-775	250-350	640-800
4.0		650-800	620-750	590-700	650-775	250-350	640-800
4.0- 6.0		650-750	620-750	590-700	650-725	250-350	640-750
4.0- 6.0		680-760	680-740	680-720	680-760	250-350	680-760
4.0		650-800	635-750	625-700	650-775	150-350	675-825
4.0		650-800	635-750	625-700	650-775	150-350	675-825
6.0		650-775	650-750	650-700	650-775	200-350	660-800
4.0		650-800	635-750	625-700	650-775	150-350	675-825
4.0		650-800	635-750	625-700	650-775	150-350	675-825
4.0		650-800	635-750	625-700	650-775	150-350	675-825
4.0		650-800	635-750	625-700	650-775	150-350	675-825
4.0- 6.0		650-740	635-725	625-725	650-800	150-350	675-825
6.0		650-775	650-750	650-700	650-775	225-350	660-775
4.0- 6.0		650-740	635-725	625-725	650-800	150-350	675-825
6.0		640-700	620-650	600-620	640-700	225-300	640-700
6.0		640-700	620-650	600-620	640-700	225-300	640-700
6.0		640-700	620-650	600-620	640-700	225-300	640-700
6.0		650-775	650-750	650-700	650-775	200-350	660-800
6.0		650-775	650-750	650-700	650-775	200-350	660-800
6.0		650-775	650-750	650-700	650-775	200-350	660-800
6.0		650-775	650-750	650-700	650-775	200-350	660-800
4.0		650-800	620-750	590-700	650-800	250-350	640-800
4.0		650-800	620-750	590-700	650-800	250-350	640-800
4.0		650-800	620-750	590-700	650-800	250-350	640-800
6.0		650-775	650-750	650-700	650-775	200-350	660-800
6.0		650-775	650-750	650-700	650-775	200-350	660-800
4.0		650-800	620-750	590-700	650-800	250-350	640-800
4.0		650-800	620-750	590-700	650-800	250-350	640-800
4.0		650-800	620-750	590-700	650-775	150-350	600-800
4.0		650-800	620-750	590-700	650-800	250-350	640-800
4.0		650-800	620-750	590-700	650-800	250-350	640-800
6.0		650-775	650-750	650-700	650-775	200-350	660-800
4.0		650-800	620-750	590-700	650-775	250-350	640-800
4.0		650-800	620-750	590-700	650-775	250-350	640-800
4.0- 6.0		650-720	640-680	600-640	650-720	225-300	650-720
4.0- 6.0		650-720	640-680	600-640	650-720	225-300	650-720
4.0- 6.0		660-710	650-700	650-700	670-720	225-300	670-720
4.0- 6.0		650-720	640-680	600-640	650-720	150-300	650-720
6.0		640-700	620-650	600-620	640-700	225-350	640-700
6.0		640-700	620-650	600-620	640-700	225-350	640-700

Grade	Filler	Sp Grav	Shrink, mils/in	Melt flow, g/10 min	Melt temp, °F	Back pres, psi	Drying temp, °F
PEKEKK		**RTP Polymers**			**RTP**		
3901	GLF	1.37	6.0			25- 100	350
3903	GLF	1.44	4.0			25- 100	350
3905	GLF	1.52	2.0			25- 100	350
3981	CF	1.33	1.0			25- 100	350
3983	CF	1.37	0.5			25- 100	350
3985	CF	1.41	0.5			25- 100	350
3999X56542-A		1.45	1.0			25- 100	350
3999X56542-B		1.43	0.5			25- 100	350
PES		**CTI Polyethersulfone**				**CTI**	
ES-10CF/15T	CF 10	1.48	2.5				250
ES-10GF/000	GLF 10	1.44	4.0				250
ES-20GF/000	GLF 20	1.51	3.0				250
ES-20GF/15T	GLF 20	1.60	3.0				250
ES-30CF/000	CF 30	1.48	0.5- 1.0				250
ES-30GF/000	GLF 30	1.60	2.0				250
ES-40GF/000	GLF 40	1.68	2.5				250
ES-7CF-7GF/15T	CF 7	1.52	2.5- 3.0				250
ES-7CF/000	CF 7	1.39	3.0				250
PES		**DSM Polyethersulfone**				**DSM**	
J-1100/20	GLF 20	1.52	3.0				220-300
J-1100/40	GLF 40	1.72	1.0				220-300
PES		**Fiberfil**			**DSM**		
J-1100/20	GLF 20	1.52	3.0				220-300
J-1100/40	GLF 40	1.72	1.0				220-300
PES		**Fiberstran**			**DSM**		
G-1100/SS/10	STS 10	1.55	5.0				220-300
PES		**RTP Polymers**			**RTP**		
1400	GLF	1.37	7.0			25- 75	300
1400.5L MG 15	GLF	1.51	6.0			25- 75	300
1401	GLF	1.44	6.0			25- 75	300
1401L	GLF	1.47	5.0			25- 75	300
1403	GLF	1.51	2.0			25- 75	300
1405	GLF	1.59	1.0			25- 75	300
1407	GLF	1.68	1.0			25- 75	300
1475	GLF	1.47	3.0			25- 75	300
1483	CF	1.44	1.5			50- 75	300
1485	CF	1.48	1.0			50- 75	300
1487	CF	1.52	0.5			50- 75	300
ESD-A-1480	CF	1.39	2.0			50- 75	300
ESD-C-1480	CF	1.41	2.0			50- 75	300
PES		**Thermocomp**			**LNP**		
JF-1004	GLF 20	1.51	2.0- 4.0			10- 100	300
JF-1004 EM LE	GLF 20	1.51	2.0- 3.0			10- 100	300
JF-1008	GLF 40	1.68	1.0- 2.0			10- 100	300
PES		**Ultrason**			**BASF**		
E1010		1.37	7.1- 8.6				270
E1010 G2	GLF 10	1.45	5.2- 5.4				270

Drying time, hr	Inj time, sec	Front temp, °F	Mid temp, °F	Rear temp, °F	Nozzle temp, °F	Mold temp, °F	Proc temp, °F
3.0		725-775	700-750	680-725		350-425	
3.0		725-775	700-750	680-725		350-425	
3.0		725-775	700-750	680-725		350-425	
3.0		725-775	700-750	680-725		350-425	
3.0		725-775	700-750	680-725		350-425	
3.0		725-775	700-750	680-725		350-425	
3.0		725-775	700-750	680-725		350-425	
3.0		725-775	700-750	680-725		350-425	
4.0		670-730	650-670	630-650	680-700	225-320	
4.0		670-730	650-670	630-650	680-700	225-320	
4.0		670-730	650-670	630-650	680-700	225-320	
4.0		670-730	650-670	630-650	680-700	225-320	
4.0		670-730	650-670	630-650	680-700	225-320	
4.0		670-730	650-670	630-650	680-700	225-320	
4.0		670-730	650-670	630-650	680-700	225-320	
4.0		670-730	650-670	630-650	680-700	225-320	
4.0		670-730	650-670	630-650	680-700	225-320	
4.0-16.0		680-710	670-720	640-660	700-720	200-350	680-720
4.0-16.0		680-710	670-720	640-660	700-720	200-350	680-720
2.0-16.0		680-710	670-720	640-660	700-720	200-350	680-720
2.0-16.0		680-710	670-720	640-660	700-720	200-350	680-720
4.0-16.0		680-710	670-720	640-660	700-720	200-350	680-720
6.0		690-725	670-700	650-680		200-325	
6.0		690-725	670-700	650-680		200-325	
6.0		690-725	670-700	650-680		200-325	
6.0		690-725	670-700	650-680		200-325	
6.0		690-725	670-700	650-680		200-325	
6.0		690-725	670-700	650-680		200-325	
6.0		690-725	670-700	650-680		200-325	
6.0		690-725	670-700	650-680		200-325	
6.0		690-700	650-680	625-650		200-325	
6.0		690-700	650-680	625-650		200-325	
6.0		690-700	650-680	625-650		200-325	
6.0		690-700	650-680	625-650		200-325	
6.0		690-700	650-680	625-650		200-325	
4.0		640-710	640-710	640-710		280-330	660-700
4.0		640-710	640-710	640-710		280-330	660-700
4.0		640-710	640-710	640-710		280-330	660-700
4.0		705-715	680-715	645-715	715-715	285-320	645-715
4.0		705-715	680-715	645-715	715-715	300-355	660-715

Grade	Filler	Sp Grav	Shrink, mils/in	Melt flow, g/10 min	Melt temp, °F	Back pres, psi	Drying temp, °F
E1010 G4	GLF 20	1.53	3.8- 5.8				270
E1010 G6	GLF 30	1.60	3.4- 6.5				270
E2010		1.37	7.1- 8.7				270
E2010 G2	GLF 10	1.45	5.7- 6.0				270
E2010 G4	GLF 20	1.53	4.5- 6.4				270
E2010 G6	GLF 30	1.60	4.0- 7.5				270
E3010		1.37	7.5- 9.1				270
KR4101	MN 30	1.62	6.0- 8.0				270
PET		**Arnite**			**DSM**		
AV2 390	GLF 50	1.78	1.0-15.0		491	43- 87	212-230
AV4 340	GLF 20	1.50	2.0-15.0		491	43- 87	212-230
PET		**Bexloy**			**Du Pont**		
K550	GLF	1.38	8.0		482		225-275
PET		**Celstran**			**Hoechst**		
PETG30-01-4	GLF 30	1.61					300
PETG40-01-4	GLF 40	1.70					300
PETG50-01-4	GLF 50	1.85					300
PETG60-01-4	GLF 60	1.91					300
PET		**DSM PET**			**DSM**		
J-1800/30	GLF 30	1.56	1.0				180-275
J-1800/45	GLF 45	1.68	1.0				180-275
J-1800/55	GLF 55	1.78	2.0				180-275
PET		**Ektar**			**Eastman**		
EN002		1.33				100- 200	300-325
PET		**Ektar FB**			**Eastman**		
EG001	GLF 30	1.53	1.0- 4.0				250-275
EG003	GLF 30	1.53	2.0				300
EG015	GLF 15	1.33	1.0- 4.0				275-300
EG045	GLF 45	1.68	1.0- 3.0				250-300
EG220	GLF 20	1.52	2.0- 5.0				300
EG711	GLF 15	1.60	2.0- 5.0				250-275
EG721	GLF 20	1.63	1.5- 4.0				250-275
EG731	GLF 30	1.71	1.0- 4.0				250-275
EG741	GLF 40	1.78	1.0- 4.0				250-275
EG903	GL 30	1.64	1.0- 4.0				250-275
EG915	GLF 15	1.59	3.0- 6.0				250-275
PET		**Electrafil**			**DSM**		
J-1800/CF/30	CF 30	1.42	1.0				180-275
PET		**Grilpet**			**EMS**		
EV-20		1.26			432		250-280
EV-30	GLF 30	1.61			489		250-280
EV-302 VO	GLF 30	1.68					250-280
EV-303 VO	GLF 30	1.73					250-280
EV-45	GLF 45	1.74					250-280
PET		**Hiloy**			**ComAlloy**		
441	GLF 30	1.56	2.0- 8.0				250-280
442	GLF 45	1.69	2.0- 9.0				250-280

Drying time, hr	Inj time, sec	Front temp, °F	Mid temp, °F	Rear temp, °F	Nozzle temp, °F	Mold temp, °F	Proc temp, °F
4.0		705-715	680-715	645-715	715-715	300-355	660-715
4.0		705-715	680-715	645-715	715-715	300-355	660-715
4.0		705-715	680-715	645-715	715-715	285-320	645-715
4.0		705-715	680-715	645-715	715-715	300-375	660-715
4.0						300-375	660-735
4.0						300-375	660-735
4.0						285-320	660-735
4.0		705-715	680-715	645-715	715-715	300-355	660-715
6.0-12.0		500-536	509-545	518-554	509-536	266-284	518-572
6.0-12.0		500-536	509-545	518-554	509-536	266-284	518-572
2.0-12.0		510-560	510-560	500-550	490-550	200-250	545-581
4.0		490	480	470	510	300	
4.0		500	490	480	520	300	
4.0		510	500	490	530	300	
4.0		510	500	500	530	300	
2.0-16.0		490-500	490-510	480-500	500-510	60-280	490-520
2.0-16.0		490-500	490-510	480-500	500-510	60-280	490-520
2.0-16.0		490-500	490-510	480-500	500-510	60-280	490-520
4.0- 6.0						60- 95	530-560
4.0- 6.0	2.0- 4.					200-250	550-590
4.0- 6.0						200-250	550-590
4.0- 6.0	2.0- 4.					200-250	540-570
4.0- 6.0	2.0- 4.					200-250	550-590
4.0- 6.0	2.0- 4.					240-290	550-590
4.0- 6.0						180-250	540-560
4.0- 6.0						180-250	540-560
4.0- 6.0						180-250	540-560
4.0- 6.0						180-250	540-560
						200-250	540-560
4.0	2.0- 4.					200-250	540-560
2.0-16.0		490-500	490-510	480-500	500-510	60-280	490-520
4.0- 6.0		510	500	490	520	265	510
4.0- 6.0		510	500	490	520	265	510
4.0- 6.0		510	500	490	520	265	510
4.0- 6.0		510	500	490	520	265	510
4.0- 6.0		510	500	490	520	265	510
3.0- 4.0		510-570	510-570	500-560	520-570	150-250	520-570
3.0- 4.0		510-570	510-570	500-560	520-570	150-250	520-570

Grade	Filler	Sp Grav	Shrink, mils/in	Melt flow, g/10 min	Melt temp, °F	Back pres, psi	Drying temp, °F
443	GLF 55	1.80	2.0- 7.0				250-280
444	GMN 35	1.61	2.0- 8.0				250-280
445	GMN 45	1.69	2.0- 8.0				250-280
PET		**Impet**			**Hoechst**		
330	GLF 30	1.58	7.0				275
340	GLF 45	1.70	6.0				275
530	GLF 30	1.70	6.0				275
540	GLF 43	1.80	6.0				275
630	GMI 35	1.60	7.0				275
PET		**Kodapak**			**Eastman**		
9921		1.33			245	100- 200	300-325
PET		**Nan Ya PET**			**Nan Ya Plastics**		
4410G6	GLF 30	1.58	2.0- 9.0	10.00 W	482	43- 58	275-293
PET		**Petlon**			**Albis**		
3530	GLF 30	1.57	3.0- 8.0		491	30- 60	
4630	GLF 30	1.66	3.0- 8.0		491	30- 60	
7530	GLF 35	1.64	3.0- 8.0		491	30- 60	
PET		**Petra**			**Allied Signal**		
110	GLF 15	1.44	5.0		473		250
130	GLF 30	1.55	3.0		473		250
130 BK-112	GL 30	1.55	3.0		473		250
130 FR	GLF 30	1.68	3.0		473		250
132	GLF 30	1.48	3.0		473		250
140	GLF 40	1.70	2.0		473		250
140 BK-112	GL 45	1.70	2.0- 4.0		473		250
149	GLF 45	1.73	2.0		473		250
230	GMN 35	1.61	3.0		473		250
230 BK-112	GMN 35	1.61	3.0		473		250
242	GMN 40	1.59	3.0		473		250
242 BK-112	GMN 40	1.59			473		250
D-110	GL 15	1.44	5.0		473		250
D-130 FR	GL 30	1.68	3.0		473		250
D-130 FR BK-112	GL 30	1.68	3.0		473		250
PET		**Plenco**			**Plenco**		
50030	GLF 30	1.58	2.0				250
50045	GLF 45	1.72	0.4				250
50125	GLF	1.83	1.0				250
50130	GLF 30	1.72	1.0				250
50240	GLF	1.79	0.5				250
50335	GLF	1.61	2.0				250
PET		**RTP Polymers**			**RTP**		
1103	GLF	1.48	2.0			25- 75	250
1105	GLF	1.56	2.0			25- 75	250
1105 FR	GLF	1.68	2.0			25- 75	250
1107 TFE 10	GLF	1.70	1.0			25- 75	250
1108	GLF	1.70	1.0			25- 75	250
1110	GLF	1.80	1.0			25- 75	250
1199X50089	GLC	1.60	3.0			25- 75	250

Drying time, hr	Inj time, sec	Front temp, °F	Mid temp, °F	Rear temp, °F	Nozzle temp, °F	Mold temp, °F	Proc temp, °F
3.0- 4.0		510-570	510-570	500-560	520-570	150-250	520-570
3.0- 4.0		510-570	510-570	500-560	520-570	150-250	520-570
3.0- 4.0		510-570	510-570	500-560	520-570	150-250	520-570
3.0		520-530	520-530	520-530	530-540	210-230	520-570
3.0		520-530	520-530	520-530	530-540	210-230	520-570
3.0		520-530	520-530	520-530	530-540	210-230	520-570
3.0		520-530	520-530	520-530	530-540	210-230	520-570
3.0		520-530	520-530	520-530	530-540	210-230	520-570
4.0- 6.0							530-560
3.0- 4.0		500-527	490-508	473-492	508-527	230-266	508-527
		480-500	465-485	460-480	480-500	195-250	495-530
		480-500	465-485	460-480	480-500	195-250	495-530
		480-500	465-485	460-480	480-500	195-250	495-530
2.0- 4.0		520-540	500-540	480-520	520-540	210-230	520-540
2.0- 4.0		540-590	520-570	500-550	540-590	210-230	540-590
2.0- 4.0		540-590	520-570	500-550	540-590	210-230	540-590
2.0- 4.0		540-570	530-560	520-550	540-570	210-230	540-570
2.0- 4.0		540-570	530-560	520-550	540-570	210-230	540-570
2.0- 4.0		540-590	520-570	500-550	540-590	210-230	540-590
2.0- 4.0		540-590	520-570	500-550	540-590	210-230	540-590
2.0- 4.0		540-590	520-570	500-550	540-590	210-230	540-590
2.0- 4.0		540-590	520-570	500-550	540-590	210-230	540-590
2.0- 4.0		540-590	520-570	500-550	540-590	210-230	540-590
2.0- 4.0		540-570	530-560	520-550	540-570	210-230	540-570
2.0- 4.0		540-570	530-560	520-550	540-570	210-230	540-570
2.0- 4.0		520-540	480-500	460-480	520-540	180-250	500-540
2.0- 4.0		540-570	530-560	500-550	540-570	180-250	540-570
2.0- 4.0		540-570	530-560	500-550	540-570	180-250	540-570
2.0- 4.0		500-550	500-550	500-550	500-550	230-250	500-550
2.0- 4.0		500-550	500-550	500-550	500-550	230-250	500-550
2.0- 4.0		500-550	500-550	500-550	500-550	230-250	500-550
2.0- 4.0		500-550	500-550	500-550	500-550	230-250	500-550
2.0- 4.0		500-550	500-550	500-550	500-550	230-250	500-550
2.0- 4.0		500-550	500-550	500-550	500-550	230-250	500-550
4.0		560-590	540-570	520-550		180-250	
4.0		560-590	540-570	520-550		180-250	
4.0		520-560	510-550	500-540		180-250	
4.0		560-590	540-570	520-550		180-250	
4.0		560-590	540-570	520-550		180-250	
4.0		560-590	540-570	520-550		180-250	
4.0		560-590	540-570	520-550		180-250	

Grade	Filler	Sp Grav	Shrink, mils/in	Melt flow, g/10 min	Melt temp, °F	Back pres, psi	Drying temp, °F
PET		**Rynite**			**Du Pont**		
415HP	GLF 15	1.39	3.0-10.0		482		225-275
430	GLF 30	1.49	2.0- 9.0		482		225-275
530	GLF 30	1.56	2.0- 9.0		489		225-275
545	GLF 45	1.69	2.0- 8.0		489		225-275
555	GLF 55	1.80	2.0- 7.0		490		225-275
935	GMI 35	1.58	4.0- 8.0		485		225-275
940	GMI 40	1.62	2.0- 7.0		482		225-275
FR-515	GLF 15	1.53	3.0-10.0		489		225-275
FR-530	GLF 30	1.67	2.0- 9.0		489		225-275
FR-543	GLF 43	1.79	2.0- 7.0		489		225-275
FR-945	GMN 45	1.84	3.0- 7.0		482		225-275
SST-35	GLF 35	1.51	2.0- 9.0		482		225-275
PET		**Selar PT**			**Du Pont**		
2251		1.26			509		
5270							
PET		**Valox**			**GE**		
9215	GL 15	1.47	4.0- 9.0			25- 50	275
9230	GL 30	1.59	3.0- 8.0			50- 100	260-280
9245	GL 45	1.72	3.0- 5.0			25- 50	275
9515	GL 15	1.59	4.0- 9.0			25- 50	275
9530	GL 30	1.72	3.0- 8.0			25- 50	275
9530M	GL 15	1.72	3.0- 8.0			25- 50	275
PET		**Voloy**			**ComAlloy**		
441	GL 30	1.67	2.0- 9.0				250-280
PETG		**Ektar**			**Eastman**		
GN001		1.27	2.0- 5.0				160
GN002		1.27	2.0- 5.0				160
GN005		1.27	2.0- 5.0				160
PETG		**Kodar**			**Eastman**		
6763		1.27					
PFA		**Teflon PFA**			**Du Pont**		
340		2.14			590		
Phenolic		**Norsophen**			**Norold**		
BMC	GLF 25	1.90	1.0				
Phenolic		**Plenco**			**Plenco**		
02000	WDF	1.39	4.0-10.0				
02300		1.40	5.0-10.0				
02308	WDF	1.42	5.0-12.0				
02311		1.41	3.0-10.0				
02369	WDF	1.38	5.0-10.0				
02408	WDF	1.38	4.0- 9.0				
02482	WDF	1.42	5.0-10.0				
02535	WDF	1.39	4.0-10.0				
02571	WDF	1.40	5.0-10.0				
02929		1.42	5.0- 9.0				
03509	FLK	1.56	3.0- 7.0				

Drying time, hr	Inj time, sec	Front temp, °F	Mid temp, °F	Rear temp, °F	Nozzle temp, °F	Mold temp, °F	Proc temp, °F
2.0-16.0	1.0- 4.	500-540	500-540	500-530	500-560	165-205	520-550
2.0-16.0	1.0- 4.	500-540	500-540	500-530	500-560	190-250	520-550
2.0-16.0	1.0- 4.	510-560	500-560	500-550	530-570	190-230	540-570
2.0-16.0	1.0- 4.	510-560	500-560	500-550	530-570	190-230	540-570
2.0-16.0	1.0- 4.	510-560	500-560	500-550	530-570	190-230	540-570
2.0-16.0	1.0- 4.	510-560	500-560	500-550	530-570	190-230	540-570
2.0-16.0	1.0- 4.	510-560	500-560	500-550	530-570	190-230	540-570
2.0-16.0	1.0- 4.	500-540	500-540	500-530	500-560	190-230	520-550
2.0-16.0	1.0- 4.	500-540	500-540	500-530	500-560	190-230	520-550
2.0-16.0	1.0- 4.	500-540	500-540	500-530	500-560	190-230	520-550
2.0-16.0	1.0- 4.	500-540	500-540	500-530	500-560	190-230	520-550
2.0-16.0	1.0- 4.	500-540	500-540	500-530	500-560	165-205	520-550
		525	500	475		80-100	527-545
		525	500	475		80-100	527-545
4.0- 6.0		520-540	520-540	500-530	520-540	100-140	520-540
3.0- 4.0		520-550	500-530	480-520	540-570	200-250	500-550
4.0- 6.0		520-540	520-540	500-530	520-540	100-140	520-540
4.0- 6.0		520-540	520-540	500-530	520-540	100-140	520-540
4.0- 6.0		520-540	520-540	500-530	520-540	100-140	520-540
4.0- 6.0		520-540	520-540	500-530	520-540	100-140	520-540
3.0- 4.0		510-570	510-570	500-560	520-570	150-250	520-570
4.0- 6.0						60-100	480-520
4.0- 6.0						60-100	480-520
6.0						60-100	480-520
	20.0	470	450	420	470	80	420-500
		700	625-650	600-630	700	300-500	650-750
						302	
						300-360	200-250
						300-360	
						300-360	
						300-360	
						300-340	
						300-360	
						300-360	
						300-360	
						300-360	
						300-360	
						300-360	

Grade	Filler	Sp Grav	Shrink, mils/in	Melt flow, g/10 min	Melt temp, °F	Back pres, psi	Drying temp, °
03597	UNS	1.55	4.0- 9.0				
04002	MN	1.58	6.0				
04300	CE	1.53	3.0- 8.0				
04301	MN	1.50	3.0- 7.0				
04304	MN	1.59	7.0				
04309	GMN	1.55	8.0				
04349	MN	1.53	3.0- 9.0				
04400	MN	1.54	9.0				
04414	FLK	1.54	4.0- 9.0				
04466	FLK	1.54	4.0- 8.0				
04485	MN	1.46	4.0- 9.0				
04527	FLK	1.58	4.0- 9.0				
04548	MN	1.48	5.0- 9.0				
04700	MN	1.64	8.0				
05115	GRP	1.77	2.0				
05482	GRK	1.42	10.0				
06582/MX582		1.59	2.0				
07002	FLK	1.37	10.0				
07080	FLK	1.39	5.0- 9.0				
07100	CE	1.40	7.0				
07200	UNS	1.36	5.0-10.0				
07202	UNS	1.37	9.0				
07203	UNS	1.39	4.0- 9.0				
07321	FLK	1.38	8.0- 9.0				
07417	FLK	1.42	8.0				
07476	FLK	1.41	5.0- 8.0				
07487		1.32	6.0-10.0				
07500	FLK	1.37	6.0-10.0				
07502	CE	1.54	8.0				
07507	CE	1.52	5.0-10.0				
07523	UNS	1.38	5.0- 8.0				
07552	GLF	1.71	2.0- 4.0				
07556	GLF	1.85	1.0- 2.0				
07579	WDF	1.38	4.0- 9.0				

Phenolic Resinoid Resinoid

Grade	Filler	Sp Grav	Shrink, mils/in	Melt flow, g/10 min	Melt temp, °F	Back pres, psi	Drying temp, °
1310	GLF	1.82	1.0- 2.0				
1312	GLF	1.89	3.0- 4.0				
1320	GLF	1.99	2.0				
1321	GLF	1.94	2.0				
1322	GLF	1.99	2.0				
1323	GLF	1.99	2.0				
1324	GLF	1.99	2.0				
2010		1.37	2.0- 4.0				
2016		1.37	2.0- 4.0				
2016P		1.37	2.0- 3.0				
2017	UNS	1.35	4.0- 5.0				
2018		1.36	2.0- 4.0				
7003S	GLF	1.84	2.0- 3.0				
7005	GLF	1.92	1.5				
7051	GLF	1.75	1.0- 2.0				
7051SS	GLF	1.75	1.0- 2.0				
7201	GLF	1.60	1.0- 3.0				

Phenolic Rogers Phenolic Rogers

Grade	Filler	Sp Grav	Shrink, mils/in	Melt flow, g/10 min	Melt temp, °F	Back pres, psi	Drying temp, °
RX 340	CE	1.43	6.0			50	

Drying time, hr	Inj time, sec	Front temp, °F	Mid temp, °F	Rear temp, °F	Nozzle temp, °F	Mold temp, °F	Proc temp, °F
						300-360	
						300-360	
						300-360	
						300-360	
						300-360	
						300-360	
						300-360	
						300-360	
						300-360	
						300-360	
						300-360	
						300-340	
						300-360	
						300-340	
						300-360	
						300-360	
						300-360	
						300-340	
						300-340	
						300	
						300-360	
						300-360	
						300-360	
						300-360	
						300-340	
						300-340	
						300-360	
						300-360	
						300-360	
						300-340	
						300-360	
						300-360	
						300-360	
						300-360	
							300-350
							300-350
							300-350
							300-350
							300-350
							300-350
							300-350
							300-350
							300-350
							300-350
							300-350
							300-350
							300-350
							300-350
							300-360
							300-360
							300-350
	8.0- 10.		165	140	210	340	235

Grade	Filler	Sp Grav	Shrink, mils/in	Melt flow, g/10 min	Melt temp, °F	Back pres, psi	Drying temp, °F
RX 342	CE	1.44	6.0			50	
RX 431	CE	1.42	5.0			50	
RX 448	CE	1.42	6.0			50	
RX 525	CE	1.43	6.0			50	
RX 611	GL	1.75	2.0			30	
RX 630	GL	1.78	1.5			30	
RX 640	GL	1.74	2.0			30	
RX 643	GL	1.74	1.5			30	
RX 647	GL	1.90	2.0			30	
RX 660	GL	1.80	2.0			30	
RX 660C	GL	1.82	2.0			30	
RX 670	GL	1.90	2.0			30	
RX 853	GL	1.80	2.0			30	
RX 865	GL	1.86	2.0			30	
RX 866	GL	1.70	2.0			30	
RX 867	GL	1.77	2.5			30	
RX 868	GL	1.77	2.0			30	
XB-22	GL	1.50	1.5			30	
XT-26	GL	1.82	2.0			30	
Plastomer		**Affinity**			**Dow**		
SM 1250		0.89		30.00			
Plastomer		**Exact**			**Exxon**		
4023		0.88		35.00	154		
4028		0.88		10.00	144		
PMMA		**Lucky PMMA**			**Lucky**		
EF 940		1.18	2.0- 6.0	3.00 I			
EG 920		1.18	2.0- 6.0	1.60 I			
EG 930		1.18	2.0- 6.0	2.20 I			
EH 910		1.18	2.0- 6.0	1.00 I			
IF 850		1.18	2.0- 6.0	12.50 I			
IF 870		1.18	2.0- 6.0	23.00 I			
IG 840		1.18	2.0- 5.0	5.80 I			
IH 830		1.18	2.0- 6.0	2.30 I			
PMMA		**ShinkoLite-P**			**Mitsubishi Ray.**		
MDP		1.19					158-167
VHP		1.19					167-185
VP		1.19					176-194
PMP		**Crystalor**			**Phillips**		
BC-1	GL	1.05			464		
CBN-020		0.83	18.0-34.0	20.00	455		
DC-6				70.00			
DC-7				27.00			
HBG-15	GL 15	0.93			464		
HBG-30	GLF 30	1.05	3.0-12.0		464		
HBG-30HP	GL 30	1.05			464		
HBG-40HP	GL 40	1.14			464		
HBN-010		0.83		10.00	464		
HBN-020		0.83	1.0-21.0	20.00	464		
PMP		**DSM Polymethylpentene**				**DSM**	
J-98/10	GLF 10	0.87	5.0			0- 100	

Drying time, hr	Inj time, sec	Front temp, °F	Mid temp, °F	Rear temp, °F	Nozzle temp, °F	Mold temp, °F	Proc temp, °F
	8.0- 10.		165	140	210	340	235
	8.0- 10.		165	140	210	340	235
	8.0- 10.		165	140	210	340	235
	8.0- 10.		165	140	210	340	235
	5.0- 7.		165	140	190	330	240
	5.0- 7.		165	140	190	330	240
	5.0- 7.		165	140	190	330	240
	5.0- 7.		165	140	190	330	240
	5.0- 7.		165	140	190	330	240
	5.0- 7.		165	140	190	330	240
	5.0- 7.		165	140	190	330	240
	5.0- 7.		165	140	190	330	240
	5.0- 7.		165	140	190	330	240
	5.0- 7.		165	140	190	330	240
	5.0- 7.		165	140	190	330	240
	5.0- 7.		165	140	190	330	240
	5.0- 7.		165	140	190	330	240
	5.0- 7.		165	140	190	330	240
	5.0- 7.		165	140	190	330	240
	0.4	350-550	350-550	350-550		50- 60	350-500
		329-365		302-347	347-374	59- 86	
		365-392		347-374	347-374	59- 86	
		410-482	410-482	410-482		104-194	
		464-518	464-518	464-518		104-194	
		428-500	428-500	428-500		104-194	
		464-518	464-518	464-518		104-194	
		392-464	392-464	392-464		104-194	
		374-464	374-464	374-464		104-194	
		410-482	410-482	410-482		104-194	
		428-500	428-500	428-500		104-194	
4.0- 6.0		374-446	374-446	374-446			
4.0- 6.0		392-491	392-491	392-491			
4.0- 6.0		410-491	410-491	410-491			
		590	560	560	590	100-190	
		590	560	560	590	100-190	
		590	560	560	590	100-190	
		590	560	560	590	100-190	
		590	560	560	590	100-190	
		590	560	560	590	100-190	
		590	560	560	590	100-190	
		590	560	560	590	100-190	
		590	560	560	590	100-190	
		590	560	560	590	100-190	
						150	560-595

Grade	Filler	Sp Grav	Shrink, mils/in	Melt flow, g/10 min	Melt temp, °F	Back pres, psi	Drying temp, °F
J-98/20	GLF 20	0.93	2.0			0- 100	
J-98/30	GLF 30	1.04	1.0			0- 100	
J-98/40	GLF 40	1.14	1.0			0- 100	
Polyamide-Imide		**Torlon**			**Amoco Perform.**		
4203L	TIO 3	1.42	10.0			50	250
4275	GRP 20	1.51	7.0			50	250
4301	GRP 12	1.46	7.0			50	250
4347	GRP 12	1.50	7.0			50	250
5030	GLF 30	1.61	4.0			50	250
7130	GRF 30	1.48	2.0			50	250
7330	CF	1.50	2.0			50	250
9040	GLF 40	1.68	2.0			50	250
Polyaryl Amide		**Ixef**			**Solvay**		
1002	GLF 30	1.43	3.0- 6.0			0- 435	176
1022	GLF 50	1.64	1.0- 4.0			0- 435	176
1501	GLF 30	1.51	2.0- 5.0			0- 435	176
1521	GLF 50	1.69	1.0- 4.0			0- 435	176
2010	GLF	1.63				0- 435	176
2030	GLF	1.73				0- 435	176
9203	GLF 30	1.32				0- 435	176
Polyarylate		**Ardel**			**Amoco Perform.**		
D-100		1.21	9.0	4.50			300
D-170		1.22	9.0	15.00			300
D-240		1.26		9.00			210
Polyarylate		**Durel**			**Hoechst**		
400		1.21	6.0			0-1000	265
410	GLF 10	1.28	3.0			0-1000	265
420	GLF 20	1.35	2.0			0-1000	265
430	GLF 30	1.44	1.0			0-1000	265
Polyarylsulfone		**CTI General**			**CTI**		
CTXC-020	CF	1.40	2.0			50- 100	300
CTXC-021	CF	1.40	3.0- 4.0			50- 100	300
Polyarylsulfone		**Radel**			**Amoco Perform.**		
A-100		1.37	6.0	12.50		50- 100	350
A-200		1.37	6.0	20.00		50- 100	350
A-300		1.37	6.0	30.00		50- 100	350
AG-210	GLF 10	1.43	5.0	12.00		50- 100	350
AG-220	GLF 20	1.51	4.0	10.00		50- 100	350
AG-230	GLF 30	1.58	3.0	10.00		50- 100	350
AG-360	GLF	1.54	3.0	16.00		50- 100	275-350
Polyarylsulfone		**RTP Polymers**			**RTP**		
1600	GLF	1.37	6.0			25- 75	300
1601	GLF	1.44	6.0			25- 75	300
1603	GLF	1.51	2.0			25- 75	300
1605	GLF	1.59	1.0			25- 75	300
1607	GLF	1.68	1.0			25- 75	300
1699X56969	CF	1.40	3.0			50- 75	300

Drying time, hr	Inj time, sec	Front temp, °F	Mid temp, °F	Rear temp, °F	Nozzle temp, °F	Mold temp, °F	Proc temp, °F
						150	560-595
						150	560-595
						150	560-595
24.0						400-425	650
24.0						400-425	650
24.0						400-425	650
24.0						400-425	650
24.0						400-425	650
24.0						400-425	650
24.0						400-425	650
24.0						400-425	650
12.0	1.0- 2.	482-536	482-536	482-536	500-554	248-302	482-536
12.0	1.0- 2.	482-536	482-536	482-536	500-554	248-302	482-536
12.0	1.0- 2.	482-536	482-536	482-536	500-554	248-302	482-536
12.0	1.0- 2.	482-536	482-536	482-536	500-554	248-302	482-536
12.0	1.0- 2.	482-536	482-536	482-536	500-554	248-302	482-536
12.0	1.0- 2.	482-536	482-536	482-536	500-554	248-302	482-536
12.0	1.0- 2.	482-536	482-536	482-536	500-554	248-302	482-536
3.0- 5.0						250-300	675-735
3.0- 5.0						250-300	675-735
7.0-10.0						150-200	520-610
6.0- 8.0		600-615	600-615	590-605	615-640	275-300	590-640
6.0- 8.0		600-615	600-615	590-605	615-640	275-300	590-640
6.0- 8.0		600-615	600-615	590-605	615-640	275-300	590-640
6.0- 8.0		600-615	600-615	590-605	615-640	275-300	590-640
4.0		690-725	660-680	650-670	690-700	275-300	
4.0		690-725	660-680	650-670	690-700	275-300	
2.5		660	660	610		280-325	650-730
2.5		660	660	610		280-325	650-725
2.5		660	660	610		280-325	650-725
2.5		660	660	610		300-325	650-725
2.5		660	660	610		300-325	650-725
2.5		660	660	610		300-325	650-725
2.5- 1.5		660	660	610		280-325	650-750
6.0		690-725	670-700	650-680		200-325	
6.0		690-725	670-700	650-680		200-325	
6.0		690-725	670-700	650-680		200-325	
6.0		690-725	670-700	650-680		200-325	
6.0		690-725	670-700	650-680		200-325	
6.0		690-700	650-680	625-650		200-325	

Grade	Filler	Sp Grav	Shrink, mils/in	Melt flow, g/10 min	Melt temp, °F	Back pres, psi	Drying temp, °F
Polybutylene		**Duraflex**			**Shell**		
DP 8510		0.90		45.00 E	194		
DP 8910		0.90			194		
Polyester		**Ektar**			**Eastman**		
K1000		1.31	2.0- 5.0			100- 200	175
Polyester		**Valox**			**GE**		
HW100		1.31				0- 50	250
HW110	MNF 10	1.36	6.0-16.0			0- 50	250
NBV051		1.30	17.0-21.0			25- 50	250
Polyester Alloy		**Comtuf**			**ComAlloy**		
410		1.20	6.0- 8.0	7.00			250-280
411	GL	1.25	4.0	7.00		50- 100	260
412	MN	1.30	6.0				250-280
Polyester, Ela		**RTP Polymers**			**RTP**		
1500-55D		1.20					
1500.5-40D	GLF 5	1.20	2.0				
1500.5-55D	GLF 5	1.23	6.0				
1501-40D	GLF 10	1.23	2.0				
1501-55D	GLF 10	1.26	5.0				
1502-40D	GLF 15	1.26	2.0				
1503-55D	GLF 20	1.34	2.0				
1505-55D	GLF 30	1.43	2.0				
1507-55D	GLF 40	1.52	2.0				
Polyester, Ela		**Stat-Kon**			**LNP**		
YC-1006	CF 30	1.33	2.0- 4.0		412	0- 50	225
Polyester, ElCo		**Ecdel**			**Eastman**		
9965		1.13		15.00 L	383-419		151
9966		1.13		10.00 L	383-419		151
9967		1.13		4.00 L	383-419		151
Polyester, ElCo		**Hytrel**			**Du Pont**		
4056		1.17		5.30 E	302	50- 80	230
4069		1.11		8.50	379	50- 80	230
4556		1.14		8.50	379	50- 80	230
5526		1.20		18.00	397	50- 80	230
5555HS		1.20		8.50	397	50- 80	230
5556		1.20		7.50	397	50- 80	230
6356		1.22		8.50	412	50- 80	230
7246		1.25		12.50	424	50- 80	230
8238		1.28		12.50	433	50- 80	230
G3548W		1.15		10.00 E	312	50- 80	230
G4074		1.18		5.20 E	338	50- 80	230
G4078W		1.18		5.30 E	338	50- 80	230
G4774		1.20		11.00 L	406	50- 80	230
G5544		1.22		10.00 L	419	50- 80	230
HTR6108		1.24		5.20 E	334	50- 80	230
HTR8068		1.43		4.60 E	336	50- 80	230
HTR8122		1.07		5.00 E	338	50- 80	230

Drying time, hr	Inj time, sec	Front temp, °F	Mid temp, °F	Rear temp, °F	Nozzle temp, °F	Mold temp, °F	Proc temp, °F
6.0						60-100	530-560
3.0- 4.0		460-490	450-480	440-470	460-490	110-140	455-490
3.0- 4.0		460-490	450-480	440-470	460-490	110-140	455-490
3.0- 4.0		460-490	450-480	440-470	460-490	110-140	455-490
3.0- 4.0		420-460	420-450	410-440	420-470	100-250	430-490
4.0		570	560	550	550-570	160-180	550-570
3.0- 4.0		420-460	420-450	410-440	420-470	100-250	430-490
		370-420	370-420	370-420		70-100	
		390-420	380-410	370-400		70-100	
		390-420	380-410	370-400		70-100	
		390-420	380-410	370-400		70-100	
		390-420	380-410	370-400		70-100	
		390-420	380-410	370-400		70-100	
		390-420	380-410	370-400		70-100	
		390-420	380-410	370-400		70-100	
		390-420	380-410	370-400		70-100	
4.0- 6.0		360-460	360-460	360-460		75-125	420-460
4.0						104-149	225-260
4.0						104-149	225-260
4.0						104-149	225-260
2.0- 3.0		375	375	329-375	356	113	356-400
2.0- 3.0		455	455	400-455	428	113	428-482
2.0- 3.0		455	455	400-455	428	113	428-482
2.0- 3.0		455	455	400-455	428	113	428-482
2.0- 3.0		455	455	400-455	428	113	428-482
2.0- 3.0		455	455	400-455	428	113	428-482
2.0- 3.0		473	473	428-473	455	113	455-500
2.0- 3.0		473	473	428-473	464	113	464-500
2.0- 3.0		473	473	428-473	464	113	464-500
2.0- 3.0		375	375	329-375	356	113	356-400
2.0- 3.0		400	400	356-400	375	113	375-425
2.0- 3.0		400	400	356-400	375	113	375-425
2.0- 3.0		473	473	428-473	455	113	455-500
2.0- 3.0		473	473	428-473	455	113	455-500
2.0- 3.0		400	400	356-400	375	113	375-410
2.0- 3.0		400	400	356-400	375	113	375-410
2.0- 3.0		400	400	356-400	375	113	375-410

Grade	Filler	Sp Grav	Shrink, mils/in	Melt flow, g/10 min	Melt temp, °F	Back pres, psi	Drying temp, °F
Polyester, ElCo		**Lomod**			**GE**		
B3000	GLF 30	1.43	2.0- 4.0	7.00 E	340	50- 75	220
FR30125A		1.43	14.0-16.0	13.00	424	50- 75	220
FR5020A		1.34	19.0-21.0	12.00	320	50- 75	220-230
FR5030A		1.39	16.0-18.0	12.00	360	50- 75	220-230
ST3090A		1.15	15.0-17.0	8.70	365	50- 75	220
ST5090A		1.13	11.0-13.0	7.40 E	293	50- 75	220
TE3040A		1.18	18.0-20.0	25.00	390	50- 75	220
TE3045A		1.20	19.0-21.0	18.00	397	50- 75	220
TE3055A		1.22	19.0-21.0	16.00	406	50- 75	220
TE3055B		1.22	19.0-21.0	16.00 L	406	50- 75	220
TE3060A		1.24	20.0-22.0	37.00	424	50- 75	
TE3070A		1.25	20.0-22.0	14.00	424	50- 75	220
TE3340A		1.18	18.0-20.0	25.00	390	50- 75	220
TE5040A		1.16	12.0-14.0	10.00 E	316	50- 75	220
TE5050A		1.21	12.0-14.0	7.00 E	331	50- 75	220
Polyester, TP		**Arnite**			**DSM**		
AV2 343	GLF 20	1.50			491		
AV2 360 S	GLF	1.73	1.5-15.0		491	43- 87	212-230
AV2 363SN	GLF 33	1.74			491		
AV2 370	GLF	1.63	2.0-15.0		491	43- 87	212-230
D04 300		1.34			486	43- 87	212-230
TV4 270	GLF 35	1.57	2.5-15.0		433	72- 145	212-230
Polyester, TP		**Bexloy**			**Du Pont**		
V-572		1.42	4.0	20.00	432		
Polyester, TP		**Celanex**			**Hoechst**		
2016		1.44	18.0-22.0			0- 50	250-280
3116	GLF 10	1.50				0- 50	250-280
3216	GLF 15	1.54				0- 50	250-280
3300	GLF 30	1.54	3.0- 5.0			0- 50	250-280
3316	GLF 30	1.66				0- 50	250-280
4305	GLF 33	1.50		7.50		0- 50	250-280
4306	GLF 30	1.50		10.00		0- 50	250-280
Polyester, TP		**CTI Polyester**			**CTI**		
PS-05CF/27CC	CF 5	1.26	5.0- 7.0				200
PS-10GF/000	GLF 10	1.38	6.0				225
PS-15GF/000	GLF 15	1.42	4.0				225
PS-20GF/000	GLF 20	1.43	3.5				225
PS-20GF/000 FR	GLF 20	1.60	3.0				225
PS-30GF/000	GLF 30	1.52	3.0				200
PS-30GF/000 FR	GLF 30	1.16	2.0				200
PS-30GM/000	GL 30	1.54	14.0				225
Polyester, TP		**Ektar**			**Eastman**		
DN003		1.23					160-170
DN101		1.22	3.0- 5.0				160-170
GN004		1.26	2.0- 5.0				160
Polyester, TP		**Plenco**			**Plenco**		
01500	MN	1.86	7.0- 9.0				
01505	CE	1.80	8.0				

Drying time, hr	Inj time, sec	Front temp, °F	Mid temp, °F	Rear temp, °F	Nozzle temp, °F	Mold temp, °F	Proc temp, °F
4.0- 6.0		370-400	350-380	340-360	370-400	60- 90	370-420
4.0- 6.0		470-500	450-490	440-470	470-500	75-120	460-500
2.0- 3.0		420-450	420-440	410-430	420-450	60- 90	410-450
2.0- 3.0		420-450	420-440	410-430	420-450	60- 90	410-450
4.0- 6.0		420-450	420-440	410-430	420-450	60- 90	410-450
4.0- 6.0		350-385	340-380	310-340	350-390	60- 90	350-390
4.0- 6.0		430-460	430-460	420-450	430-460	70-120	410-460
4.0- 6.0		440-480	430-470	420-460	440-480	70-120	440-480
4.0- 6.0		450-470	440-470	430-450	450-470	70-120	440-480
4.0- 6.0		450-470	440-470	430-450	450-470	70-120	440-480
		460-490	450-480	440-460	460-490	70-120	460-490
4.0- 6.0		460-500	450-480	440-470	460-500	70-120	460-500
4.0- 6.0		430-460	430-460	420-450	430-460	70-120	410-460
4.0- 6.0		360-390	350-395	340-360	360-390	60- 90	370-420
4.0- 6.0		370-400	350-380	340-360	370-400	60- 90	370-420
						266-284	509-563
6.0-12.0		491-509	500-518	509-527	491-509	266-284	509-527
						266-284	459-513
6.0-12.0		500-536	509-545	518-554	509-536	266-284	518-572
6.0-12.0		527-572	527-572	527-572	527-572	50-122	518-572
6.0-12.0		464		455	473	86-194	464-518
	5.0- 15.	460-510	460-510	430-480	470-520	80-200	460-500
3.0- 4.0		470	460	450	480	150-200	450-500
3.0- 4.0		470	460	450	480	150-200	450-500
3.0- 4.0		470	460	450	480	150-200	450-500
3.0- 4.0		470	460	450	480	150-200	450-500
3.0- 4.0		470	460	450	480	150-200	450-500
3.0- 4.0		470	460	450	480	150-200	450-500
3.0- 4.0		470	460	450	480	150-200	450-500
						150-200	460-480
						200-250	480-510
						200-250	475-510
						200-250	470-510
						200	440-480
						200-250	450-480
						200	450-480
						250	480-520
						60-100	530-560
6.0						60-100	530-560
4.0- 6.0						60-100	480-520
	5.0- 8.					335	
	5.0- 8.					335	

Grade	Filler	Sp Grav	Shrink, mils/in	Melt flow, g/10 min	Melt temp, °F	Back pres, psi	Drying temp, °F
01506	CE	1.68	4.0- 5.0				
01510	MN	2.05	4.0- 6.0				
01528	GL	1.97	3.0- 4.0				
01530	GMN	1.98	2.0- 4.0				
01535	GLF	1.92	4.0				
01580	GMN	1.76	2.0- 5.0				
01581	GMN	1.80	3.0- 5.0				
Polyester, TP		**Rynite**			**Du Pont**		
CR503 BK570		1.67	3.5- 5.3				250-275
Polyester, TP		**Selar PT**			**Du Pont**		
4232		1.26			491		250-275
4234		1.28			491		250-275
Polyester, TP		**Stat-Kon**			**LNP**		
PDX-W-92154		1.28	15.0-20.0		430	50- 100	230
Polyester, TP		**Texapol Polyester**			**Texapol**		
8400		1.31	17.0	19.00 T	441		210-250
8410 HS		1.31	17.0	23.00 T	441		210-250
8410		1.31	18.0	23.00 T	441		210-250
8411 FR		1.41	25.0	25.00 T	441		210-250
GB 8400-30	GLB 30	1.53	16.0	9.00 T	433		210-250
GF 8400 7.5 FR	GLF 8	1.45	11.0	10.00 T	441		210-250
GF 8400-15 FR	GLF 15	1.49	7.0	10.00 T	441		210-250
GF 8400-15	GLF 15	1.41	4.0	12.00 T	433		210-250
GF 8400-30 FR	GLF 30	1.65	3.0	5.00 T	441		210-250
GF 8400-30	GLF 30	1.53	3.0	11.00 T	437		210-250
GF 8400-40	GLF 40	1.63	2.0	10.00 T	437		210-250
GMF 8400-45 FR	GMN 45	1.88	2.0	6.00 T	441		210-250
Polyester, TP		**Texin**			**Miles**		
345-D		1.22	8.0			0- 200	180-220
355-D		1.22	8.0			0- 200	180-220
360-D		1.23	8.0			0- 200	180-220
445-D		1.22	8.0			0- 200	180-220
455-D		1.21	8.0			0- 200	180-220
458-D		1.22	8.0			0- 200	180-220
470-D		1.24	8.0			0- 200	180-220
480-A		1.20	8.0			0- 200	180-220
591-A		1.22	8.0			0- 200	180-220
688-A		1.26	8.0			0- 200	180-220
Polyester, TP		**Valox**			**GE**		
210HP		1.31	6.0-16.0			0- 50	250
220HP		1.31	6.0-16.0			0- 50	250
225HP		1.31	6.0-16.0			0- 50	250
230HP		1.31	6.0-16.0			0- 50	250
235HP		1.31	6.0-16.0			0- 50	250
260HP		1.31	6.0-16.0			0- 50	250
265HP		1.31	6.0-16.0			0- 50	250
280HP		1.31	6.0-16.0			0- 50	250
285HP		1.31	6.0-16.0			0- 50	250
302		1.31	6.0-16.0			0- 50	250
304		1.31	6.0-16.0			0- 50	250

Drying time, hr	Inj time, sec	Front temp, °F	Mid temp, °F	Rear temp, °F	Nozzle temp, °F	Mold temp, °F	Proc temp, °F
	5.0- 8.					335	
	5.0- 8.					335	
	5.0- 8.					335	
	5.0- 8.					335	
	5.0- 8.					335	
	5.0- 8.					335	
	5.0- 8.					335	
2.0- 3.0						100-250	555-610
6.0- 8.0							
6.0- 8.0							
4.0- 6.0		400-450	400-450	400-450		50-120	430-445
3.0- 4.0		440	440	440	465	150-200	
3.0- 4.0		440	440	440	465	150-200	
3.0- 4.0		440	440	440	465	150-200	
3.0- 4.0		440	435	435	460	150-200	
3.0- 4.0		450	450	460	475	150-200	
3.0- 4.0		440	435	435	460	150-200	
3.0- 4.0		440	435	435	460	150-200	
3.0- 4.0		460	460	470	475	150-200	
3.0- 4.0		440	435	435	460	150-200	
3.0- 4.0		460	460	470	475	150-200	
3.0- 4.0		460	460	470	475	150-200	
3.0- 4.0		440	435	435	460	150-200	
1.0- 3.0		360-410	360-400	360-390	370-415	60-110	400
1.0- 3.0		390-430	380-420	380-410	400-440	60-110	410
1.0- 3.0		390-430	380-420	380-410	400-440	60-110	425
1.0- 3.0		360-410	360-400	360-390	370-415	60-110	400
1.0- 3.0		390-430	380-420	380-410	400-440	60-110	410
1.0- 3.0		390-430	380-420	380-410	400-440	60-110	425
1.0- 3.0		420-460	415-460	410-455	425-465	60-110	445
1.0- 3.0		360-410	360-400	360-390	370-415	60-110	395
1.0- 3.0		360-410	360-400	360-390	370-415	60-110	395
1.0- 3.0		360-410	360-400	360-390	370-415	60-110	385
3.0- 4.0		455-480	450-470	440-460	460-480	60-150	460-480
3.0- 4.0		455-480	450-470	440-460	460-480	60-150	460-480
3.0- 4.0		455-480	450-470	440-460	460-480	60-150	460-480
3.0- 4.0		455-480	450-470	440-460	460-480	60-150	460-480
3.0- 4.0		455-480	450-470	440-460	460-480	60-150	460-480
3.0- 4.0		455-480	450-470	440-460	460-480	60-150	460-480
3.0- 4.0		455-480	450-470	440-460	460-480	60-150	460-480
3.0- 4.0		455-480	450-470	440-460	460-480	60-150	460-480
3.0- 4.0		455-480	450-470	440-460	460-480	60-150	460-480
3.0- 4.0		460-490	450-480	440-470	460-490	110-140	455-490
3.0- 4.0		460-490	450-480	440-470	460-490	110-140	455-490

Grade	Filler	Sp Grav	Shrink, mils/in	Melt flow, g/10 min	Melt temp, °F	Back pres, psi	Drying temp, °F
310		1.31	6.0-16.0			25- 50	250-260
310G		1.31	6.0-16.0			25- 50	250-260
310R		1.31	6.0-16.0			25- 50	250-260
317		1.31	6.0-16.0			0- 50	250
9731	GL 30	1.57	3.0- 8.0			50- 100	250-275
Polyester, TP		**Voloy**			**ComAlloy**		
462	UNS	1.61	2.0- 5.0				250-280
Polyester, TP		**Vybex**			**Ferro**		
22003 NA	GLF 30	1.54	3.0				230
Polyester, TS		**BMC**			**BMC**		
100		1.94	1.0- 4.0				
102		1.94	1.0- 4.0				
200	GLF	1.94	1.0- 4.0				
Polyester, TS		**Cosmic Polyester**			**Cosmic Plastics**		
3D36	GLF	2.00	1.0- 4.0				
Polyester, TS		**Cyglas**			**Cytec**		
2274	GLF	2.02	2.5- 4.0				
2693	GLF	2.00	2.0- 3.0				
4501	GLF	2.08	1.0- 2.5				
4660	GLF	1.92	2.0- 3.0				
4743	GLF	1.98	1.2- 1.6				
4783	UNS	1.74	8.0-10.0				
4943	GLF	1.99	3.0- 4.0				
501	GLF	1.93					
508	GLF	1.82	1.5- 2.5				
5209	GLF	1.90	1.0- 2.0				
5307	GLF	2.03	2.5- 4.0				
5307LS	GLF	2.00	1.0- 2.5				
5436	GLF	2.04	3.0- 4.0				
5547	GLF	1.86	2.5- 3.5				
5592	GLF	1.87	2.0- 3.0				
5610	GLF	1.93	1.0				
5631	GLF	2.02	1.0- 2.0				
600	GLF	2.02	2.5- 4.0				
600LS	GLF	1.95	0.5- 1.5				
604	GLF	2.06	3.0- 4.0				
605	GLF	1.95	2.5- 4.0				
605L	GLF	1.95	2.5- 4.0				
605LS	GLF	1.89	1.0- 2.2				
610	GLF	1.93	2.5- 4.5				
615	GLF	1.88	2.0- 3.5				
620	GLF	1.77	2.5- 3.5				
620M	GLF	1.86	1.5- 2.5				
620X	GLF	1.79	2.5- 3.5				
680	GLF	1.71	1.0- 2.0				
685	GLF	1.75	1.5- 2.5				
Polyether		**Texin**			**Miles**		
970-D		1.18	8.0			0- 200	180-220
985-A		1.12	8.0			0- 200	180-220
990-A		1.13	8.0			0- 200	180-220

Drying time, hr	Inj time, sec	Front temp, °F	Mid temp, °F	Rear temp, °F	Nozzle temp, °F	Mold temp, °F	Proc temp, °F
3.0- 4.0		460-490	450-480	440-470	460-490	110-140	455-490
3.0- 4.0		460-490	450-480	440-470	460-490	110-140	455-490
3.0- 4.0		460-490	450-480	440-470	460-490	110-140	455-490
3.0- 4.0		460-490	450-480	440-470	460-490	110-140	455-490
4.0- 6.0		560-580	550-570	540-560	560-580	200-250	550-580
3.0- 4.0		500-560	500-560	490-530	510-560	150-250	500-560
						60-180	440-500
	3.0	90-110	90-110	90-110		315-350	
	3.0	90-110	90-110	90-110		315-350	
	3.0	90-110	90-110	90-110		315-350	
							275-330
							280-330
							280-330
							330-350
							280-330
							280-330
							280-330
							280-330
							280-330
							280-330
							280-330
							300-350
							300-350
							300-350
							280-330
							280-330
							300-330
							320-340
							280-330
							280-330
							280-330
							280-330
							280-330
							280-330
							280-330
							280-330
							280-330
							280-330
							280-330
							280-330
							280-330
1.0- 3.0		420-460	415-460	410-455	425-465	60-110	445
1.0- 3.0		360-410	360-400	360-390	370-415	60-110	395
1.0- 3.0		360-410	360-400	360-390	370-415	60-110	395

Grade	Filler	Sp Grav	Shrink, mils/in	Melt flow, g/10 min	Melt temp, °F	Back pres, psi	Drying temp, °F
Polyimide (TP)		**Aurum**			**Mitsui Toatsu**		
400		1.33					180-200
450		1.33	8.3				180-200
500		1.33					180-200
JAF3040	KEV	1.45					180-200
JCN3030	CF	1.43	2.1				180-200
JCN6030	CF	1.45	2.1				180-200
JCN6530 (Am.)	CF	1.44	2.1				180-200
JCN6530 (Cry.)	CF	1.47					180-200
JGN3030	GLF	1.56	4.4				180-200
JNF3010	PTF	1.38					180-200
JNF3020	PTF	1.44					180-200
JQF3025	PTF	1.48					180-200
JRF3025	GRF	1.40					180-200
JRN3015	GRF	1.34					180-200
Polyolefin		**Ferrene**			**Ferro**		
CPE40JG	MN 40	1.29	15.0	0.25		20- 50	200
LPE30HO	MN 30	1.18	15.0	5.00		20- 50	200
RPE30HW	GLF 30	1.19	4.0	2.00		20- 50	200
RPE40HW	GLF 40	1.29	3.0	2.00		20- 50	200
Polyolefin		**Ferrocon**			**Ferro**		
EPP99GA01		1.00	15.0	1.00		20- 50	200
EPP99GA02		1.00	15.0	0.40		20- 50	200
Polyolefin		**Hostalloy**			**Hoechst**		
731		0.95	20.0-30.0			0- 150	
Polyolefin		**Lupol**			**Lucky**		
GP-2100	GLF 10	0.97	4.0- 6.0			0- 568	158-194
GP-2200	GLF 20	1.03	4.0- 6.0			0- 568	158-194
GP-2201F	GLF 20	1.34	4.0- 6.0			0- 568	158-194
GP-2300	GLF 30	1.15	3.0- 5.0			0- 568	158-194
GP-3100	MN 10	0.96	10.0-16.0			0- 568	158-194
GP-3100H	MN 10	0.96	10.0-16.0			0- 568	158-194
GP-3102	MN 10	0.96	10.0-16.0			0- 568	158-194
GP-3151F	MN 15	1.32	8.0-14.0			0- 568	158-194
GP-3200	MN 20	1.06	8.0-14.0			0- 568	158-194
GP-3200H	MN 20	1.06	8.0-14.0			0- 568	158-194
GP-3201F	MN 20	1.25	8.0-14.0			0- 568	158-194
GP-3202	MN 20	1.06	8.0-14.0			0- 568	158-194
GP-3300	MN 30	1.13	7.0-12.0			0- 568	158-194
GP-3300H	MN 30	1.13	7.0-12.0			0- 568	158-194
GP-3302	MN 30	1.13	7.0-12.0			0- 568	158-194
GP-3400	MN 40	1.22	6.0-10.0			0- 568	158-194
HF-3208	MN 20	1.04	8.0-16.0			0- 568	158-194
HF-3308	MN 30	1.03	8.0-16.0			0- 568	158-194
HG-3100	MN 10	0.96	10.0-16.0			0- 568	158-194
HG-3200	MN 20	1.06	8.0-14.0			0- 568	158-194
HI-2202	GLF 20	1.02	5.0- 8.0			0- 568	158-194
Polyolefin		**Safe-FR**			**UVtec**		
Safe-FR		1.40	7.0-10.0	30.00 F			

Drying time, hr	Inj time, sec	Front temp, °F	Mid temp, °F	Rear temp, °F	Nozzle temp, °F	Mold temp, °F	Proc temp, °F
5.0-10.0						338-392	734-788
5.0-10.0						338-392	734-788
5.0-10.0						338-392	734-788
5.0-10.0						338-392	734-788
5.0-10.0						338-392	734-788
5.0-10.0						338-392	734-788
5.0-10.0						338-392	734-788
5.0-10.0						338-392	734-788
5.0-10.0						338-392	734-788
5.0-10.0						338-392	734-788
5.0-10.0						338-392	734-788
5.0-10.0						338-392	734-788
5.0-10.0						338-392	734-788
5.0-10.0						338-392	734-788
2.0		420-480	400-480	390-430	420-480	80-150	
2.0		420-480	400-480	390-430	420-480	80-150	
2.0		420-480	400-480	390-430	420-480	80-150	
2.0		420-480	400-480	390-430	420-480	80-150	
2.0		430-500	420-500	400-450	420-500	80-150	
2.0		430-500	420-500	400-450	420-500	80-150	
	2.0- 4.	380-500	380-500	380-400	380-500	70-120	380-450
2.0- 3.0		410-446	392-446	392-428	410-464	104-194	410-464
2.0- 3.0		410-446	392-446	392-428	410-464	104-194	410-464
2.0- 3.0		410-446	392-446	392-428	410-464	104-194	410-464
2.0- 3.0		410-446	392-446	392-428	410-464	104-194	410-464
2.0- 3.0		410-446	392-446	392-428	410-464	104-194	410-464
2.0- 3.0		410-446	392-446	392-428	410-464	104-194	410-464
2.0- 3.0		410-446	392-446	392-428	410-464	104-194	410-464
2.0- 3.0		410-446	392-446	392-428	410-464	104-194	410-464
2.0- 3.0		410-446	392-446	392-428	410-464	104-194	410-464
2.0- 3.0		410-446	392-446	392-428	410-464	104-194	410-464
2.0- 3.0		410-446	392-446	392-428	410-464	104-194	410-464
2.0- 3.0		410-446	392-446	392-428	410-464	104-194	410-464
2.0- 3.0		410-446	392-446	392-428	410-464	104-194	410-464
2.0- 3.0		410-446	392-446	392-428	410-464	104-194	410-464
2.0- 3.0		410-446	392-446	392-428	410-464	104-194	410-464
2.0- 3.0		410-446	392-446	392-428	410-464	104-194	410-464
2.0- 3.0		410-446	392-446	392-428	410-464	104-194	410-464
2.0- 3.0		410-446	392-446	392-428	410-464	104-194	410-464
2.0- 3.0		410-446	392-446	392-428	410-464	104-194	410-464
2.0- 3.0		410-446	392-446	392-428	410-464	104-194	410-464
2.0- 3.0		410-446	392-446	392-428	410-464	104-194	410-464
		450-550	430-490	400-450	460-530	100-150	450-520

Grade	Filler	Sp Grav	Shrink, mils/in	Melt flow, g/10 min	Melt temp, °F	Back pres, psi	Drying temp, °F
Polyolefin		**Zeonex**			**Nippon Zeon**		
280		1.01	4.0- 6.0				
280S		1.01	4.0- 6.0	15.00 W			
Polyolefin Ela		**Engage**			**Dow**		
CL8001		0.87		0.50			
CL8002		0.87		1.00			
EG8100		0.87		1.00			
EG8150		0.87		0.50			
EG8200		0.87		5.00			
POM		**Ultraform**			**BASF**		
E3320X		1.41	19.0-20.0	1.00 E	327-335	50- 100	230
H2320 006		1.41	19.0-20.0	3.50 E	327-335	50- 100	230
N2200 G2	GLF 10	1.47		5.50 E	327-335	50- 100	230
N2200 G3	GLF 15	1.51		5.50 E	327-335	50- 100	230
N2200 G4	GLF 20	1.54	8.0-15.0	5.50 E	327-335	50- 100	230
N2200 G5	GLF 25	1.58	7.0-14.0	5.50 E	327-335	50- 100	230
N2310P		1.41	19.0-20.0	9.00 E	327-335	50- 100	230
N2320		1.41	19.0-20.0	9.00	327-335	50- 100	230
N2320 003		1.41	19.0-20.0	9.00 E	327-335	50- 100	230
N2320Y	MOS	1.42	19.0-20.0	9.00 E	327-335	50- 100	230
N2325U		1.41	19.0-20.0	9.00 E	327-335	50- 100	230
N2350		1.41	19.0-20.0	9.00 E	327-335	50- 100	230
N2520XL	CBL	1.39	19.0-20.0	9.00 E	327-335	50- 100	230
N2640 Z2	PU 10	1.38	18.0-19.0	8.00 E	327-335	50- 100	230
N2640 Z4	PU 20	1.36	18.0-19.0	6.50 E	327-335	50- 100	230
N2640 Z5	PU 25	1.34	17.0-18.0	6.00 E	327-335	50- 100	230
N2640 Z6	PU 30	1.33	17.0-18.0	5.00 E	327-335	50- 100	230
N2720XM210	CSO 10	1.49	18.0-20.0	6.00 E	327-335	50- 100	230
N2720XM43	CSO 20	1.56	18.0-19.0	6.00 E	327-335	50- 100	230
N2720XM63	CSO 30	1.64	17.0-18.0	5.50 E	327-335	50- 100	230
N2770K	CH	1.42	19.0-20.0	9.00 E	327-335	50- 100	230
N2771		1.42	2.0	9.00 E	327-335	50- 100	230
S2320		1.41	19.0-20.0	13.00 E	327-335	50- 100	230
S2320 003		1.41	19.0-20.0	13.00 E	327-335	50- 100	230
S2320 009		1.41	19.0-20.0	13.00 E	327-335	50- 100	230
W2320 003		1.41	19.0	30.00 E	327-335	50- 100	230
W2330 003		1.41	19.0	30.00 E	327-335	50- 100	230
W2640XZ2	PU 10	1.38	19.0	26.00 E	327-335	50- 100	230
Z2320 003		1.41	19.0	50.00 E	327-335	50- 100	230
Z2330 003		1.41	19.0	50.00 E	327-335	50- 100	230
PP		**Acctuf**			**Amoco Chemical**		
3143			10.0-25.0	2.50 L			
3234			10.0-25.0	6.00 L			
3243			10.0-25.0	5.00 L			
3541			10.0-25.0	20.00 L			
3543			10.0-25.0	20.00 L			
3638			10.0-25.0	24.00 L			
PP		**Albis Polypropylene**				**Albis**	
A12H-GF20	GLF 20	1.03		5.50			
A12H-GF20C	GLF 20	1.03		5.00			
A12H-GF30	GLF 30	1.12		6.00			

Drying time, hr	Inj time, sec	Front temp, °F	Mid temp, °F	Rear temp, °F	Nozzle temp, °F	Mold temp, °F	Proc temp, °F
		536-554	518-518	464-482	518-536	257-266	
		536-554	518-518	464-482	518-536	257-266	
		350-550	350-550	350-550		50	
		350-550	350-550	350-550		50	
		350-550	350-550	350-550		50	
		350-550	350-550	350-550		50	
		350-550	350-550	350-550		50	
2.0						140-210	375-430
2.0						140-210	375-430
2.0						180-230	375-430
2.0						180-230	375-430
2.0						180-230	375-430
2.0						180-230	375-430
2.0						140-210	375-445
2.0						140-210	375-445
2.0						140-210	375-445
2.0						140-210	375-445
2.0						140-210	375-430
2.0						140-210	375-430
2.0						140-180	375-430
2.0						140-180	375-430
2.0						140-180	375-430
2.0						140-180	375-420
2.0						140-180	375-420
2.0						140-210	375-430
2.0						180-230	375-430
2.0						180-230	375-430
2.0						140-210	375-445
2.0						140-210	375-445
2.0						140-210	375-445
2.0						140-210	375-445
2.0						140-210	375-445
2.0						140-180	375-445
2.0						140-180	335-445
2.0						140-210	375-430
2.0						140-180	375-445
2.0						140-180	375-445
		400-525	400-525	400-525	430-575	60-150	
		400-525	400-525	400-525	430-575	60-150	
		400-525	400-525	400-525	430-575	60-150	
		400-525	400-525	400-525	430-575	60-150	
		400-525	400-525	400-525	430-575	60-150	
		400-525	400-525	400-525	430-575	60-150	
						70-140	390-540
						70-140	390-540
						70-140	390-540

Grade	Filler	Sp Grav	Shrink, mils/in	Melt flow, g/10 min	Melt temp, °F	Back pres, psi	Drying temp, °F
A12H-GK20/GF20C	GLF 20	1.20		6.00			
A12H-VO		1.39		1.00			
PP		**Amoco Polypropylene**				**Amoco Chem.**	
7232		0.91	10.0-25.0	12.00 L		50- 100	
7236		0.91	10.0-25.0	12.00		50- 100	
7734		0.91	10.0-25.0	28.00 L		50- 100	
7744		0.90	15.0-20.0	28.00 L		50- 100	
7938		0.91	10.0-25.0	35.00 L		50- 100	
PP		**Aqualoy**			**ComAlloy**		
125	GLF 20	1.04	4.0	9.00			180-200
135	UNS	1.17	5.0	6.00			180-200
145	UNS	1.25	3.0	9.00			180-200
PP		**Celstran**			**Hoechst**		
PPG30-01-4	GLF 30	1.12					200
PPG30-02-4	GLF 30	1.12					270
PPG40-01-4	GLF 40	1.21					200
PPG40-02-4	GLF 40	1.21					270
PPG50-01-4	GLF 50	1.33					200
PPG50-02-4	GLF 50	1.33					270
PP		**Comshield**			**ComAlloy**		
101		0.98	15.0	4.00			180-200
105		1.08	10.0	3.50			180-200
106	CBL	1.22	8.0	3.50			180-200
110	UNS	1.22	4.0	3.50			180-200
160	CBL	1.20	4.0	3.50			180-200
PP		**Comtuf**			**ComAlloy**		
101	MN	1.07	15.0	6.00			180-200
102	MN	1.06	14.0	6.00			180-200
103	GL 15	0.98	7.0	5.00			180-200
104	GL 25	1.07	6.0	4.00			180-200
105	GL	1.19	5.0	3.00			180-200
PP		**CTI Polypropylene**			**CTI**		
PP-10GF/000	GLF 10	0.96	6.0				
PP-20CC/000	CAC 20	1.04	16.0				
PP-20GF/000	GLF 20	1.04	4.0				
PP-20GF/000	GLF 20	1.04	4.0				
PP-30GF/000	GLF 30	1.12	3.0				
PP-40TC/000	TAL 40	1.23	10.0				
PPX-10GF/000 HC	GLF 10	0.96	6.0				
PPX-15GF/000 HC	GLF 15	1.00	5.0				
PPX-20GF/000 HC	GLF 20	1.04	4.0				
PPX-30GF/000 HC	GLF 30	1.12	3.0				
PPX-35GF/000 HC	GLF 35	1.16	1.0- 2.0				
PPX-40GF/000 HC	GLF 40	1.21	2.0				
PP		**DSM Polypropylene**			**DSM**		
F-60/20	GLF 20	0.75	3.0			0- 50	165
J-60/10/E3	GLF 10	0.95	6.0				
J-60/10/E4	GLF 10	0.95	6.0				
J-60/20	GLF 20	1.04	4.0				

Drying time, hr	Inj time, sec	Front temp, °F	Mid temp, °F	Rear temp, °F	Nozzle temp, °F	Mold temp, °F	Proc temp, °F
						70-140	390-540
						70-140	390-430
		400-570	400-570	400-570			
		400-570	400-570	400-570			
		400-570	400-570	400-570			
		400-570	400-570	400-570			
		400-570	400-570	400-570			
1.0- 2.0		440	450	460	440-460	50-150	440-460
1.0- 2.0		460	470	480	460-480	50-150	460-480
1.0- 2.0		460	470	480	460-480	50-150	460-480
2.0		370-390	360-380	350-370	370-390	140-160	370-390
4.0		570-590	560-580	550-570	550-570	290-310	570-590
2.0		380-400	370-390	360-380	380-400	140-160	380-400
4.0		570-590	560-580	550-570	550-570	290-310	570-590
2.0		390-410	380-400	370-390	390-410	140-160	390-410
4.0		570-590	560-580	550-570	550-570	290-310	570-590
1.0- 2.0		420	430	440	420-440	50-150	430-450
1.0- 2.0		440	450	460	440-460	50-150	440-460
1.0- 2.0		440	450	460	440-460	50-150	440-460
1.0- 2.0		460	470	480	460-480	50-150	460-480
1.0- 2.0		460	470	480	460-480	50-150	460-480
1.0- 2.0		420	430	440	420-440	50-150	430-450
1.0- 2.0		420	430	440	420-440	50-150	430-450
1.0- 2.0		420	430	440	420-440	50-150	430-450
1.0- 2.0		440	450	460	440-460	50-150	440-460
1.0- 2.0		460	470	480	460-480	50-150	460-480
		475-525	445-475	400-445	420-440	70-120	
		475-525	445-475	400-445	420-440	70-120	
		475-525	445-475	400-445	420-440	70-120	
		475-525	445-475	400-445	420-440	70-120	
		475-525	445-475	400-445	420-440	70-120	
		475-525	445-475	400-445	420-440	70-120	
		475-525	445-475	400-445	420-440	70-120	
		475-525	445-475	400-445	420-440	70-120	
		475-525	445-475	400-445	420-440	70-120	
		475-525	445-475	400-445	420-440	70-120	
		475-525	445-475	400-445	420-440	70-120	
		475-525	445-475	400-445	420-440	70-120	
2.0						20-130	410-500
		360-390	400-440	390-410	360-380	100-160	390-550
		360-390	400-440	390-410	360-380	100-160	390-550
		360-390	400-440	390-410	360-380	100-160	390-550

Grade	Filler	Sp Grav	Shrink, mils/in	Melt flow, g/10 min	Melt temp, °F	Back pres, psi	Drying temp, °F
J-60/20/E	GLF 20	1.04	3.0				
J-60/20/E3	GLF 20	1.30	4.0				
J-60/20/E4	GLF 20	1.04	2.0				
J-60/20/E8	GLF 20	1.04	3.0				
J-60/30	GLF 30	1.13	4.0				
J-60/30/E	GLF 30	1.13	3.0				
J-60/30/E3	GLF 30	1.13	4.0				
J-60/30/E4	GLF 30	1.13	2.0				
J-60/30/E8	GLF 30	1.13	2.0				
J-60/40	GLF 40	1.22	3.0				
J-60/40/E	GLF 40	1.22	2.0				
J-60/40/E3	GLF 40	1.22	3.0				
J-60/40/E4	GLF 40	1.22	1.0				
J-60/40/E8	GLF 40	1.22	2.0				
J-61/10	GLF 10	0.98	4.0				
J-61/20	GLF 20	1.04	2.5				
J-61/30	GLF 30	1.11	1.4				
J-61/40	GLF 40	1.21	0.8				
J-62/10/E4	GLF 10	0.96	5.0				
J-62/20/E4	GLF 20	1.04	3.0				
J-62/30/E4	GLF 30	1.13	2.0				
J-62/40/E4	GLF 40	1.20	1.0				
PP-60/CC/20	CAC 20	1.06	12.0				210
PP-60/CC/40	CAC 40	1.25	10.0				210
PP-60/TC/20	TAL 20	1.07	10.0				210
PP-60/TC/40	TAL 40	1.25	7.0				210
PP-61/CC/20	CAC 20	1.05	12.0				210
PP-61/CC/40	CAC 40	1.24	9.0				210
PP-61/TC/20	TAL 20	1.05	10.0				210
PP-61/TC/40	TAL 40	1.24	7.0				210
PP-62/CC/20	CAC 20	1.06	13.0				210
PP-62/CC/40	CAC 40	1.25	12.0				210
PP-62/MI/40	MI 40	1.23	7.0				210

PP — Ektar FB — Eastman

Grade	Filler	Sp Grav	Shrink, mils/in	Melt flow, g/10 min	Melt temp, °F	Back pres, psi	Drying temp, °F
PG001	GLF 10	0.97	3.0- 4.0				
PG002	GLF 20	1.03	2.0- 3.0				
PG003	GLF 30	1.12	2.0- 3.0				

PP — Electrafil — DSM

Grade	Filler	Sp Grav	Shrink, mils/in	Melt flow, g/10 min	Melt temp, °F	Back pres, psi	Drying temp, °F
J-60/CF/30	CF 30	1.06	1.0				
JM-61/CF/10	CF 10	0.93	3.0				
PP-60/SD		0.95	17.0				170

PP — Escalloy — ComAlloy

Grade	Filler	Sp Grav	Shrink, mils/in	Melt flow, g/10 min	Melt temp, °F	Back pres, psi	Drying temp, °F
104	UNS 25	1.07	6.0	4.00			180-200
124	GL 40	1.22	4.0	6.00			180-200
129	UNS 40	1.22	5.0	6.00			180-200

PP — Ferrex — Ferro

Grade	Filler	Sp Grav	Shrink, mils/in	Melt flow, g/10 min	Melt temp, °F	Back pres, psi	Drying temp, °F
GPP10CC	MN 10	0.97	13.0	8.00		20- 50	200
GPP10CK	MN 10	0.97	13.0	8.00		20- 50	200
GPP10CS	MN 10	0.97	13.0	8.00		20- 50	200
GPP20CC	MN 20	1.05	12.0	8.00		20- 50	200
GPP20CF	MN 20	1.05	12.0	20.00		20- 50	200
GPP20CF HB	MN 22	1.05	10.0-14.0	14.50 L		20- 50	200

Drying time, hr	Inj time, sec	Front temp, °F	Mid temp, °F	Rear temp, °F	Nozzle temp, °F	Mold temp, °F	Proc temp, °F
		360-390	400-440	390-410	360-380	100-160	390-550
		360-390	400-440	390-410	360-380	100-160	390-550
		360-390	400-440	390-410	360-380	100-160	390-550
		360-390	400-440	390-410	360-380	100-160	390-550
		360-390	400-440	390-410	360-380	100-160	390-550
		360-390	400-440	390-410	360-380	100-160	390-550
		360-390	400-440	390-410	360-380	100-160	390-550
		360-390	400-440	390-410	360-380	100-160	390-550
		360-390	400-440	390-410	360-380	100-160	390-550
		360-390	400-440	390-410	360-380	100-160	390-550
		360-390	400-440	390-410	360-380	100-160	390-550
		360-390	400-440	390-410	360-380	100-160	390-550
		360-390	400-440	390-410	360-380	100-160	390-550
		360-390	400-440	390-410	360-380	100-160	390-550
		360-390	400-440	390-410	360-380	100-160	390-550
		360-390	400-440	390-410	360-380	100-160	390-550
		360-390	400-440	390-410	360-380	100-160	390-550
		360-390	400-440	390-410	360-380	100-160	390-550
		360-390	400-440	390-410	360-380	100-160	390-550
		360-390	400-440	390-410	360-380	100-160	390-550
		360-390	400-440	390-410	360-380	100-160	390-550
		360-390	400-440	390-410	360-380	100-160	390-550
1.0		390-440	400-470	350-450	380-430	90-120	400-450
1.0		390-440	400-470	350-450	380-430	90-120	400-450
1.0		390-440	400-470	350-450	380-430	90-120	400-450
1.0		390-440	400-470	350-450	380-430	90-120	400-450
1.0		390-440	400-470	350-450	380-430	90-120	400-450
1.0		390-440	400-470	350-450	380-430	90-120	400-450
1.0		390-440	400-470	350-450	380-430	90-120	400-450
1.0		390-440	400-470	350-450	380-430	90-120	400-450
1.0		390-440	400-470	350-450	380-430	90-120	400-450
1.0		390-440	400-470	350-450	380-430	90-120	400-450
1.0		390-440	400-470	350-450	380-430	90-120	400-450
						80-120	425-475
						80-120	425-475
						80-120	425-475
		360-390	400-440	390-410	360-380	100-160	390-550
		360-390	400-440	390-410	360-380	100-160	390-550
2.0	0.5- 1.	360	370	350		80-100	360
1.0- 2.0		440	450	460	440-460	50-150	440-460
1.0- 2.0		460	470	480	460-480	50-150	460-480
1.0- 2.0		460	470	480	460-480	50-150	460-480
2.0- 3.0		410-420	400-410	390-400	420-430	115-140	
2.0- 3.0		410-420	400-410	390-400	420-430	115-140	
2.0- 3.0		410-420	400-410	390-400	420-430	115-140	
2.0- 3.0		410-420	400-410	390-400	420-430	115-140	
2.0- 3.0		410-420	400-410	390-400	420-430	115-140	
2.0- 3.0		410-420	400-410	390-400	420-430	115-140	

Grade	Filler	Sp Grav	Shrink, mils/in	Melt flow, g/10 min	Melt temp, °F	Back pres, psi	Drying temp, °F
GPP20CF NA	MN 23	1.06	10.0-14.0	21.00 L		20- 50	200
GPP20CF UL	MN 23	1.09	13.0-16.0	21.00 L		20- 50	200
GPP20CK	MN 20	1.05	12.0	8.00		20- 50	200
GPP20CK HB-BK	MN 20	1.06	12.0-15.0	8.00 L		20- 50	200
GPP20CN	MN 20	1.06	12.0-16.0	22.00 L		20- 50	200
GPP20CS	MN 21	1.06	13.0-16.0	10.00 L		20- 50	200
GPP30CC	MN 30	1.14	11.0	8.00		20- 50	200
GPP30CK	MN 30	1.14	11.0	8.00		20- 50	200
GPP35CF UL	MN 36	1.24	11.0-14.0	22.00 L		20- 50	200
GPP35CN	MN 36	1.20	12.0-14.0	21.00 L		20- 50	200
GPP35CN UL	MN 35	1.20	11.0-16.0	21.00 L		20- 50	200
GPP35CS UL	MN 36	1.21	10.0-13.0	8.50 L		20- 50	200
GPP40CC	MN 40	1.24	9.0	8.00		20- 50	200
GPP40CF	MN 40	1.24	9.0	20.00		20- 50	200
GPP40CK	MN 40	1.24	10.0	8.00		20- 50	200
GPP40CN	MN 40	1.24	10.0	20.00		20- 50	200
ULLPP6020 HS HG	MN 20	1.05	12.0	20.00		20- 50	200

PP — Ferro Polypropylene — Ferro

Grade	Filler	Sp Grav	Shrink, mils/in	Melt flow, g/10 min	Melt temp, °F	Back pres, psi	Drying temp, °F
CPP20GH	MN 26	1.06	11.0-14.0	20.00 L		50- 100	200
CPP25GF	MN 23	1.08	10.0-15.0	18.50 L		50- 100	200
CPP30GF	MN 30	1.15	10.0	8.00		50- 100	200
CPP30GH	MN 25	1.11	6.0-14.0	19.00 L		50- 100	200
CPP40GF	MN 44	1.30	8.0-13.0	10.00 L		50- 100	200
CPP45GF	MN 45	1.29	7.0	20.00		50- 100	200
CPP45GH	MN 46	1.33	8.0-12.0	19.00 L		50- 100	200
HPP30GR	GMN 30	1.11	1.0- 5.0	5.50 L		50- 100	200
HPP35GP	GMN 39	1.20	6.0-10.0	10.00 L		50- 100	200
HPP40GR	GMN 37	1.23	2.0- 5.0	3.00 L		50- 100	200
HPP45GR	GMN 45	1.26	4.0	1.00		50- 100	200
LPP10BC	CAC 10	0.97	13.0	8.00		50- 100	200
LPP10BK	CAC 10	0.97	13.0	8.00		50- 100	200
LPP10BS	CAC 10	0.97	13.0	4.00		50- 100	200
LPP20BC	CAC 23	1.07	13.0-15.0	12.00 L		50- 100	200
LPP20BK	CAC 20	1.05	12.0-14.0	10.00 L		50- 100	200
LPP20BN HB	CAC 21	1.08	11.0-13.0	27.00 L		50- 100	200
LPP20BS	CAC 20	1.05	12.0	4.00		50- 100	200
LPP30BC	CAC 30	1.14	11.0	8.00		50- 100	200
LPP30BK	CAC 30	1.14	11.0	8.00		50- 100	200
LPP30BS	CAC 30	1.14	11.0	8.00		50- 100	200
LPP40BA	CAC 43	1.28	12.0-14.0	3.00 L		50- 100	200
LPP40BC	CAC 42	1.26	9.0-13.0	10.00 L		50- 100	200
LPP40BK	CAC 40	1.22	10.0-13.0	10.00 L		50- 100	200
LPP40BN	CAC 40	1.22		15.00 L		50- 100	200
LPP40BS	CAC 40	1.24	10.0	8.00		50- 100	200
MPP12FU	MI 12	1.00	9.0	3.00		50- 100	200
MPP20FA	MI 20	1.05	8.0	3.00		50- 100	200
MPP20FJ	MI 20	1.05	8.0	3.00		50- 100	200
MPP20FU	MI 20	1.06	9.0-12.0	8.00 L		50- 100	200
MPP40FA	MI 40	1.25	6.0	3.00		50- 100	200
MPP40FJ	MI 40	1.24	5.0- 7.0	12.00 L		50- 100	200
MPP40FU	MI 40	1.25	6.0	3.00		50- 100	200
RPP10DA	GLF 10	0.97	6.0	2.00		20- 50	200
RPP10EA	GLF 10	0.97	5.0	3.00		20- 50	200
RPP10ER	GLF 10	0.97	6.0	3.00		20- 50	200
RPP13DA	GLF 13	1.00	5.0	2.00		20- 50	200

Drying time, hr	Inj time, sec	Front temp, °F	Mid temp, °F	Rear temp, °F	Nozzle temp, °F	Mold temp, °F	Proc temp, °F
2.0- 3.0		410-420	400-410	390-400	420-430	115-140	
2.0- 3.0		410-420	400-410	390-400	420-430	115-140	
2.0- 3.0		410-420	400-410	390-400	420-430	115-140	
2.0- 3.0		410-420	400-410	390-400	420-430	115-140	
2.0- 3.0		410-420	400-410	390-400	420-430	115-140	
2.0- 3.0		410-420	400-410	390-400	420-430	115-140	
2.0- 3.0		410-420	400-410	390-400	420-430	115-140	
2.0- 3.0		410-420	400-410	390-400	420-430	115-140	
2.0- 3.0		410-420	400-410	390-400	420-430	115-140	
2.0- 3.0		410-420	400-410	390-400	420-430	115-140	
2.0- 3.0		410-420	400-410	390-400	420-430	115-140	
2.0- 3.0		410-420	400-410	390-400	420-430	115-140	
2.0- 3.0		410-420	400-410	390-400	420-430	115-140	
2.0- 3.0		410-420	400-410	390-400	420-430	115-140	
2.0- 3.0		410-420	400-410	390-400	420-430	115-140	
2.0- 3.0		410-420	400-410	390-400	420-430	115-140	
2.0		380-420	380-420	360-400	380-420	40-100	
2.0- 3.0		410-415	400-410	395-400	415-425	110-125	
2.0- 3.0		410-415	400-410	395-400	415-425	110-125	
2.0- 3.0		410-415	400-410	395-400	415-425	110-125	
2.0- 3.0		410-415	400-410	395-400	415-425	110-125	
2.0- 3.0		410-415	400-410	395-400	415-425	110-125	
2.0- 3.0		410-415	400-410	395-400	415-425	110-125	
2.0- 3.0		410-415	400-410	395-400	415-425	110-125	
2.0- 3.0		420-425	410-420	400-415	425-440	110-135	
2.0- 3.0		420-425	410-420	400-415	425-440	110-135	
2.0- 3.0		420-425	410-420	400-415	425-440	110-135	
2.0- 3.0		420-425	410-420	400-415	425-440	110-135	
2.0- 3.0		410-415	400-410	395-400	415-425	110-125	
2.0- 3.0		410-415	400-410	395-400	415-425	110-125	
2.0- 3.0		410-415	400-410	395-400	415-425	110-125	
2.0- 3.0		410-415	400-410	395-400	415-425	110-125	
2.0- 3.0		410-415	400-410	395-400	415-425	110-125	
2.0- 3.0		410-415	400-410	395-400	415-425	110-125	
2.0- 3.0		410-415	400-410	395-400	415-425	110-125	
2.0- 3.0		410-415	400-410	395-400	415-425	110-125	
2.0- 3.0		410-415	400-410	395-400	415-425	110-125	
2.0- 3.0		410-415	400-410	395-400	415-425	110-125	
2.0- 3.0		410-415	400-410	395-400	415-425	110-125	
2.0- 3.0		410-415	400-410	395-400	415-425	110-125	
2.0- 3.0		410-415	400-410	395-400	415-425	110-125	
2.0- 3.0		410-415	400-410	395-400	415-425	110-125	
2.0- 3.0		410-415	400-410	395-400	415-425	110-125	
2.0- 3.0		420-425	410-420	400-415	425-440	110-135	
2.0- 3.0		420-425	410-420	400-415	425-440	110-135	
2.0- 3.0		420-425	410-420	400-415	425-440	110-135	
2.0- 3.0		420-425	410-420	400-415	425-440	110-135	
2.0- 3.0		420-425	410-420	400-415	425-440	110-135	
2.0- 3.0		420-425	410-420	400-415	425-440	110-135	
2.0- 3.0		420-425	410-420	400-415	425-440	110-135	
2.0- 3.0		440-465	430-460	420-450	440-470	110-135	
2.0- 3.0		440-465	430-460	420-450	440-470	110-135	
2.0- 3.0		440-465	430-460	420-450	440-470	110-135	
2.0- 3.0		440-465	430-460	420-450	440-470	110-135	

Grade	Filler	Sp Grav	Shrink, mils/in	Melt flow, g/10 min	Melt temp, °F	Back pres, psi	Drying temp, °F
RPP20DA	GLF 20	1.04	2.0- 4.0	5.00 L		20- 50	200
RPP20EA	GLF 20	1.04	3.0	3.00		20- 50	200
RPP20EA HB	GLF 20		3.0- 6.0	5.25 L		20- 50	200
RPP20EA06HB-BK	GLF 20	1.04	2.0- 4.0	5.00 L		20- 50	200
RPP20ER	GLF 20	1.04	4.0	3.00		20- 50	200
RPP25DA	GLF 25	1.08	4.0	2.00		20- 50	200
RPP30DA	GLF 30	1.13	4.0	2.00		20- 50	200
RPP30EA	GLF 30	1.15	3.0	2.00		20- 50	200
RPP40DA	GLF 40	1.22	3.0	2.00		20- 50	200
RPP40EA	GLF 40	1.21	2.0- 3.0	7.50 L		20- 50	200
TPP10AC	TAL 10	0.98	12.0	4.00		20- 50	200
TPP10AN	TAL 10	0.98	12.0	4.00		20- 50	200
TPP10AR	TAL 10	0.98	12.0	4.00		20- 50	200
TPP20AC	TAL 17	1.02		4.00 L		20- 50	200
TPP20AE UL	TAL 21	1.06	10.0-14.0	15.00 L		20- 50	200
TPP20AJ	TAL 20	1.02	9.0-11.0	23.00 L		20- 50	200
TPP20AN	TAL 23	1.07	8.0-11.0	7.00 L		20- 50	200
TPP20AR	TAL 21	1.05	12.0-15.0	0.70 L		20- 50	200
TPP20AT	TAL 20	1.06	11.0-15.0	5.00 L		20- 50	200
TPP30AC	TAL 30	1.15	8.0	4.00		20- 50	200
TPP40AC	TAL 35	1.15		4.50 L		20- 50	200
TPP40AD UL	TAL 42	1.26	7.0-13.0	5.00 L		20- 50	200
TPP40AN HB	TAL 42	1.24	7.0-11.0	8.00 L		20- 50	200

PP Ferropak Ferro

Grade	Filler	Sp Grav	Shrink, mils/in	Melt flow, g/10 min	Melt temp, °F	Back pres, psi	Drying temp, °F
GPP20YE	MN 21	1.07	13.0-16.0	10.00 L		20- 50	200
TPP20WL	TAL 20	1.07	11.0-18.0	3.00 L		20- 50	200
TPP40WA	TAL 38	1.24	11.0-13.0	0.80 L		20- 50	200

PP Fiberfil DSM

Grade	Filler	Sp Grav	Shrink, mils/in	Melt flow, g/10 min	Melt temp, °F	Back pres, psi	Drying temp, °F
J-60/20	GLF 20	1.04	4.0- 5.0				
J-60/20/E8	GLF 20	1.04	3.0				
J-60/30	GLF 30	1.13	4.0- 5.0				
J-60/30/E8	GLF 30	1.13	2.0				
J-60/40	GLF 40	1.22	3.0- 4.0				
J-60/40/E8	GLF 40	1.22	2.0				
J-61/10	GLF 10	0.98	4.0- 5.3				
J-61/10/E8	GLF 10	0.97	3.0				
J-61/20	GLF 20	1.04	2.5- 3.3				
J-61/20/E8	GLF 20	1.04	2.0				
J-61/30	GLF 30	1.11	1.4- 2.6				
J-61/30/E8	GLF 30	1.12	1.0				
J-61/40	GLF 40	1.21	0.8- 2.4				
J-61/40/E8	GLF 40	1.20	1.0				
J-62/10/E8	GLF 10	0.97	7.0				
J-62/20/E8	GLF 20	1.04	3.0				
J-62/30/E8	GLF 30	1.12	2.0				
J-62/40/E8	GLF 40	1.20	1.0				
PP-60/CC/20	CAC 20	1.06	12.0				210
PP-60/CC/40	CAC 40	1.24	10.0				210
PP-60/TC/20	TAL 20	1.06	10.0-12.0				210
PP-60/TC/40	TAL 40	1.24	7.0- 9.0				210
PP-61/CC/20	CAC 20	1.05	12.0				210
PP-61/CC/40	CAC 40	1.24	9.0				210
PP-61/TC/20	TAL 20	1.05	10.0				210
PP-61/TC/40	TAL 40	1.24	7.0				210

Drying time, hr	Inj time, sec	Front temp, °F	Mid temp, °F	Rear temp, °F	Nozzle temp, °F	Mold temp, °F	Proc temp, °F
2.0- 3.0		440-465	430-460	420-450	440-470	110-135	
2.0- 3.0		440-465	430-460	420-450	440-470	110-135	
2.0- 3.0		440-465	430-460	420-450	440-470	110-135	
2.0- 3.0		440-465	430-460	420-450	440-470	110-135	
2.0- 3.0		440-465	430-460	420-450	440-470	110-135	
2.0- 3.0		440-465	430-460	420-450	440-470	110-135	
2.0- 3.0		440-465	430-460	420-450	440-470	110-135	
2.0- 3.0		440-465	430-460	420-450	440-470	110-135	
2.0- 3.0		440-465	430-460	420-450	440-470	110-135	
2.0- 3.0		440-465	430-460	420-450	440-470	110-135	
2.0- 3.0		415-420	410-415	400-410	420-425	110-130	
2.0- 3.0		415-420	410-415	400-410	420-425	110-130	
2.0- 3.0		415-420	410-415	400-410	420-425	110-130	
2.0- 3.0		415-420	410-415	400-410	420-425	110-130	
2.0- 3.0		415-420	410-415	400-410	420-425	110-130	
2.0- 3.0		415-420	410-415	400-410	420-425	110-130	
2.0- 3.0		415-420	410-415	400-410	420-425	110-130	
2.0- 3.0		415-420	410-415	400-410	420-425	110-130	
2.0- 3.0		415-420	410-415	400-410	420-425	110-130	
2.0- 3.0		415-420	410-415	400-410	420-425	110-130	
2.0- 3.0		415-420	410-415	400-410	420-425	110-130	
2.0- 3.0		415-420	410-415	400-410	420-425	110-130	
2.0- 3.0		415-420	410-415	400-410	420-425	110-130	
2.0- 3.0		410-420	400-410	390-400	420-430	115-140	
2.0- 3.0		415-420	410-415	400-410	420-425	110-130	
2.0- 3.0		415-420	410-415	400-410	420-425	110-130	
		360-390	400-440	390-410	360-380	100-160	390-550
		360-390	400-440	390-410	360-380	100-160	390-550
		360-390	400-440	390-410	360-380	100-160	390-550
		360-390	400-440	390-410	360-380	100-160	390-550
		360-390	400-440	390-410	360-380	100-160	390-550
		360-390	400-440	390-410	360-380	100-160	390-550
		360-390	400-440	390-410	360-380	100-160	390-550
		360-390	400-440	390-410	360-380	100-160	390-550
		360-390	400-440	390-410	360-380	100-160	390-550
		360-390	400-440	390-410	360-380	100-160	390-550
		360-390	400-440	390-410	360-380	100-160	390-550
		360-390	400-440	390-410	360-380	100-160	390-550
		360-390	400-440	390-410	360-380	100-160	390-550
		360-390	400-440	390-410	360-380	100-160	390-550
		360-390	400-440	390-410	360-380	100-160	390-550
		360-390	400-440	390-410	360-380	100-160	390-550
		360-390	400-440	390-410	360-380	100-160	390-550
		360-390	400-440	390-410	360-380	100-160	390-550
2.0		390-440	400-470	350-450	380-430	90-120	400-450
2.0		390-440	400-470	350-450	380-430	90-120	400-450
2.0		390-440	400-470	350-450	380-430	90-120	400-450
2.0		390-440	400-470	350-450	380-430	90-120	400-450
2.0		390-440	400-470	350-450	380-430	90-120	400-450
2.0		390-440	400-470	350-450	380-430	90-120	400-450
2.0		390-440	400-470	350-450	380-430	90-120	400-450
2.0		390-440	400-470	350-450	380-430	90-120	400-450

Grade	Filler	Sp Grav	Shrink, mils/in	Melt flow, g/10 min	Melt temp, °F	Back pres, psi	Drying temp, °F
PP-62/CC/20	CAC 20	1.06	13.0				210
PP-62/CC/40	CAC 40	1.25	12.0-14.0				210
PP-62/MI/40	MN 40	1.23	7.0- 8.0				210
PP		**Fiberfil V0**			**DSM**		
J-60/10/FR	GLF 30	1.32	5.0			25- 50	160
J-60/20/FR	GLF 30	1.40	2.0			25- 50	160
J-60/30/FR	GLF 30	1.47	1.0			25- 50	160
M-1774/BK-4992	GLF	1.49	1.0				165
PP-60/FR		1.42	13.0			0- 50	160
PP-60/V2		0.94	15.0			50- 100	210
PP-60/V0/NH		1.03	15.0			50	150
PP		**Fina Polypropylene**				**Fina**	
3260WZ		0.90		12.00 L	330		
3260Z		0.90		12.00 L	330		
3422		0.90		5.00 L	330		
3429		0.90		4.50 L	330		
3621WZ		0.90		12.00 L	330		
3622		0.90		12.00 L	330		
3622MZ		0.90		12.00 L	330		
3720Z		0.90		20.00 L	330		
3721WZ		0.90		20.00 L	330		
3724Z		0.90		20.00 L	330		
3761		0.90		18.00 L	330		
3763		0.90		25.00 L	330		
3824		0.90		30.00 L	330		
3824TZ		0.90		30.00 L	330		
3824WZ		0.90		30.00 L	330		
3824Z		0.90		30.00 L	330		
3825		0.90		30.00 L	330		
3825WZ		0.90		30.00 L	330		
3825Z		0.90		30.00 L	330		
3826		0.90		36.00 L	330		
3829		0.90		30.00 L	330		
3840		0.90		30.00 L	330		
7425		0.90		4.00 L	293		
7522MZ		0.90		10.00 L	293		
7622MZ		0.90		11.00 L	293		
7640		0.90		20.00 L	293		
7822MZ		0.90		30.00 L	293		
7825		0.90		30.00 L	293		
7840		0.90		30.00 L	293		
PP		**HiFax**			**Himont**		
AB-6103		1.80	9.0-11.0	13.00			
AB-7023		1.20	11.0-13.0	20.00			
CA45		0.90	13.0-15.0	8.00			
CA53		0.89	12.0-14.0	8.70			
CA53AC		0.89	12.0-14.0	8.70			
ETA-3081		0.92	10.0-12.0	4.00			
ETA-3081HS		0.90	14.0-16.0	6.40			
ETA-3163		0.96	9.5-11.5	5.00			
ETA-3183		0.98	9.0-11.0	4.50			
RTA-3263		1.10	8.0-10.0	7.50			
RTA-3263E		1.11	7.5- 9.5	0.40			

Drying time, hr	Inj time, sec	Front temp, °F	Mid temp, °F	Rear temp, °F	Nozzle temp, °F	Mold temp, °F	Proc temp, °F
2.0		390-440	400-470	350-450	380-430	90-120	400-450
2.0		390-440	400-470	350-450	380-430	90-120	400-450
2.0		390-440	400-470	350-450	380-430	90-120	400-450
2.0		390-440	400-470	350-450	380-430	90-120	400-450
2.0		390-440	400-470	350-450	380-430	90-120	400-450
2.0		390-440	400-470	350-450	380-430	90-120	400-450
1.0		390	410	390	400	90	400
2.0		390	400	370	400	90	400
2.0		370-430	380-440	350-410	370-430	90-120	380-420
2.0		390-430	390-430	380-420	380-420	100-150	380-430
							390-450
							390-450
							390-450
							390-450
							390-450
							390-450
							390-450
							390-450
							390-450
							390-450
							400-500
							400-500
							390-450
							390-450
							390-450
							390-450
							390-450
							390-450
							390-450
							390-450
							350-450
							390-450
							390-450
							390-450
							390-450
							390-450
							390-450
							390-450
							390-450
						80-100	430-470
						80-100	430-470
						80-100	430-470
						80-100	430-470
						80-100	430-470
						80-100	430-470
						80-100	430-470
						80-100	430-470
						80-100	430-470
						80-100	430-470
						80-100	430-470

Grade	Filler	Sp Grav	Shrink, mils/in	Melt flow, g/10 min	Melt temp, °F	Back pres, psi	Drying temp, °F
PP		**Hiloy**			**ComAlloy**		
101	GLF 10	0.98	8.0	9.00			180-200
102	GLF 20	1.04	7.0	9.00			180-200
103	GLF 30	1.13	6.0	9.00			180-200
104	GLF 40	1.22	5.0	9.00			180-200
111	GLF 10	0.98	5.0	9.00			180-200
112	GLF 20	1.04	4.0	9.00			180-200
113	GLF 30	1.13	4.0	9.00			180-200
114	GLF 40	1.22	3.0	9.00			180-200
119	GMN 40	1.22	4.0	9.00			180-200
120	UNS	1.22	5.0	6.00			180-200
121	GL 10	0.97	7.0	8.00			180-200
122	GL 20	1.04	6.0	8.00			180-200
123	GL	1.12	5.0	6.00			180-200
124	GL 40	1.22	4.0	6.00			180-200
129	UNS	1.22	5.0	6.00			180-200
130	UNS	1.22	5.0	6.00			180-200
PP		**Hostacom**			**Hoechst**		
G2 N01	GLF 20	1.05		1.70 L	327-333		
G2 N02	GLF 20	1.05		1.00 L	327-333		
G2 N03	GLB 20	1.03		1.30 L	327-333		
G2 U01	GLF 20	1.05		13.00 L	327-333		
G2 U02	GLF 20	1.05		15.00 L	327-333		
G3 N01	GLF 30	1.05		1.00 L	327-333		
G3 U01	GLF 30	1.05		15.00 L	327-333		
M1 U01	MNF 10	0.97		16.00 L	327-333		
M2 N01	MNF 20	1.04		2.00 L	327-333		
M2 N02	MNF 20	1.04		3.00 L	327-333		
M2 U01	MNF 20	1.04		16.00 L	327-333		
M4 N01	MNF 40	1.21		2.50 L	327-333		
M4 U01	MNF 40	1.21		15.00 L	327-333		
M4 U02	MNF 40	1.21		15.00 L	327-333		
M4 U03	MNF 40	1.24		13.00 L	327-333		
PP		**Marlex**			**Phillips**		
HGL-050-01		0.91		5.00			
HGL-120-01		0.91		12.00			
HGL-200		0.91		20.00			
HGL-350		0.91		35.00			
HGN-020-05		0.91		2.10			
HGN-020-06		0.91		1.60			
HGN-200		0.91		20.00			
HGX-010		0.91		1.20			
HGX-010-01		0.91		1.00			
HGX-030		0.91		3.00			
HGX-330		0.91		33.00			
HGY-040		0.91		4.00			
HGZ-120-02		0.91		12.00			
HGZ-120-04		0.91		12.00			
HGZ-200		0.91		20.00			
HGZ-350		0.91		35.00			
HLM-020		0.91		2.00			
HLN-120-01		0.91		12.00			
HLN-200		0.91		20.00			

Drying time, hr	Inj time, sec	Front temp, °F	Mid temp, °F	Rear temp, °F	Nozzle temp, °F	Mold temp, °F	Proc temp, °F
1.0- 2.0		420	430	440	420-440	50-150	430-450
1.0- 2.0		440	450	460	440-460	50-150	440-460
1.0- 2.0		440	450	460	440-460	50-150	440-460
1.0- 2.0		460	470	480	460-480	50-150	460-480
1.0- 2.0		420	430	440	420-440	50-150	430-450
1.0- 2.0		440	450	460	440-460	50-150	440-460
1.0- 2.0		440	450	460	440-460	50-150	440-460
1.0- 2.0		460	470	480	460-480	50-150	460-480
1.0- 2.0		460	470	480	460-480	50-150	460-480
1.0- 2.0		460	470	480	460-480	50-150	460-480
1.0- 2.0		420	430	440	420-440	50-150	430-450
1.0- 2.0		440	450	460	440-460	50-150	440-460
1.0- 2.0		440	450	460	440-460	50-150	440-460
1.0- 2.0		460	470	480	460-480	50-150	460-480
1.0- 2.0		460	470	480	460-480	50-150	460-480
1.0- 2.0		460	470	480	460-480	50-150	460-480
						68-194	392-500
						68-194	392-500
						68-194	464-500
						68-194	392-500
						68-194	392-500
						68-194	392-500
						68-194	392-500
						68-194	464-500
						68-194	464-500
						68-194	464-500
						68-194	464-500
						68-194	464-500
						68-194	464-500
						68-194	464-500
						68-194	464-500
						60-100	375-575
						60-100	375-575
						60-100	375-575
						60-100	375-575
						60-100	375-575
						60-100	375-575
						60-100	375-575
						60-100	375-575
						60-100	375-575
						60-100	375-575
						60-100	375-575
						60-100	375-575
						60-100	375-575
						60-100	375-575
						60-100	375-575
						60-100	375-575
						60-100	375-575
						60-100	375-575
						60-100	375-575

Grade	Filler	Sp Grav	Shrink, mils/in	Melt flow, g/10 min	Melt temp, °F	Back pres, psi	Drying temp, °F
HLN-200-01		0.91		20.00			
HLN-350		0.91		35.00			
HLV-040		0.91		4.30			
HLV-050		0.91		5.30			
HNS-080		0.91		8.00			
HNZ-020		0.91		2.00			
HXV-035		0.91		3.50			
RLX-020		0.90		1.80			
RMN-020		0.90		2.00			
RMX-020		0.90		2.00			

PP Multi-Pro Multibase

Grade	Filler	Sp Grav	Shrink, mils/in	Melt flow, g/10 min	Melt temp, °F	Back pres, psi	Drying temp, °F
0612 CW	CAC 6	0.94	16.0	12.00 L		20- 50	200
1004 R	GLF 10	0.97	6.0	4.00		20- 50	200
1008 T	TAL 10	0.95	12.0	8.00		20- 50	200
1010 T	TAL 10	0.97	13.0	11.00		20- 50	200
1012 C	CAC 12	1.00	14.0	12.00 L		20- 50	200
1020 C	CAC 10	0.99	14.0	20.00		20- 50	200
1505 R	GLF 15	1.04	6.0	5.00		20- 50	200
1508 M	MI 15	1.02	10.0	8.50		20- 50	200
1512 RU	GLF 15	1.04	6.0	12.00		20- 50	200
2003 TI	TAL 20	1.04	11.0	3.00		20- 50	200
2004 R	GLF 20	1.05	5.0	4.00		20- 50	200
2005 M	MI 20	1.05	8.0	5.00		20- 50	200
2005 RCI	GLF 20	1.05	5.0	5.00		20- 50	200
2006 T	TAL 20	1.04	8.0	6.00		20- 50	200
2010 M	MI 20	1.05	9.0	10.00		20- 50	200
2015 CW	CAC 20	1.04	13.0	15.00		20- 50	200
2018 CU	CAC 20	1.04	12.0	18.00 L		20- 50	200
2508 C	CAC 25	1.08	12.0	8.00 L		20- 50	200
2510 XG	MN 25	1.24	9.0	10.00		20- 50	200
2510 XGW	MN 27	1.14		14.00		20- 50	200
2514 CUW	CAC 25	1.08	9.0	14.00		20- 50	200
2514 XU	MN 25	1.10	10.0	14.00		20- 50	200
2516 CU	CAC 25	1.08	11.0	16.50		20- 50	200
2518 CU	CAC 25	1.08	9.0	18.00		20- 50	200
2518 TU	TAL 25	1.09	8.0	18.00		20- 50	200
2820 XUG	MN 28	1.14	9.0	20.00		20- 50	200
3003 R	GLF 30	1.13	5.0	3.00		20- 50	200
3003 RC	GLF 30	1.13	4.0	3.00 L		20- 50	200
3011 CHI	CAC 30	1.15	11.0	11.00 L		20- 50	200
3012 RC	GL 30	1.11	3.0	7.00 L		20- 50	200
3014 CUW	CAC 30	1.15	10.0	14.00 L		20- 50	200
3014 XUGW	MN	1.14	10.0	14.00		20- 50	200
3015 CUA	CAC 30	1.14	11.0	15.00		20- 50	200
3018 TU	TAL 30	1.12	8.5	18.00		20- 50	200
3018 XUG	MN 30	1.15	10.0	18.00		20- 50	200
3019 CXUW	CAC 30	1.14	10.0	19.00		20- 50	200
4000	CAC 40	1.24	10.0	0.90		20- 50	200
4002 C	CAC 40	1.21	9.0	2.00 L		20- 50	200
4003 C	CAC 40	1.24	10.0	3.00 L		20- 50	200
4003 R	GLF 40	1.24	4.0	3.00		20- 50	200
4003 RC	GL 40	1.24	4.0	3.00 L		20- 50	200
4005 M	MI 40	1.25	6.0	5.00		20- 50	200
4005 MI	MI 40	1.25	6.0	5.00		20- 50	200
4006 TH	TAL 40	1.24	8.0	6.00		20- 50	200

Drying time, hr	Inj time, sec	Front temp, °F	Mid temp, °F	Rear temp, °F	Nozzle temp, °F	Mold temp, °F	Proc temp, °F
						60-100	375-575
						60-100	375-575
						60-100	375-575
						60-100	375-575
						60-100	375-575
						60-100	375-575
						60-100	375-575
						60-100	375-575
						60-100	375-575
						60-100	375-575
2.0		390-450	390-450	370-420	380-440	40-150	
2.0		420-500	400-500	390-500	420-480	40-150	
2.0		400-450	400-450	390-430	380-440	40-150	
2.0		400-450	400-450	390-430	380-440	40-150	
2.0		390-450	390-450	370-420	380-440	40-150	
2.0		390-450	390-450	370-420	380-440	40-150	
2.0		420-500	400-500	390-500	420-480	40-150	
2.0		430-500	420-500	420-500	420-480	40-150	
2.0		420-500	400-500	390-500	420-480	40-150	
2.0		400-450	400-450	390-430	380-440	40-150	
2.0		420-500	400-500	390-500	420-480	40-150	
2.0		430-500	420-500	420-500	420-480	40-150	
2.0		420-500	400-500	390-500	420-480	40-150	
2.0		400-450	400-450	390-430	380-440	40-150	
2.0		430-500	420-500	420-500	420-480	40-150	
2.0		390-450	390-450	370-420	380-440	40-150	
2.0		390-450	390-450	370-420	380-440	40-150	
2.0		390-450	390-450	370-420	380-440	40-150	
2.0		390-450	380-450	380-430	380-440	40-150	
2.0		390-450	380-450	380-430	380-440	40-150	
2.0		390-450	390-450	370-420	380-440	40-150	
2.0		390-450	380-450	380-430	380-440	40-150	
2.0		390-450	390-450	370-420	380-440	40-150	
2.0		390-450	390-450	370-420	380-440	40-150	
2.0		400-450	400-450	390-430	380-440	40-150	
2.0		390-450	380-450	380-430	380-440	40-150	
2.0		420-500	400-500	390-500	420-480	40-150	
2.0		420-500	400-500	390-500	420-480	40-150	
2.0		390-450	390-450	370-420	380-440	40-150	
2.0		420-500	400-500	390-500	420-480	40-150	
2.0		390-450	390-450	370-420	380-440	40-150	
2.0		390-450	380-450	380-430	380-440	40-150	
2.0		390-450	390-450	370-420	380-440	40-150	
2.0		400-450	400-450	390-430	380-440	40-150	
2.0		390-450	380-450	380-430	380-440	40-150	
2.0		390-450	390-450	370-420	380-440	40-150	
2.0		390-450	390-450	370-420	380-440	40-150	
2.0		390-450	390-450	370-420	380-440	40-150	
2.0		390-450	390-450	370-420	380-440	40-150	
2.0		420-500	400-500	390-500	420-480	40-150	
2.0		420-500	400-500	390-500	420-480	40-150	
2.0		430-500	420-500	420-500	420-480	40-150	
2.0		430-500	420-500	420-500	420-480	40-150	
2.0		400-450	400-450	390-430	380-440	40-150	

Grade	Filler	Sp Grav	Shrink, mils/in	Melt flow, g/10 min	Melt temp, °F	Back pres, psi	Drying temp, °F
4007 XGS	MN 40	1.23	70.0	7.00		20- 50	200
4008 RC	GL 40	1.22	2.0	8.50		20- 50	200
4008 T	TAL 40	1.23	9.0	8.00		20- 50	200
4008 XB	MN 40	1.25	8.0	8.00		20- 50	200
4009 T	TAL 40	1.24	9.0	9.00		20- 50	200
4010 M	MI 40	1.25	6.0	10.00		20- 50	200
4012 TH	TAL 40	1.24	8.0	12.00		20- 50	200
4012 X Platinum	GL 42	1.26	50.0	14.00		20- 50	200
4013 XS	MN 40	1.25	8.0	13.00		20- 50	200
4015 RS	GLF 40	1.24	5.0	15.00		20- 50	200
4015 THW	TAL 40	1.25	9.0	15.00		20- 50	200
4017 C	CAC 40	1.23	10.0	17.00		20- 50	200
4022 RS	GLF 40	1.23	50.0	22.00		20- 50	200
4518 CUW	CAC 45	1.35	9.0	18.00		20- 50	200
5014 CUW	CAC	1.08	11.0	14.00		20- 50	200
5102 MR	MI 50	1.34	5.0	12.00		20- 50	200

PP Nortuff Polymerland

Grade	Filler	Sp Grav	Shrink, mils/in	Melt flow, g/10 min	Melt temp, °F	Back pres, psi	Drying temp, °F
NFA 4400TO	MN	1.26	10.0-14.0	1.70			
NFC 1700MO	MN	1.17	9.0-12.0	7.00			
NFC 1800TO	MN	1.23	9.0-12.0	4.00			
NFC 2400CO		1.04	4.0- 6.0	7.00			
NFC 2440CF	UNS	1.23	3.0- 5.0	5.00			
NFC 2600CO	UNS	1.12	1.0- 3.0	6.00			
NFC 4000FR		0.98	10.0-14.0	12.00			
NFC 4400CO		1.20	3.0- 6.0	5.00			
NFC 4600CO	UNS	1.48	1.0- 4.0	5.00			
NFH 2800CO	UNS	1.26	1.0- 2.0	5.00			
RC 1800-KV		1.24		7.30			
RX 3241-MK	MN 30	1.12		14.00 L			

PP Nyloy Nytex Compos.

Grade	Filler	Sp Grav	Shrink, mils/in	Melt flow, g/10 min	Melt temp, °F	Back pres, psi	Drying temp, °F
PG-0020	GLF 20	1.04	5.0				
PG-0030	GLF 30	1.12	4.0				
PG-0230	GLF 30	1.12	4.0				

PP PermaStat RTP

Grade	Filler	Sp Grav	Shrink, mils/in	Melt flow, g/10 min	Melt temp, °F	Back pres, psi	Drying temp, °F
100		0.95	17.0-20.0				
100FR							
100LE							
200		1.11	15.0-23.0				

PP Petrothene Quantum

Grade	Filler	Sp Grav	Shrink, mils/in	Melt flow, g/10 min	Melt temp, °F	Back pres, psi	Drying temp, °F
PP 1510-PC		0.90		0.70			
PP 1602-WF				2.24			
PP 1610-PF				2.50			

PP Polyfort A. Schulman

Grade	Filler	Sp Grav	Shrink, mils/in	Melt flow, g/10 min	Melt temp, °F	Back pres, psi	Drying temp, °F
FPP 1285	GLF 20	1.03				50- 250	
FPP 1603	GLF 10	0.97				50- 250	
FPP 1604	GLF 30	1.12				50- 250	
FPP 1807	GLF 40	1.04				50- 250	
PP 1531		0.89				50- 250	
PP 1549		0.89				50- 250	
PP 828		0.89				50- 250	
PP 829		0.89				50- 250	

Drying time, hr	Inj time, sec	Front temp, °F	Mid temp, °F	Rear temp, °F	Nozzle temp, °F	Mold temp, °F	Proc temp, °F
2.0		390-450	380-450	380-430	380-440	40-150	
2.0		420-500	400-500	390-500	420-480	40-150	
2.0		400-450	400-450	390-430	380-440	40-150	
2.0		390-450	380-450	380-430	380-440	40-150	
2.0		400-450	400-450	390-430	380-440	40-150	
2.0		430-500	420-500	420-500	420-480	40-150	
2.0		400-450	400-450	390-430	380-440	40-150	
2.0		420-500	400-500	390-500	420-480	40-150	
2.0		390-450	380-450	380-430	380-440	40-150	
2.0		420-500	400-500	390-500	420-480	40-150	
2.0		400-450	400-450	390-430	380-440	40-150	
2.0		390-450	390-450	370-420	380-440	40-150	
2.0		420-500	400-500	390-500	420-480	40-150	
2.0		390-450	390-450	370-420	380-440	40-150	
2.0		390-450	390-450	370-420	380-440	40-150	
2.0		430-500	420-500	420-500	420-480	40-150	
						120-140	390-450
						120-140	390-450
						120-140	390-450
						120-140	390-450
						120-140	390-450
						120-140	390-450
						120-140	390-450
						120-140	390-450
						120-140	390-450
						120-140	390-450
						110-125	410-450
						110-125	410-450
							445
							445
							445
		360-400	360-400	360-400		90-150	
		360-400	360-400	360-400		90-150	
		360-400	360-400	360-400		90-150	
		460-525	460-525	460-525		150-200	
	15.0	435	425	400	450	125	
	15.0	435	425	400	450	125	
	15.0	435	425	400	450	125	
		425		375		70-120	385-475
		425		375		70-120	385-475
		425		375		70-120	385-475
		425		375		70-120	385-475
		425		375		70-120	385-475
		425		375		70-120	385-475
		425		375		70-120	385-475
		425		375		70-120	385-475

Grade	Filler	Sp Grav	Shrink, mils/in	Melt flow, g/10 min	Melt temp, °F	Back pres, psi	Drying temp, °F
PP		**Pro-fax**			**Himont**		
6331NW		0.90		12.00		50- 300	
SR-011				21.00		50- 300	
PP		**Rexene**			**Rexene**		
PP 11S12A		0.90		12.00			
PP 14S30V		0.90		30.00			
PP 14S6A		0.90		6.00			
PP 17B30V		0.90		30.00			
PP 17S12A		0.90		12.00			
PP 17S4A		0.90		4.00			
PP 18S3A		0.90		3.00			
PP 18S4A		0.90		4.00			
PP		**RTP Polymers**			**RTP**		
100		0.91	25.0			25- 50	180
100 CC HS		0.91	25.0			25- 50	180
100 FR		1.00	20.0			50	180
100 FR A		1.00	20.0			50	180
100 GB 10	GLB 10	0.97	23.0			50	180
100 GB 20	GLB 20	1.04	20.0			50	180
100 GB 30	GLB 30	1.12	18.0			50	180
100 GB 40	GLB 40	1.22	16.0			50	180
100 HB	TAL	0.91	15.0			50	180
100 HS		0.91	25.0			25- 50	180
100 SP		0.73	7.0			25- 50	180
100Z		0.91	25.0			25- 50	180
101	GLF 10	0.98	9.0			50	180
101 FR	GLF	1.45	5.0			50	180
101 HB	GLF	0.98	7.0			50	180
101 M 20	GLF 10	1.14	7.0			50	180
101 SP	GLF 10	0.77	8.0			50	180
101CC	GLF 10	0.98	9.0			50	180
101Z	GLF 10	0.98	9.0			50	180
101Z	GLF 10	0.98	7.0			50	180
102 HB	GLF	1.00	5.0			50	180
102 M 25	GLF 15	1.25	5.0			50	180
103	GLF 20	1.05	5.0			50	180
103 AV	GLF 20	0.84	5.0			50	180
103 FR	GLF	1.46	2.0			50	180
103 HB	GLF	1.05	4.0			50	180
103 SP	GLF 20	0.84	5.0			50	180
103CC	GLF 20	1.05	5.0			50	180
103CC HB	GLF	1.05	4.0			50	180
103CC HI HB	GLF	1.05	4.0			50	180
103CC M 10	GMN	1.13	3.0			50	180
103Z	GLF 20	1.05	5.0			50	180
103Z	GLF 20	1.05	4.0			50	180
104 HB	GLF	1.08	4.0			50	180
105	GLF 30	1.13	5.0			50	180
105 HB	GLF	1.13	3.0			50	180
105 SP	GLF 30	0.90	5.0			50	180
105CC	GLF 30	1.13	5.0			50	180
105CC FR	GLF	1.47	2.0			50	180
105CC HB	GLF	1.13	3.0			50	180

Drying time, hr	Inj time, sec	Front temp, °F	Mid temp, °F	Rear temp, °F	Nozzle temp, °F	Mold temp, °F	Proc temp, °F
						70-120	400-445
						70-120	385-420
		470-580	475-525	475-500		40-220	400-600
		470-580	475-525	475-500		40-220	400-600
		470-580	475-525	475-500		40-220	400-600
		470-580	475-525	475-500		40-220	400-600
		470-580	475-525	475-500		40-220	400-600
		470-580	475-525	475-500		40-220	400-600
		470-580	475-525	475-500		40-220	400-600
		470-580	475-525	475-500		40-220	400-600
2.0		470-525	460-515	450-505		90-150	
2.0		470-525	460-515	450-505		90-150	
2.0		370-430	350-400	330-350		100-170	
2.0		370-430	350-400	330-350		100-170	
2.0		420-450	410-440	400-430		70-150	
2.0		420-450	410-440	400-430		70-150	
2.0		420-450	410-440	400-430		70-150	
2.0		420-450	410-440	400-430		70-150	
2.0		400-440	390-430	380-420		70-120	
2.0		470-525	460-515	450-505		90-150	
2.0		470-525	460-515	450-505		90-150	
2.0		470-525	460-515	450-505		90-150	
2.0		420-450	410-440	400-430		70-150	
2.0		370-430	350-400	330-350		100-170	
2.0		420-450	410-440	400-430		70-150	
2.0		420-450	410-440	400-430		70-150	
2.0		420-450	410-440	400-430		70-150	
2.0		420-450	410-440	400-430		70-150	
2.0		420-450	410-440	400-430		70-150	
2.0		420-450	410-440	400-430		70-150	
2.0		420-450	410-440	400-430		70-150	
2.0		420-450	410-440	400-430		70-150	
2.0		420-450	410-440	400-430		70-150	
2.0		420-450	410-440	400-430		70-150	
2.0		370-430	350-400	330-350		100-170	
2.0		420-450	410-440	400-430		70-150	
2.0		420-450	410-440	400-430		70-150	
2.0		420-450	410-440	400-430		70-150	
2.0		420-450	410-440	400-430		70-150	
2.0		420-450	410-440	400-430		70-150	
2.0		420-450	410-440	400-430		70-150	
2.0		420-450	410-440	400-430		70-150	
2.0		420-450	410-440	400-430		70-150	
2.0		420-450	410-440	400-430		70-150	
2.0		420-450	410-440	400-430		70-150	
2.0		420-450	410-440	400-430		70-150	
2.0		420-450	410-440	400-430		70-150	
2.0		420-450	410-440	400-430		70-150	
2.0		370-430	350-400	330-350		100-170	
2.0		420-450	410-440	400-430		70-150	

Grade	Filler	Sp Grav	Shrink, mils/in	Melt flow, g/10 min	Melt temp, °F	Back pres, psi	Drying temp, °F
105CC T	GLF	1.13	3.0			50	180
105Z	GLF 30	1.13	5.0			50	180
105Z	GLF 30	1.13	4.0			50	180
106 HB	GLF	1.17	3.0			50	180
107	GLF 40	1.23	4.0			50	180
107 HB	GLF	1.23	2.0			50	180
107CC	GLF 40	1.23	4.0			50	180
107CC HB	GLF	1.23	2.0			50	180
107Z	GLF 40	1.23	4.0			50	180
107Z	GLF 40	1.23	3.0			50	180
127	TAL 40	1.25	12.0			50	180
127 HB	TAL	1.25	8.0			50	180
127 HI	TAL 40	1.23	10.0			50	180
128	TAL 20	1.05	20.0			50	180
128 HI	TAL 20	1.04	18.0			50	180
131	TAL 10	0.98	25.0			50	180
131 HI	TAL 10	0.98	20.0			50	180
132	TAL 30	1.14	16.0			50	180
136	TAL 35	1.27	12.0			50	180
136 HB	MN	1.27	8.0			50	180
140	CAC 40	1.24	14.0			50	180
140 HB	CAC	1.24	10.0			50	180
141	CAC 20	1.05	20.0			50	180
141 HI	CAC 10	1.05	20.0			50	180
142	CAC 10	0.98	25.0			50	180
143	CAC 30	1.14	18.0			50	180
148	MI 40	1.26	9.0			50	180
149	MI 25	1.10	16.0			50	180
149	MI 25	1.10	16.0				
149 HB	MI	1.10	10.0			50	180
150		1.25	12.0			50	180
150 HF		1.25	12.0			50	180
150 LF		1.25	12.0			50	180
152		0.94	20.0			50	180
152 HF		0.94	20.0			50	180
152 LF		0.94	20.0			50	180
175	GLF	1.12	6.0			50	180
175X	GLF	1.12	6.0			50	180
175X HB	GLB	1.12	5.0			50	180
177	GLF	1.24	6.0			50	180
178	GLF	1.13	5.0			50	180
178 HB	GLB	1.13	4.0			50	180
178X	GLF	1.13	5.0			50	180
199X50505	GLB	1.25	10.0			50	180
199X53840	CAC	1.19	14.0			50	180
ESD-AS-100		0.90	15.0-20.0				
PP		**Shell PP**			**Shell**		
5A95				10.00 L		50- 200	
5A97				3.90 L		50- 200	
5D45				0.65 L		50- 200	
5D98				3.40 L		50- 200	
5E66				8.50 L		50- 200	
5E98				3.40 L		50- 200	
6D20				1.90 L	298	50- 200	
6D21				8.00 L	288	50- 200	

Drying time, hr	Inj time, sec	Front temp, °F	Mid temp, °F	Rear temp, °F	Nozzle temp, °F	Mold temp, °F	Proc temp, °F
2.0		420-450	410-440	400-430		70-150	
2.0		420-450	410-440	400-430		70-150	
2.0		420-450	410-440	400-430		70-150	
2.0		420-450	410-440	400-430		70-150	
2.0		420-450	410-440	400-430		70-150	
2.0		420-450	410-440	400-430		70-150	
2.0		420-450	410-440	400-430		70-150	
2.0		420-450	410-440	400-430		70-150	
2.0		420-450	410-440	400-430		70-150	
2.0		420-450	410-440	400-430		70-150	
2.0		400-440	390-430	380-420		70-120	
2.0		400-440	390-430	380-420		70-120	
2.0		400-440	390-430	380-420		70-120	
2.0		400-440	390-430	380-420		70-120	
2.0		400-440	390-430	380-420		70-120	
2.0		400-440	390-430	380-420		70-120	
2.0		400-440	390-430	380-420		70-120	
2.0		400-440	390-430	380-420		70-120	
2.0		400-440	390-430	380-420		70-120	
2.0		400-440	390-430	380-420		70-120	
2.0		400-440	390-430	380-420		70-120	
2.0		400-440	390-430	380-420		70-120	
2.0		400-440	390-430	380-420		70-120	
2.0		400-440	390-430	380-420		70-120	
2.0		400-440	390-430	380-420		70-120	
2.0		400-440	390-430	380-420		70-120	
2.0		400-440	390-430	380-420		70-120	
2.0		400-440	390-430	380-420		70-120	
		450-525	450-525	450-525		90-150	
2.0		400-440	390-430	380-420		70-120	
2.0		370-430	350-400	330-350		100-170	
2.0		370-430	350-400	330-350		100-170	
2.0		370-430	350-400	330-350		100-170	
2.0		370-430	350-400	330-350		100-170	
2.0		370-430	350-400	330-350		100-170	
2.0		370-430	350-400	330-350		100-170	
2.0		420-450	410-440	400-430		70-150	
2.0		420-450	410-440	400-430		70-150	
2.0		420-450	410-440	400-430		70-150	
2.0		420-450	410-440	400-430		70-150	
2.0		420-450	410-440	400-430		70-150	
2.0		420-450	410-440	400-430		70-150	
2.0		420-450	410-440	400-430		70-150	
2.0		420-450	410-440	400-430		70-150	
2.0		400-440	390-430	380-420		70-120	
		425-525	425-525	425-525		90-150	
		400	400	390	400	60-150	
		400	400	390	400	60-150	
		400	400	390	400	60-150	
		400	400	390	400	60-150	
		400	400	390	400	60-150	
		400	400	390	400	60-150	
		400	400	390	400	60-150	
		400	400	390	400	60-150	

Grade	Filler	Sp Grav	Shrink, mils/in	Melt flow, g/10 min	Melt temp, °F	Back pres, psi	Drying temp, °F
6D81				5.00 L	273	50- 200	
6D82				7.00 L	273	50- 200	
6E20				1.90 L	298	50- 200	
6E21				8.00 L	288	50- 200	
6E77				3.50 L	316	50- 200	
7129				0.60 L		50- 200	
7328				2.00 L		50- 200	
7623				8.00 L		50- 200	
7635				8.50 L		50- 200	
7C54H				12.00 L		50- 200	
PP		**Tempalloy**			**ComAlloy**		
114	GL 40	1.19	5.0	3.00			180-200
120	UNS	1.22	5.0	6.00			180-200
PP		**Tenite**			**Eastman**		
4L31	TAL	0.97		4.50 L	350-430		
PP		**Thermofil Polypropylene**				**Thermofil**	
P6-20FG-0100	GLC 20	1.04	4.0	2.50 L		50	160
P6-20FG-0103	GLC 20	1.04	4.0	2.50		50	160
P6-20FG-0600	GLC 20	1.04	4.0	2.50 L		50	160
P6-20FG-0734	GLC 20	1.04	4.0	3.00 L		50	160
P6-20FG-1600	GLC 20	1.04	4.0	4.00 L		50	160
P6-20FG-2100	GLC 20	1.04	4.0	5.00 L		50	160
P6-23FG-1600	GLC 23	1.07	4.0	5.00 L		50	160
P6-23FG-2600	GLC 23	1.07	3.0	5.00 L		50	160
P6-30FG-0100	GLC 30	1.13	3.0	2.50 L		50	160
P6-30FG-0103	GLC 30	1.13	3.0	2.50		50	160
P6-30FG-0600	GLC 30	1.13	3.0	2.50 L		50	160
P6-30FG-0684	GLC 30	1.13	3.0	2.50 L		50	160
P6-30FG-0686	GLC 30	1.13	3.0	5.00		50	160
P6-30FG-1100	GLC 30	1.13	3.0	4.00 L		50	160
P6-30FG-1804	GLC 30	1.12	4.0	4.00 L		50	160
P6-30FG-2100	GLC 30	1.12	4.0	5.00 L		50	160
P6-35FG-0100	GLC 35	1.17	3.0	3.00 L		50	160
P6-40FM-Y430	GMN 40	1.23	3.0	5.00 L		50	160
PP		**Voloy**			**ComAlloy**		
100		1.33	12.0	9.00			180-200
102		1.24	10.0	8.00			180-200
110		0.96	15.0	3.00			160
112	GL 20	1.33	5.0	3.50			180-200
113	GL 30	1.38	4.0	3.50			180-200
120		0.94	15.0	3.00			160
122		1.18	10.0	2.00			160
PP		**WPP PP**			**Washington Penn**		
PP-H2TF-4C	TAL 40	1.24		5.00		150- 300	
PP Copolymer		**Cefor**			**Shell**		
SRD4-101					316		
SRD4-127					289		
PP Copolymer		**Comtuf**			**ComAlloy**		
106	GMN 30	1.11	6.0	9.00			180-200

Drying time, hr	Inj time, sec	Front temp, °F	Mid temp, °F	Rear temp, °F	Nozzle temp, °F	Mold temp, °F	Proc temp, °F
		400	400	390	400	60-150	
		400	400	390	400	60-150	
		400	400	390	400	60-150	
		400	400	390	400	60-150	
		400	400	390	400	60-150	
		400	400	390	400	60-150	
		400	400	390	400	60-150	
		400-440	400-440	390-430	400-440	60-150	
		400-440	400-440	390-430	400-440	60-150	
		400-440	400-440	390-430	400-440	60-150	
1.0- 2.0		460	470	480	460-480	50-150	460-480
1.0- 2.0		460	470	480	460-480	50-150	460-480
							375-450
2.0		420-450	400-420	380-400	430-460	80-150	
2.0		420-450	400-420	380-400	430-460	80-150	
2.0		420-450	400-420	380-400	430-460	80-150	
2.0		420-450	400-420	380-400	430-460	80-150	
2.0		420-450	400-420	380-400	430-460	80-150	
2.0		420-450	400-420	380-400	430-460	80-150	
2.0		420-450	400-420	380-400	430-460	80-150	
2.0		420-450	400-420	380-400	430-460	80-150	
2.0		420-450	400-420	380-400	430-460	80-150	
2.0		420-450	400-420	380-400	430-460	80-150	
2.0		420-450	400-420	380-400	430-460	80-150	
2.0		420-450	400-420	380-400	430-460	80-150	
2.0		420-450	400-420	380-400	430-460	80-150	
2.0		420-450	400-420	380-400	430-460	80-150	
2.0		420-450	400-420	380-400	430-460	80-150	
2.0		420-450	400-420	380-400	430-460	80-150	
2.0		420-450	400-420	380-400	430-460	80-150	
2.0		420-450	400-420	380-400	430-460	80-150	
1.0- 2.0		420	430	440	420-440	50-150	430-450
1.0- 2.0		420	430	440	420-440	50-150	430-450
1.0- 2.0		410	400	390	400	50-120	400
1.0- 2.0		440	450	460	440-460	50-150	440-460
1.0- 2.0		460	470	480	460-480	50-150	460-480
1.0- 2.0		410	400	390	400	50-120	400
1.0 2.0		410	400	390	400	50-120	400
		435-445	430-440	425-435	440-450	80-120	440-460
1.0- 2.0		460	470	480	460-480	50-150	460-480

Grade	Filler	Sp Grav	Shrink, mils/in	Melt flow, g/10 min	Melt temp, °F	Back pres, psi	Drying temp, °F
PP Copolymer		**CTI Polypropylene**			**CTI**		
PPX-20GF/20GB/000	GLF 20	1.21	1.0- 2.0				
PPX-30GF/000 HI	GLF 30	1.12	2.0				
PP Copolymer		**Electrafil**			**DSM**		
PP-61/EC	CP	1.00	15.0			50	150
PP Copolymer		**Escorene**			**Exxon**		
PD 7292		0.90		4.00 L			
PD 7292N		0.90		4.00 L			
PD 7313		0.90		8.00 L			
PD 8092		0.90		2.00 L			
PD 8102		0.90		4.00 L			
PP 7032		0.90		4.00 L			
PP 9074MED		0.88	19.6	24.00 L			
PP 9374MED		0.90	19.0	12.00 L			
PP Copolymer		**Petrothene**			**Quantum**		
PP 1402-GF		0.90		2.00			
PP 1402-NF		0.90		2.00			
PP 1404-ZJ				4.50			
PP 1415-HC				0.80			
PP 1510-HC				0.70			
PP 1510-LC				0.70			
PP 7200-AF				2.00			
PP 7200-GF				2.00			
PP 7200-MF				2.00			
PP 7300-KF				2.00			
PP 7300-MF				2.00			
PP 8310-GO			15.0-25.0	10.00			
PP 8310-KO			15.0-25.0	10.00			
PP 8380-KW				30.00			
PP 8380-ZW				30.00			
PP 8402-HO			15.0-25.0	8.00			
PP 8402-TO			15.0-25.0	8.00			
PP 8403-HO			15.0-25.0	8.00			
PP 8404-HJ				4.50			
PP 8404-ZJ			15.0-25.0	4.50			
PP 8410-ZR			15.0-25.0	11.00			
PP 8411-NR				10.00			
PP 8411-ZR			15.0-25.0	10.00			
PP 8412-HK			15.0-25.0	7.00			
PP 8412-TK			15.0-25.0	7.00			
PP 8420-HK			15.0-25.0	6.00			
PP 8430-ZR				11.00			
PP 8462-HR			15.0-25.0	12.00			
PP 8470-HU			15.0-25.0	20.00			
PP 8470-ZU			15.0-25.0	20.00			
PP 8502-HK			15.0-25.0	5.00			
PP 8602-HJ			15.0-25.0	2.50			
PP 8752-HF			15.0-25.0	2.00			
PP 8755-HK			15.0-25.0	5.00			
PP 8762-HR			15.0-25.0	12.00			
PP 8770-HU			15.0-25.0	20.00			
PP 8775-HU			15.0-25.0	20.00			

Drying time, hr	Inj time, sec	Front temp, °F	Mid temp, °F	Rear temp, °F	Nozzle temp, °F	Mold temp, °F	Proc temp, °F
		475-525	445-475	400-445	420-440	70-120	
		475-525	445-475	400-445	420-440	70-120	
2.0		430-450	440-460	420-440	430-450	80-150	430-450
							420-500
							420-500
							420-500
							430-520
							420-500
							420-500
						50	428
						50	428
	15.0	435	425	400	450	125	
	15.0	435	425	400	450	125	
	15.0	435	425	400	450	125	
	15.0	435	425	400	450	125	
	15.0	435	425	400	450	125	
	15.0	435	425	400	450	125	
	15.0	435	425	400	450	125	
	15.0	435	425	400	450	125	
	15.0	435	425	400	450	125	
	15.0	435	425	400	450	125	
	15.0	435	425	400	450	125	
	15.0	435	425	400	450	125	410-440
	15.0	435	425	400	450	125	410-440
	15.0	435	425	400	450	125	410-450
	15.0	435	425	400	450	125	410-450
	15.0	435	425	400	450	125	430-470
	15.0	435	425	400	450	125	430-470
	15.0	435	425	400	450	125	
	15.0	435	425	400	450	125	
	15.0	435	425	400	450	125	
	15.0	435	425	400	450	125	410-440
	15.0	435	425	400	450	125	410-440
	15.0	435	425	400	450	125	410-440
	15.0	435	425	400	450	125	
	15.0	435	425	400	450	125	
	15.0	435	425	400	450	125	430-470
	15.0	435	425	400	450	125	410-440
	15.0	435	425	400	450	125	410-440
	15.0	435	425	400	450	125	410-440
	15.0	435	425	400	450	125	410-440
	15.0	435	425	400	450	125	430-470
	15.0	435	425	400	450	125	430-470
	15.0	435	425	400	450	125	430-470
	15.0	435	425	400	450	125	430-470
	15.0	435	425	400	450	125	410-440
	15.0	435	425	400	450	125	410-440
	15.0	435	425	400	450	125	410-440

Grade	Filler	Sp Grav	Shrink, mils/in	Melt flow, g/10 min	Melt temp, °F	Back pres, psi	Drying temp, °F
PP 8802-HO			15.0-25.0	8.00			
PP 8815-ZR			15.0-25.0	12.00			
PP 8820-HU			15.0-25.0	20.00			
PP TR040				0.80			
PP Copolymer		**Polyfort**			**A. Schulman**		
FPP 1182	TAL 20	1.02				50- 250	
FPP 1223	CAC 10	0.97				50- 250	
FPP 1227	CAC 40	1.22				50- 250	
FPP 1245	CAC 20	1.02				50- 250	
FPP 1272	TAL 40	1.22				50- 250	
FPP 1345	CAC 30	1.13				50- 250	
FPP 1383	TAL 30	1.13				50- 250	
FPP 1488	MI 20	1.02				50- 250	
FPP 1512	TAL 10	0.97				50- 250	
FPP 1609	MI 10	0.97				50- 250	
FPP 1611	MI 30	1.13				50- 250	
FPP 1612	MI 40	1.22				50- 250	
FPP 1651	MI 10	0.97				50- 250	
FPP 1652	MI 20	1.02				50- 250	
FPP 1653	MI 30	1.13				50- 250	
FPP 1654	MI 40	1.22				50- 250	
PP Copolymer		**Pro-fax**			**Himont**		
7523		0.90		4.00		50- 300	
7531		0.90		4.00		50- 300	
7624		0.90		2.00		50- 300	
8523		0.90		4.00		50- 300	
8623		0.90		2.00		50- 300	
SA-747M		0.90		11.00		50- 300	
SA-994		0.90		11.00		50- 300	
SB-642		0.90		20.00		50- 300	
SB-751		0.89		30.00		50- 300	
SB-787		0.90		20.00			
SB-821		0.90		11.00		50- 300	
SB-823		0.90		21.00			
SB-912		0.90		6.00		50- 300	
SD-101W		0.90		4.00		50- 300	
SD-102		0.90		4.00		50- 300	
SD-206		0.90		9.00		50- 300	
SD-242		0.90		35.00		50- 300	
SD-374				15.00		50- 300	
PP Copolymer		**Rexene**			**Rexene**		
PP 14B12A		0.90		12.00			
PP 14B12ACS253		0.90		12.00			
PP 14B5A		0.90		5.00			
PP 14S2A		0.90		2.00			
PP 14S4A		0.90		4.00			
PP 17C7A		0.90	0.9	7.00			
PP 17C9ACS258		0.90		9.00			
PP 17S20V		0.90		20.00			
PP 17S7ACS228		0.91		7.00			
PP 18C3A		0.90		3.00			
PP 18S10A		0.90		10.00			
PP 18S2A		0.90		2.00			

Drying time, hr	Inj time, sec	Front temp, °F	Mid temp, °F	Rear temp, °F	Nozzle temp, °F	Mold temp, °F	Proc temp, °F
	15.0	435	425	400	450	125	430-470
	15.0	435	425	400	450	125	410-440
	15.0	435	425	400	450	125	410-440
	15.0	435	425	400	450	125	
		425		375		70-120	385-475
		425		375		70-120	385-475
		425		375		70-120	385-475
		425		375		70-120	385-475
		425		375		70-120	385-475
		425		375		70-120	385-475
		425		375		70-120	385-475
		425		375		70-120	385-475
		425		375		70-120	385-475
		425		375		70-120	385-475
		425		375		70-120	385-475
		425		375		70-120	385-475
		425		375		70-120	385-475
		425		375		70-120	385-475
		425		375		70-120	385-475
		425		375		70-120	385-475
						70-120	445-475
						70-120	445-475
						70-120	475-525
						70-120	445-475
						70-120	475-525
						70-120	400-445
						70-120	400-445
						70-120	400-445
						70-120	385-420
		400-430	380-410	360-400	400-430	50- 80	380-410
						70-120	400-445
		400-430	380-410	360-400	400-430	50- 80	380-410
						70-120	445-475
						70-120	445-475
						70-120	445-475
						70-120	445-475
						70-120	385-420
						70-120	400-445
		470-580	475-525	475-500		40-220	400-600
		470-580	475-525	475-500		40-220	400-600
		470-580	475-525	475-500		40-220	400-600
		470-580	475-525	475-500		40-220	400-600
		470-580	475-525	475-500		40-220	400-600
		470-580	475-525	475-500		40-220	400-600
		470-580	475-525	475-500		40-220	400-600
		470-580	475-525	475-500		40-220	400-600
		470-580	475-525	475-500		40-220	400-600
		470-580	475-525	475-500		40-220	400-600
		470-580	475-525	475-500		40-220	400-600
		470-580	475-525	475-500		40-220	400-600

Grade	Filler	Sp Grav	Shrink, mils/in	Melt flow, g/10 min	Melt temp, °F	Back pres, psi	Drying temp, °F
PP 18S6A		0.90		6.00			
PP Copolymer		**RTP Polymers**			**RTP**		
150 HI		1.25	12.0			50	180
181 HI	CF	0.97	2.0			50	180
199X54218	GMN	1.23	3.0			50	180
199X56365	MN	1.37	8.0			50	180
ESD-A-100		0.97	17.0				
ESD-A-100 FR		1.11	16.0				
ESD-A-102	GLF	1.09	4.0				
ESD-A-180	CF	0.94	2.0				
ESD-C-100		0.97	17.0				
ESD-C-100 FR		1.11	16.0				
ESD-C-102	GLF	1.09	4.0				
ESD-C-180	CF	0.99	2.0				
PP Copolymer		**Shell PP**			**Shell**		
6A01K				10.00 L		50- 200	
6A20				2.00 L	297	50- 200	
6A20S				2.00 L	297	50- 200	
6C20				1.90 L	300	50- 200	
6C20S				1.90 L	300	50- 200	
6C44				7.00 L	280	50- 200	
6D20S				1.90 L		50- 200	
7C05N				15.00 L		50- 200	
7C06				1.50 L		50- 200	
7C06N				1.50 L		50- 200	
7C06U				1.50 L		50- 200	
7C12N				22.00 L		50- 200	
7C12Z				22.00 L		50- 200	
7C49				4.00 L		50- 200	
7C49S				4.00 L		50- 200	
7C50		0.90	12.0-25.0	8.00 L		50- 200	
7C50H				8.00 L		50- 200	
7C55H				1.80 L		50- 200	
7C56		0.90	12.0-25.0	8.00 L		50- 200	
7C60		0.90	12.0-25.0	20.00 L		50- 200	
7C62H				8.00 L		50- 200	
DS7C06W				1.50 L		50- 200	
DS7C07				6.50 L		50- 200	
DS7C42				12.00 L		50- 200	
DS7C49S				4.00 L		50- 200	
DS7C51HW				8.00 L		50- 200	
DS7C56				8.00 L		50- 200	
DS7C56W				8.00 L		50- 200	
DS7C57H				0.60 L		50- 200	
DS7C58				28.00 L		50- 200	
DS7C60W				20.00 L		50- 200	
E7C47S				5.00 L		50- 200	
SRC7-442				35.00 L		50- 200	
SRC7-443				35.00 L		50- 200	
SRC7-446				35.00 L		50- 200	
SRC7-461				4.00 L		50- 200	
SRD6-354				12.00 L		50- 200	
SRD6-376				20.00 L		50- 200	
WRS7-430				4.00 L		50- 200	

Drying time, hr	Inj time, sec	Front temp, °F	Mid temp, °F	Rear temp, °F	Nozzle temp, °F	Mold temp, °F	Proc temp, °F
		470-580	475-525	475-500		40-220	400-600
2.0		370-430	350-400	330-350		100-170	
2.0		440-460	430-450	410-440		100-170	
2.0		420-450	410-440	400-430		70-150	
2.0		370-430	350-400	330-350		100-170	
		425-525	425-525	425-525		90-150	
		425-525	425-525	425-525		90-150	
		425-525	425-525	425-525		90-150	
		400-500	400-500	400-500		115-175	
		425-525	425-525	425-525		90-150	
		425-525	425-525	425-525		90-150	
		425-525	425-525	425-525		90-150	
		400-500	400-500	400-500		115-175	
		400-440	400-440	390-430	400-440	60-150	430-455
		400	400	390	400	60-150	
		400	400	390	400	60-150	
		400	400	390	400	60-150	
		400	400	390	400	60-150	
		400	400	390	400	60-150	
		400	400	390	400	60-150	
		400	400	390	400	60-150	
		400-440	400-440	390-430	400-440	60-150	
		400	400	390	400	60-150	
		400-440	400-440	390-430	400-440	60-150	
		400-440	400-440	390-430	400-440	60-150	
		400-440	400-440	390-430	400-440	60-150	
		400-440	400-440	390-430	400-440	60-150	
		400-440	400-440	390-430	400-440	60-150	
		400	400	390	400	60-150	
		400-440	400-440	390-430	400-440	60-150	
		400-440	400-440	390-430	400-440	60-150	
		400-440	400-440	390-430	400-440	60-150	
		400-440	400-440	390-430	400-440	60-150	
		400	400	390	400	60-150	
		400	400	390	400	60-150	
		400	400	390	400	60-150	
		400	400	390	400	60-150	
		400	400	390	400	60-150	
		400	400	390	400	60-150	
		400	400	390	400	60-150	
		400	400	390	400	60-150	
		400	400	390	400	60-150	
		400	400	390	400	60-150	
		400	400	390	400	60-150	
		400	400	390	400	60-150	
		400	400	390	400	60-150	
		400	400	390	400	60-150	
		400	400	390	400	60-150	
		400	400	390	400	60-150	
		400	400	390	400	60-150	
		400	400	390	400	60-150	
		400	400	390	400	60-150	

Grade	Filler	Sp Grav	Shrink, mils/in	Melt flow, g/10 min	Melt temp, °F	Back pres, psi	Drying temp, °F
WRS7-431				8.00 L		50- 200	
WRS7-432				12.00 L		50- 200	
WRS7-440				20.00 L		50- 200	

PP Homopolymer Albis Polypropylene Albis Canada

Grade	Filler	Sp Grav	Shrink, mils/in	Melt flow, g/10 min	Melt temp, °F	Back pres, psi	Drying temp, °F
A12H-TV20	TAL 20	1.06		2.00			
A12H-TV40	TAL 40	1.24		1.50			

PP Homopolymer Amoco Polypropylene Amoco Chem.

Grade	Filler	Sp Grav	Shrink, mils/in	Melt flow, g/10 min	Melt temp, °F	Back pres, psi	Drying temp, °F
1016		0.91	15.0-20.0	5.00 L		50- 100	
1046		0.91	15.0-20.0	5.00 L		50- 100	
1246		0.90	15.0-20.0	20.00 L		50- 100	
4017		0.91	15.0-20.0	8.00 L		50- 100	
4018		0.91	15.0-20.0	12.00 L		50- 100	
4036		0.91	15.0-20.0	5.00 L		50- 100	
4039		0.91	15.0-20.0	18.00 L		50- 100	
6534		0.91	10.0-25.0	5.00 L		50- 100	
7234		0.91	10.0-25.0	12.00 L		50- 100	
7239		0.91	10.0-25.0	12.00 L		50- 100	
7634		0.91	10.0-25.0	23.00 L		50- 100	
7644		0.90	15.0-20.0	23.00 L		50- 100	
7934		0.91	10.0-25.0	35.00 L		50- 100	
7944		0.90	15.0-25.0	35.00 L		50- 100	

PP Homopolymer Escorene Exxon

Grade	Filler	Sp Grav	Shrink, mils/in	Melt flow, g/10 min	Melt temp, °F	Back pres, psi	Drying temp, °F
PP 1403F	TAL 40	1.24		6.50 L			

PP Homopolymer Fiberstran DSM

Grade	Filler	Sp Grav	Shrink, mils/in	Melt flow, g/10 min	Melt temp, °F	Back pres, psi	Drying temp, °F
G-60/20	GLF 20	1.04	3.0				
G-60/30	GLF 30	1.13	3.0				

PP Homopolymer Hiloy ComAlloy

Grade	Filler	Sp Grav	Shrink, mils/in	Melt flow, g/10 min	Melt temp, °F	Back pres, psi	Drying temp, °F
109	UNS	1.22	5.0	6.00			180-200
110	UNS	1.22	5.0	6.00			180-200

PP Homopolymer Petrothene Quantum

Grade	Filler	Sp Grav	Shrink, mils/in	Melt flow, g/10 min	Melt temp, °F	Back pres, psi	Drying temp, °F
PP 1002-NF		0.90		2.00			
PP 1004-GK		0.90		4.00			
PP 2004-MR				12.00			
PP 2085-GW				35.00			
PP 8000-GK			15.0-25.0	5.00			
PP 8001-LK			15.0-25.0	5.00			
PP 8004-MR			15.0-25.0	12.00			
PP 8004-ZR			15.0-25.0	12.00			
PP 8005-AR			15.0-25.0	12.00			
PP 8020-AU			15.0-25.0	20.00			
PP 8020-GU			15.0-25.0	20.00			
PP 8020-ZU			15.0-25.0	20.00			
PP 8070-ZU			15.0-25.0	20.00			
PP 8080-AW			15.0-25.0	35.00			
PP 8080-GW			15.0-25.0	35.00			
PP 8080-ZW				35.00			

PP Homopolymer Polyfort A. Schulman

Grade	Filler	Sp Grav	Shrink, mils/in	Melt flow, g/10 min	Melt temp, °F	Back pres, psi	Drying temp, °F
FPP 1006	TAL 20	1.02				50- 250	
FPP 1007	TAL 40	1.22				50- 250	

Drying time, hr	Inj time, sec	Front temp, °F	Mid temp, °F	Rear temp, °F	Nozzle temp, °F	Mold temp, °F	Proc temp, °F
		400	400	390	400	60-150	
		400	400	390	400	60-150	
		400	400	390	400	60-150	
						70-140	390-540
						70-140	390-540
		400-570	400-570	400-570			
		400-570	400-570	400-570			
		400-570	400-570	400-570			
		400-570	400-570	400-570			
		400-570	400-570	400-570			
		400-570	400-570	400-570			
		400-570	400-570	400-570			
		400-570	400-570	400-570			
		400-570	400-570	400-570			
		400-570	400-570	400-570			
		400-570	400-570	400-570			
		400-570	400-570	400-570			
		400-570	400-570	400-570			
		400-570	400-570	400-570			
							420-500
		360-390	400-440	390-410	360-380	100-160	390-550
		360-390	400-440	390-410	360-380	100-160	390-550
1.0- 2.0		460	470	480	460-480	50-150	460-480
1.0- 2.0		460	470	480	460-480	50-150	460-480
	15.0	435	425	400	450	125	
	15.0	435	425	400	450	125	
	15.0	435	425	400	450	125	
	15.0	435	425	400	450	125	
	15.0	435	425	400	450	125	430-470
	15.0	435	425	400	450	125	430-470
	15.0	435	425	400	450	125	410-440
	15.0	435	425	400	450	125	410-440
	15.0	435	425	400	450	125	410-440
	15.0	435	425	400	450	125	410-440
	15.0	435	425	400	450	125	410-440
	15.0	435	425	400	450	125	410-440
	15.0	435	425	400	450	125	
	15.0	435	425	400	450	125	380-420
	15.0	435	425	400	450	125	380-420
	15.0	435	425	400	450	125	380-420
		425		375		70-120	385-475
		425		375		70-120	385-475

Grade	Filler	Sp Grav	Shrink, mils/in	Melt flow, g/10 min	Melt temp, °F	Back pres, psi	Drying temp, °F
FPP 1080	GLF 20	1.04				50- 250	
FPP 1082	GLF 10	0.96				50- 250	
FPP 1200	CAC 40	1.22				50- 250	
FPP 1208	GLF 30	1.13				50- 250	
FPP 1236	GLF 20	1.04				50- 250	
FPP 1237	CAC 20	1.02				50- 250	
FPP 1239	GLF 30	1.13				50- 250	
FPP 1244	MI 40	1.22				50- 250	
FPP 1257	MI 40	1.22				50- 250	
FPP 1309	MI 10	0.97				50- 250	
FPP 1310	MI 20	1.02				50- 250	
FPP 1337	MI 20	1.02				50- 250	
FPP 1494	MI 30	1.13				50- 250	
FPP 1564	CAC 30	1.13				50- 250	
FPP 1578	CAC 10	0.97				50- 250	
FPP 1602	GLF 10	0.96				50- 250	
FPP 1607	TAL 30	1.13				50- 250	
FPP 1608	MI 10	0.97				50- 250	
FPP 1650	MI 30	1.13				50- 250	
FPP 1666	TAL 10	0.97				50- 250	
FPP 1763	GLF 40	1.18				50- 250	
FPP 1829	GLF 40	1.18				50- 250	

PP Homopolymer Pro-fax Himont

Grade	Filler	Sp Grav	Shrink, mils/in	Melt flow, g/10 min	Melt temp, °F	Back pres, psi	Drying temp, °F
6323		0.90		12.00		50- 300	
6523		0.90		4.00		50- 300	
6524		0.90		4.00		50- 300	
PD-199		0.90		1.50		50- 300	
PD-451		0.93		4.00		50- 300	
PD-626		0.90		12.00		50- 300	
PD-701		0.90		35.00		50- 300	
PF-041N		0.90	12.0	22.00			
PF-511				20.00		50- 300	
PF-531				20.00		50- 300	

PP Homopolymer Rexene Rexene

Grade	Filler	Sp Grav	Shrink, mils/in	Melt flow, g/10 min	Melt temp, °F	Back pres, psi	Drying temp, °F
PP 11B1A		0.90		1.00			
PP 11S30V		0.90		30.00			

PP Homopolymer Shell PP Shell

Grade	Filler	Sp Grav	Shrink, mils/in	Melt flow, g/10 min	Melt temp, °F	Back pres, psi	Drying temp, °F
5550				5.00 L		50- 200	
5610				8.00 L		50- 200	
5824S				12.00 L		50- 200	
5864				12.00 L		50- 200	
5A02				3.20 L		50- 200	
5A08		0.90	12.0-25.0	2.90 L	334	50- 200	
5A09		0.91	12.0-25.0	40.00 L	334	50- 200	
5A10				1.40 L		50- 200	
5A18Z				12.00 L		50- 200	
5A71		0.90	12.0-25.0	8.00 L	334	50- 200	
5A72		0.90	12.0-25.0	12.00 L	334	50- 200	
5A73		0.90	12.0-25.0	15.00 L	334	50- 200	
5A89		0.91	12.0-25.0	35.00 L	334	50- 200	
5A91		0.90	12.0-25.0	15.00 L	334	50- 200	
5C02		0.90	12.0-25.0	3.50 L	334	50- 200	
5C04S				20.00 L		50- 200	

Drying time, hr	Inj time, sec	Front temp, °F	Mid temp, °F	Rear temp, °F	Nozzle temp, °F	Mold temp, °F	Proc temp, °F
		425		375		70-120	385-475
		425		375		70-120	385-475
		425		375		70-120	385-475
		425		375		70-120	385-475
		425		375		70-120	385-475
		425		375		70-120	385-475
		425		375		70-120	385-475
		425		375		70-120	385-475
		425		375		70-120	385-475
		425		375		70-120	385-475
		425		375		70-120	385-475
		425		375		70-120	385-475
		425		375		70-120	385-475
		425		375		70-120	385-475
		425		375		70-120	385-475
		425		375		70-120	385-475
		425		375		70-120	385-475
		425		375		70-120	385-475
		425		375		70-120	385-475
		425		375		70-120	385-475
		425		375		70-120	385-475
		425		375		70-120	385-475
						70-120	400-445
						70-120	445-475
						70-120	445-475
						70-120	475-525
						70-120	445-475
						70-120	400-445
						70-120	385-420
		400-430	380-410	360-400	400-430	50- 80	380-410
						70-120	400-445
						70-120	400-445
		470-580	475-525	475-500		40-220	400-600
		470-580	475-525	475-500		40-220	400-600
		400-440	400-440	390-430	400-440	60-150	
		400	400	390	400	60-150	
		400-440	400-440	390-430	400-440	60-150	
		400-440	400-440	390-430	400-440	60-150	
		400	400	390	400	60-150	
		400	400	390	400	60-150	
		400	400	390	400	60-150	
		400	400	390	400	60-150	
		400-440	400-440	390-430	400-440	60-150	
		400	400	390	400	60-150	
		400	400	390	400	60-150	
		400	400	390	400	60-150	
		400	400	390	400	60-150	
		400	400	390	400	60-150	
		400	400	390	400	60-150	
		400-440	400-440	390-430	400-440	60-150	

Grade	Filler	Sp Grav	Shrink, mils/in	Melt flow, g/10 min	Melt temp, °F	Back pres, psi	Drying temp, °F
5C04Z				20.00 L		50- 200	
5C06L		0.90	12.0-25.0	10.00 L	334	50- 200	
5C08				2.90 L		50- 200	
5C12				11.00 L		50- 200	
5C12S				11.00 L		50- 200	
5C13				12.00 L		50- 200	
5C14				8.00 L		50- 200	
5C19U				3.00 L		50- 200	
5C55S				2.00 L		50- 200	
5C55Z				2.00 L		50- 200	
5C64				4.20 L		50- 200	
5C78				35.00 L		50- 200	
5C97				3.90 L		50- 200	
5C99				3.00 L		50- 200	
DS5A20				20.00 L		50- 200	
DS5A20N				20.00 L		50- 200	
DS5A20S				20.00 L		50- 200	
DS5A20Z				20.00 L		50- 200	
DS5D45				0.65 L		50- 200	
DX5A15				5.00 L		50- 200	
DX5A15H				5.00 L		50- 200	
DX5A15N				5.00 L		50- 200	
DX5A15S				5.00 L		50- 200	
DX5A15Z				5.00 L		50- 200	
DX5A21				8.00 L		50- 200	
DX5A22H				16.00 L		50- 200	
DX5A23				12.00 L		50- 200	
DX5A23N				12.00 L		50- 200	
DX5A23S				12.00 L		50- 200	
DX5A23Z				12.00 L		50- 200	
HM 6100				1.40 L		50- 200	
JM 6100				2.00 L		50- 200	
PDC 1123				20.00 L		50- 200	
PDC 1155				35.00 L		50- 200	
PDC 1156				35.00 L		50- 200	
PDC 1166				16.00 L		50- 200	
PDC 1174				20.00 L		50- 200	
PDC 1175				20.00 L		50- 200	
PDC 1176				20.00 L		50- 200	
PM 6100				5.00 L		50- 200	
RM 6100				8.00 L		50- 200	
SM 6100				12.00 L		50- 200	
SM 6703				12.00 L		50- 200	
SRD5-1119				0.60 L		50- 200	
SRD5-1120				0.60 L		50- 200	
VM 6100				20.00 L		50- 200	
VM 6300				20.00 L		50- 200	
VM 6307				20.00 L		50- 200	

PP Homopolymer Thermofil Polypropylene Thermofil

Grade	Filler	Sp Grav	Shrink, mils/in	Melt flow, g/10 min	Melt temp, °F	Back pres, psi	Drying temp, °F
P6-30FG-1600	GLC 30	1.13	3.0	3.50 L		50	160
P6-30FG-2600	GLC 30	1.13	3.0	5.00 L		50	160

PPA Amodel Amoco Perform.

Grade	Filler	Sp Grav	Shrink, mils/in	Melt flow, g/10 min	Melt temp, °F	Back pres, psi	Drying temp, °F
A-1115 HS	GLF 15	1.26	6.0- 7.0		590		175
A-1133 HS	GLF 33	1.43	2.0- 6.0		590		175

Drying time, hr	Inj time, sec	Front temp, °F	Mid temp, °F	Rear temp, °F	Nozzle temp, °F	Mold temp, °F	Proc temp, °F
		400-440	400-440	390-430	400-440	60-150	
		400	400	390	400	60-150	
		400	400	390	400	60-150	
		400-440	400-440	390-430	400-440	60-150	
		400-440	400-440	390-430	400-440	60-150	
		400	400	390	400	60-150	
		400-440	400-440	390-430	400-440	60-150	
		400	400	390	400	60-150	
		400	400	390	400	60-150	
		400	400	390	400	60-150	
		400	400	390	400	60-150	
		400	400	390	400	60-150	
		400	400	390	400	60-150	
		400	400	390	400	60-150	
		400	400	390	400	60-150	
		400	400	390	400	60-150	
		400	400	390	400	60-150	
		400	400	390	400	60-150	
		400	400	390	400	60-150	
		400	400	390	400	60-150	
		400	400	390	400	60-150	
		400	400	390	400	60-150	
		400	400	390	400	60-150	
		400	400	390	400	60-150	
		400	400	390	400	60-150	
		400	400	390	400	60-150	
		400	400	390	400	60-150	
		400	400	390	400	60-150	
		400	400	390	400	60-150	
		400	400	390	400	60-150	
		400	400	390	400	60-150	
		400	400	390	400	60-150	
		400	400	390	400	60-150	
		400	400	390	400	60-150	
		400	400	390	400	60-150	
		400	400	390	400	60-150	
		400	400	390	400	60-150	
		400	400	390	400	60-150	
		400	400	390	400	60-150	
		400	400	390	400	60-150	
		400	400	390	400	60-150	
		400	400	390	400	60-150	
		400	400	390	400	60-150	
		400	400	390	400	60-150	
		400	400	390	400	60-150	
		400	400	390	400	60-150	
		400	400	390	400	60-150	
		400	400	390	400	60-150	
2.0		420-450	400-420	380-400	430-460	80-150	
2.0		420-450	400-420	380-400	430-460	80-150	
16.0		585-625	585-625	585-625		150-330	610-650
16.0		585-625	585-625	585-625		150-330	610-650

Grade	Filler	Sp Grav	Shrink, mils/in	Melt flow, g/10 min	Melt temp, °F	Back pres, psi	Drying temp, °F
A-1145 HS	GLF 45	1.56	2.0- 3.0		590		175
A-1230 HS	MN 30	1.45	11.0		590		175
A-1240 HN	MN 40	1.54	8.0		590		175
A-1340 HS	GMN 40	1.54	4.0- 7.0		590		175
AD-1000		1.17	15.0-20.0		590		175
AF-1133 V0	GLF 33	1.71	2.0- 4.0		590		175
ET-1000		1.13	15.0-20.0		590		175
PPA		**PermaStat**			**RTP**		
4000		1.21	16.0-22.0				
4005	GLF 30	1.43	2.5- 3.5				
PPA		**Thermocomp**			**LNP**		
UC-1006	CBL 30	1.33	1.0- 3.0		590	0- 50	175
UF-1006	GLF 30	1.43	2.0- 4.0		590	0- 50	175
UL-4020	PTF 10	1.26	16.0-22.0		590	0- 50	175
PPC		**Lexan**			**GE**		
PPC4501		1.20	7.0- 8.0			50- 200	265
PPC4504		1.20	7.0- 8.0	2.00		50- 200	265
PPC4701		1.20	8.0-10.0			50- 200	265
PPC4704		1.20	8.0-10.0	1.00		50- 200	265
RA2000		1.20	5.0- 7.0	25.00 O		50- 100	250
PPE		**Iupiace**			**Mitsubishi Gas**		
AX5010		1.06	5.0- 7.0			1137-2132	
AX5015		1.06	5.0- 7.0			1137-2132	
AX5026		1.06	5.0- 7.0			1137-2132	
NX-9000		1.11	12.0-13.0			1137-2132	
PPE		**Prevex**			**GE**		
PMA		1.06	6.0- 7.0			50- 100	220
PQA		1.06	6.0- 7.0			50- 100	220
VFA		1.09	5.0- 7.0			50- 100	180
VFAX		1.09	5.0- 7.0			50- 100	180
VGA		1.10	6.0- 7.0			50- 100	190
VJA		1.10	5.0- 7.0			50- 100	210
VKA		1.09	6.0- 7.0			50- 100	220
VQA		1.06	6.0- 8.0			50- 100	220
W20		1.06	6.0- 7.0			50- 100	180-200
W50		1.06	6.0- 7.0			50- 100	200-220
PPE		**RTP Polymers**			**RTP**		
2499X51452A		1.57	2.0- 3.0				
PPE		**Shuman PPE**			**Shuman Plastics**		
210		1.09		1.30			180-220
211		1.10		1.50			180-220
PPO		**CTI Modified PPO**			**CTI**		
MP-10CF-4CC/15T	CF 14	1.22	1.5- 2.0			0- 100	200-250
MP-10MCF	CF 10	1.10	1.5- 2.0			0- 100	200-250
MP-24CC	CF 24	1.11	1.5- 2.0			0- 100	200-250
MP-25NCF	GRF 25	1.25	1.0			0- 100	200-250
MP-30CF	CF 30	1.12	0.9- 1.5			0- 100	200-250
MP-30GF	GLF 30	1.28	1.0			0- 100	200-250

Drying time, hr	Inj time, sec	Front temp, °F	Mid temp, °F	Rear temp, °F	Nozzle temp, °F	Mold temp, °F	Proc temp, °F
16.0		585-625	585-625	585-625		150-330	610-650
16.0		585-625	585-625	585-625		150-330	610-650
16.0		585-625	585-625	585-625		150-330	610-650
16.0		585-625	585-625	585-625		150-330	610-650
16.0		585-625	585-625	585-625		150-330	610-650
16.0		585-625	585-625	585-625		150-330	610-650
16.0		585-625	585-625	585-625		150-330	610-650
		530-580	530-580	530-580		200-300	
		530-580	530-580	530-580		200-300	
10.0-16.0		585-660	585-660	585-660		275	600-625
10.0-16.0		585-660	585-660	585-660		275	600-625
10.0-16.0		585-660	585-660	585-660		275	600-625
3.0- 5.0		620-690	575-630	575-630	630-690	180-210	630-690
3.0- 5.0		620-690	575-630	575-630	630-690	180-210	630-690
3.0- 5.0		640-710	595-650	595-650	650-710	180-210	650-710
3.0- 5.0		640-710	595-650	595-650	650-710	180-210	650-710
3.0- 4.0		500-520	480-500	470-500	490-520	140-180	500-540
						122-176	500-554
						122-176	482-536
						122-176	500-554
						158-248	500-590
2.0- 4.0		500-550	490-540	430-480	500-550	160-180	500-550
2.0- 4.0		540-580	520-570	460-510	540-580	180-230	540-580
2.0- 4.0		425-500	415-490	405-460	425-500	150-180	425-500
2.0- 4.0		425-500	415-490	405-460	425-500	150-180	425-500
2.0- 4.0		450-520	440-510	420-470	450-520	150-180	450-520
2.0- 4.0		480-540	460-510	430-470	480-540	150-170	480-540
2.0- 4.0		490-540	470-530	430-480	490-540	160-180	490-540
2.0- 4.0		540-580	520-570	460-510	540-580	180-230	540-580
2.0- 4.0		490-540	490-540	430-480	490-540	150-200	500-550
2.0- 4.0		515-565	515-565	455-505	515-565	170-220	525-575
		500-550	500-550	500-550		150-250	
2.0- 4.0							430-500
2.0- 4.0							430-500
		540-580	530-570	530-560	550-590	190-220	
		540-580	530-570	530-560	550-590	190-220	
		540-580	530-570	530-560	550-590	190-220	
		540-580	530-570	530-560	550-590	190-220	
		540-580	530-570	530-560	550-590	190-220	
		540-580	530-570	530-560	550-590	190-220	

Grade	Filler	Sp Grav	Shrink, mils/in	Melt flow, g/10 min	Melt temp, °F	Back pres, psi	Drying temp, °F
MP-40CF	CF 40	1.25	0.2- 0.2			0- 100	200-250
PPO		**DSM Polyphenylene Oxide**				**DSM**	
J-1700/20	GLF 20	1.21					160-180
J-1700/30	GLF 30	1.31					160-180
J-1700/40	GLF 40	1.40					160-180
PPO		**Fiberfil**			**DSM**		
J-1700/20	GLF 20	1.21					150-180
J-1700/30	GLF 30	1.31					150-180
J-1700/40	GLF 40	1.40					150-180
PPO		**Fiberfil V0**			**DSM**		
J-1700/20/V0/BK	GLF 20	1.26					160-180
J-1700/20/V0	GLF 20	1.30	1.6				160-180
J-1700/30/V0/BK	GLF 30	1.34					160-180
J-1700/30/V0	GLF 30	1.39	0.9				160-180
PPO		**Fiberstran**			**DSM**		
G-1704/SS/5	STS 5	1.13	7.0				160-180
PPO		**Noryl**			**GE**		
731		1.06	5.0- 7.0			50- 100	220-230
731H		1.06	5.0- 7.0			50- 100	220-230
CRT200		1.09	5.0- 7.0			50- 100	205-215
EM5101		1.06	5.0- 7.0			50- 100	210-225
EM5102		1.06	5.0- 7.0			50- 100	210-225
EM5103		1.06	5.0- 7.0			50- 100	210-225
EM5104		1.06	5.0- 7.0			50- 100	215-230
EM6100		1.04	5.0- 7.0			50- 100	205-220
EM6101		1.05	5.0- 7.0			50- 100	210-225
EM7100		1.04	5.0- 7.0			50- 100	180-190
EM7301	GLF 10	1.12	4.0- 6.0			50- 100	210-225
EM7304	GLF 16	1.16	3.0- 5.0			50- 100	210-225
EZ250		1.07	5.0- 7.0			50- 100	200-220
FMC1010	CF 10	0.92				50- 100	180-200
FMC3008A	CGM 30	1.02					200-220
FN215		0.99	6.0- 8.0				
GFN1	GL 10	1.14	2.0- 5.0			50- 100	200
GFN2	GL 20	1.20	2.0- 5.0			50- 100	240-250
GFN3	GL 30	1.28	1.0- 4.0			50- 100	240-250
HM1500	GL 15	1.25	3.0			25- 100	200-230
HM3020 (Black)	GMN 30	1.31	2.5			25- 100	200-230
HM3020 (Grey)	GMN 30	1.31	2.5			25- 100	200-230
HM4025	GMN 40	1.43	1.5			25- 100	200-230
HMC1010	CF 10	1.15	0.5- 1.5			50- 100	180-200
HMC3008A	CGM 30	1.28	0.5- 1.5				200-220
HP41		1.06	5.0- 7.0			50- 100	220
HS1000	UNS	1.21	5.0- 7.0			50- 100	210-220
HS1000X	MNF	1.23	5.0- 7.0			50- 100	210-220
HS2000	MNF	1.23	5.0- 7.0			50- 100	220-230
HS2000E		1.23	5.0- 7.0			50- 100	220-230
N190		1.10	5.0- 7.0			50- 100	170-190
N190H		1.08	5.0- 7.0			50- 100	170-190
N190HX		1.10	5.0- 7.0			50- 100	170-190
N190X		1.12	5.0- 7.0			50- 100	170-190

Drying time, hr	Inj time, sec	Front temp, °F	Mid temp, °F	Rear temp, °F	Nozzle temp, °F	Mold temp, °F	Proc temp, °F
		540-580	530-570	530-560	550-590	190-220	
2.0-16.0		520-575	530-575	500-550	530-575	140-210	530-575
2.0-16.0		520-575	530-575	500-550	530-575	140-210	530-575
2.0-16.0		520-575	530-575	500-550	530-575	140-210	530-575
2.0-16.0		520-575	530-575	500-550	530-575	140-210	530-575
2.0-16.0		520-575	530-575	500-550	530-575	140-210	530-575
2.0-16.0		520-575	530-575	500-550	530-575	140-210	530-575
2.0-16.0		520-575	530-575	500-550	530-575	140-210	530-575
2.0-16.0		520-575	530-575	500-550	530-575	140-210	530-575
2.0-16.0		520-575	530-575	500-550	530-575	140-210	530-575
2.0-16.0		520-575	530-575	500-550	530-575	140-210	530-575
2.0-16.0		520-575	530-575	500-550	530-575	140-210	530-575
2.0- 4.0		540-570	520-560	500-540	530-570	160-200	540-590
2.0- 4.0		540-570	520-560	500-540	530-570	160-200	540-590
2.0- 4.0		490-540	470-500	450-480	500-520	150-180	450-540
2.0- 4.0		570-580	550-570	530-550	550-580	120-180	550-580
2.0- 4.0		570-580	550-570	530-550	550-580	120-180	550-580
2.0- 4.0		570-590	550-580	540-560	560-590	120-180	560-590
2.0- 4.0		560-580	570-590	560-590	540-560	120-180	560-590
2.0- 4.0		520-550	520-550	520-560	510-540	120-180	520-560
2.0- 4.0		520-550	520-550	520-560	510-540	120-180	520-560
2.0- 4.0		500-520	510-530	510-540	490-510	120-160	490-540
2.0- 4.0		510-540	520-550	520-570	500-530	110-150	520-570
2.0- 4.0		520-550	510-540	500-530	520-570	110-150	520-570
2.0- 4.0		560-580	540-560	530-540	570-580	100-200	530-580
2.0- 4.0		500-540	480-500	460-480	510-540	150-200	460-540
2.0- 4.0		520-570	520-550	500-540	520-570	150-190	520-570
2.0- 4.0		550-590	540-580	540-570	550-590	160-200	550-590
2.0- 4.0		540-580	530-570	530-560	550-590	190-220	550-620
2.0- 4.0		540-580	530-570	530-560	550-590	190-220	550-620
2.0- 4.0		535-570	520-555	500-535	535-570	160-220	535-570
2.0- 4.0		560-600	540-580	520-560	560-600	160-220	560-600
2.0- 4.0		560-600	540-580	520-560	560-600	160-220	560-600
2.0- 4.0		560-600	540-580	520-560	560-600	160-200	560-600
2.0- 4.0		500-540	480-500	460-480	510-540	150-200	460-540
2.0- 4.0		520-570	520-550	500-540	520-570	150-190	520-570
2.0- 4.0		520-570	520-570	500-540	540-590	160-190	540-590
2.0- 4.0		490-520	470-500	450-480	500-530	150-180	480-540
2.0- 4.0		490-520	470-500	450-480	500-530	150-180	480-540
2.0- 4.0		540-580	540-570	520-560	550-590	160-200	540-590
2.0- 4.0		540-580	540-570	520-560	550-590	160-200	540-590
2.0- 4.0		490-550	470-490	450-470	500-520	150-180	450-540
2.0- 4.0		490-550	470-490	450-470	500-520	150-180	450-540
2.0- 4.0		490-525	470-490	450-470	500-525	150-180	450-525
2.0- 4.0		490-525	470-490	450-470	500-525	150-180	450-525

Grade	Filler	Sp Grav	Shrink, mils/in	Melt flow, g/10 min	Melt temp, °F	Back pres, psi	Drying temp, °F
N225		1.09	5.0- 7.0			50- 100	220-230
N2251	GL 20	1.21	2.0- 4.0				220-230
N225X		1.11	5.0- 7.0			50- 100	220-230
N300		1.09	5.0- 7.0			50- 100	220-230
PC180		1.07	5.0- 7.0			50- 100	210-220
PC180X		1.11	5.0- 7.0			50- 100	170-190
PN235		1.05	5.0- 7.0			50- 100	190-205
PX0722		1.04	5.0- 7.0			50- 100	190-205
PX0844		1.06	5.0- 7.0			50- 100	210-225
PX0888		1.04	5.0- 7.0			50- 100	210-225
PX1127		1.06	5.0- 7.0				220-230
PX1222		1.04	5.0- 7.0			50- 100	210-225
PX1265		1.06	5.0- 7.0			50- 100	210-225
PX1269		1.12	5.0- 7.0				220-225
PX1390		1.06	5.0- 7.0			50- 100	210-230
PX1391		1.07	5.0- 7.0			50- 100	210-230
PX1394		1.10	5.0- 7.0				225-240
PX1701		1.06	5.0- 7.0				180-190
PX1703		1.06	5.0- 7.0			50- 100	160-180
PX4058		1.06				50- 100	210
PX4207		1.06	5.0- 7.0			50- 100	170-190
PX4608		1.06	5.0- 7.0			50- 100	
PX4685		1.10	5.0- 7.0			50- 100	190-205
PX5327	CF 5	1.14	2.0- 6.0			50- 100	180-200
RFN30	GL 30	1.27	2.0- 5.0			50- 100	240-250
RFN420	UNS	1.21	2.0- 4.0			50- 100	240-250
RFN430	UNS	1.27	1.0- 3.0			50- 100	240-250
SE1		1.08	5.0- 7.0			50- 100	225
SE100		1.08	5.0- 7.0			50- 100	210-220
SE100H		1.10	5.0- 7.0			50- 100	210-220
SE100HX		1.09	5.0- 7.0			50- 100	210-220
SE100X		1.10	5.0- 7.0			50- 100	210-220
SE1GFN1	GL 10	1.16	3.0- 5.0			50- 100	240-250
SE1GFN2	GL 20	1.23	2.0- 5.0			50- 100	240-250
SE1GFN3	GL 30	1.31	1.0- 4.0			50- 100	240-250
SE1X		1.10	5.0- 7.0			50- 100	225
SPN410		1.06	5.0- 7.0			50- 100	220
SPN410H		1.06	5.0- 7.0			50- 100	220
SPN420		1.07	5.0- 7.0			50- 100	220
SPN422L		1.08	5.0- 7.0			50- 100	220
STN15		1.04				50- 60	210-220
UV180		1.08	5.0- 7.0			50- 100	200-215

PPO Noryl GTX GE

Grade	Filler	Sp Grav	Shrink, mils/in	Melt flow, g/10 min	Melt temp, °F	Back pres, psi	Drying temp, °F
GTX600		1.10	7.0-10.0			50- 100	210
GTX810	GLF 10	1.16	4.0- 7.0			50- 100	225
GTX820	GLF 20	1.24	4.0- 7.0			50- 100	225
GTX830	GLF 30	1.33	4.0- 7.0			50- 100	225
GTX901		1.08	8.0-11.0			50- 100	225
GTX902		1.08	8.0-11.0			50- 100	225
GTX904		1.08	8.0-11.0			50- 100	225
GTX909		1.13	13.0-17.0			50- 100	200
GTX910		1.10	11.0-13.0			50- 100	225
GTX917		1.10	11.0-13.0			50- 100	225
PX5231		1.08				50- 100	225

Drying time, hr	Inj time, sec	Front temp, °F	Mid temp, °F	Rear temp, °F	Nozzle temp, °F	Mold temp, °F	Proc temp, °F
2.0- 4.0		510-530	500-520	490-510	500-540	160-200	500-550
2.0- 4.0		540-580	530-570	530-560	550-590	190-220	550-620
2.0- 4.0		510-530	500-520	490-510	500-540	160-200	500-550
2.0- 4.0		560-590	550-580	540-570	560-590	190-220	550-620
2.0- 4.0		490-520	470-490	450-470	500-520	150-180	480-540
2.0- 4.0		490-540	470-510	450-480	490-540	150-180	490-540
2.0- 4.0		530-560	540-570	530-570	520-540	130-190	530-570
2.0- 4.0		540-560	520-540	510-530	530-560	120-180	530-560
2.0- 4.0		560-580	540-570	530-570	550-580	120-180	550-580
2.0- 4.0		570-580	550-570	530-550	550-580	120-180	550-580
2.0- 4.0						160-200	540-590
2.0- 4.0		560-580	540-570	530-550	550-580	120-180	550-580
2.0- 4.0		570-590	550-580	540-560	560-590	120-180	560-590
2.0- 4.0						160-200	550-580
2.0- 4.0		570-590	550-580	540-560	560-590	120-180	560-590
2.0- 4.0		580-600	560-580	540-560	570-600	120-180	570-600
2.0- 4.0						190-200	520-580
2.0- 4.0		510-530	500-520	490-510	510-540	120-160	490-540
2.0- 4.0		510-530	500-520	490-510	500-540	150-180	480-540
2.0- 4.0		535-555	525-545	515-535	525-565	120-180	525-575
2.0- 4.0		510-530	500-520	490-510	510-540	120-160	490-540
		520-570	520-570	520-560	520-570	160-180	530-580
2.0- 4.0		540-560	520-540	510-530	530-560	120-180	530-560
2.0- 4.0		500-540	480-500	460-480	510-540	150-200	460-540
2.0- 4.0		540-580	530-570	530-560	550-590	190-220	550-620
2.0- 4.0		535-590	530-580	525-570	540-600	190-220	550-600
2.0- 4.0		550-610	540-600	530-580	550-620	190-220	550-620
2.0- 4.0		540-570	520-560	500-540	550-590	160-200	540-590
2.0- 4.0		490-520	470-490	450-470	500-520	150-180	480-540
2.0- 4.0		490-520	470-490	450-470	500-520	150-180	480-540
2.0- 4.0		490-520	470-490	450-470	500-520	150-180	480-540
2.0- 4.0		490-520	470-490	450-470	500-520	150-180	480-540
2.0- 4.0		540-580	530-570	530-560	550-590	190-220	550-620
2.0- 4.0		540-580	530-570	530-560	550-590	190-220	550-620
2.0- 4.0		540-580	530-570	530-560	550-590	190-220	550-620
2.0- 4.0		540-570	520-560	500-540	550-590	160-200	540-590
2.0- 4.0		520-570	520-570	500-540	540-590	160-190	540-590
2.0- 4.0		520-570	520-570	500-540	540-590	160-190	540-590
2.0- 4.0		540-580	540-570	535-565	540-590	160-200	540-590
2.0- 4.0		540-580	540-570	535-565	540-590	160-200	540-590
3.0		530-570	510-550	490-530	550-590	110-140	550-590
2.0- 4.0		490-520	470-490	450-470	500-520	150-180	450-525
4.0		545-585	530-565	510-550	550-585	140-210	545-585
2.0- 4.0		530-575	520-570	500-550	540-575	170-250	530-575
2.0- 4.0		530-575	520-570	500-550	540-575	170-250	530-575
2.0- 4.0		530-575	520-570	500-550	540-575	170-250	530-575
2.0- 4.0		530-575	520-570	500-550	530-575	150-200	530-575
2.0- 4.0		530-575	520-570	500-550	530-575	150-200	530-575
2.0- 4.0		530-575	520-570	500-550	530-575	150-200	530-575
2.0- 4.0		520-560	510-550	500-540	520-560	150-200	520-560
2.0- 4.0		530-575	520-570	500-550	540-575	170-230	530-575
2.0- 4.0		560-600	550-580	520-580	560-600	170-230	560-600
2.0- 4.0		530-575	520-570	500-550	540-575	170-230	530-575

Grade	Filler	Sp Grav	Shrink, mils/in	Melt flow, g/10 min	Melt temp, °F	Back pres, psi	Drying temp, °F
PPO Alloy		**Noryl**			**GE**		
MX4717				6.00 0		40- 60	220
PPS		**Celstran**			**Hoechst**		
PPSC40-01-4	CF 40	1.46					270
PPSG30-01-4	GLF 30	1.52					270
PPSG40-01-4	GLF 40	1.62					270
PPSG50-01-4	GLF 50	1.72					270
PPSG60-01-4	GLF 60	1.84					270
PPSK35-01-2	KEV 35	1.35					270
PPSK35-01-4	KEV 35	1.35					270
PPS		**Celstran S**			**Hoechst**		
PPS6-01-4	STS 6	0.95	5.0- 6.0				200
PPS		**CTI General**			**CTI**		
SF-000/30T	PTF 30	1.51	7.0				300
SF-20CF/000	CF 20	1.40	1.5				300
SF-30CF/000	CF 30	1.43	1.0				300
SF-30CF/15T	CF 30	1.52	1.0				300
SF-30GF/000	GLF 30	1.56	2.0				300
SF-30GF/15T	GLF 30	1.67	2.0				300
SF-40CF/000	CF 40	1.46	0.5				300
SF-40GF/000	GLF 40	1.65	1.0				300
SF-47GF/000	GLF 47	1.72	0.5- 1.0				300
TPX-80CNI	CBI 80	2.92	6.0				300
PPS		**CTI PPS**			**CTI**		
SF-20CF	CF 20	1.38	1.5				300
SF-20CF/000	CF 20	1.40	3.5				300
SF-30CF	CF 30	1.41	1.0				300
SF-30CF/15T	CF 30	1.53	1.0				300
SF-30GF	GLF 30	1.55	2.0- 3.0				300
SF-30GF/15T	GLF 30	1.69	2.0				300
SF-40CF	CF 40	1.45	0.1				300
SF-40GF	GLF 40	1.65	2.0				300
PPS		**DSM PPS**			**DSM**		
J-1300/10	GLF 10	1.39	3.0				250-325
J-1300/20	GLF 20	1.47	2.0				250-325
J-1300/25/MF/40	GLF 25	1.94	1.0				250-325
J-1300/30	GLF 30	1.56	2.0				250-325
J-1305/10	GLF 10	1.42	4.0				250-325
J-1305/20	GLF 20	1.44	3.0				250-325
J-1305/30	GLF 30	1.54	2.0				250-325
J-1305/30/TF/15	GLF 30	1.66	1.0				250-325
J-1305/40	GLF 40	1.68	1.0				250-325
PPS		**Fiberfil**			**DSM**		
J-1300/10	GLF 10	1.39	3.0- 5.0				250-325
J-1300/20	GLF 20	1.47	2.0- 3.0				250-325
J-1300/25/MF/40	GLF 25	1.95	1.0- 2.0				250-325
J-1300/30	GLF 30	1.56	2.0				250-325

Drying time, hr	Inj time, sec	Front temp, °F	Mid temp, °F	Rear temp, °F	Nozzle temp, °F	Mold temp, °F	Proc temp, °F
4.0		490-565	460-545	440-530	500-575	130-180	500-575
4.0		590-610	580-600	570-590	590-610	290-310	590-610
4.0		570-590	560-580	550-570	550-570	290-310	570-590
4.0		580-600	570-590	560-580	560-580	290-310	580-600
4.0		590-610	580-600	570-590	570-590	290-310	590-610
4.0		600	590	580	610	300	
4.0		600-620	600-620	590-610	600-620	290-310	600-620
4.0		620	610	600	635	300	
2.0		390	380	370	390	150	
2.0		630-650	600-640	580-620	600-620	275-300	
2.0		630-650	600-640	580-620	600-620	275-300	
2.0		630-650	600-640	580-620	600-620	275-300	
2.0		630-650	600-640	580-620	600-620	275-300	
2.0		630-650	600-640	580-620	600-620	275-300	
2.0		630-650	600-640	580-620	600-620	275-300	
2.0		630-650	600-640	580-620	600-620	275-300	
2.0		630-650	600-640	580-620	600-620	275-300	
2.0		630-650	600-640	580-620	600-620	275-300	
2.0		630-650	600-640	580-620	600-620	275-300	
2.0		630-650	600-640	580-620	600-620	275-300	
2.0		630-650	600-640	580-620	600-620	275-300	
2.0		630-650	600-640	580-620	600-620	275-300	
2.0		630-650	600-640	580-620	600-620	275-300	
2.0		630-650	600-640	580-620	600-620	275-300	
2.0		630-650	600-640	580-620	600-620	275-300	
2.0		630-650	600-640	580-620	600-620	275-300	
2.0		630-650	600-640	580-620	600-620	275-300	
4.0-16.0		590-630	600-650	550-580	600-630	100-400	550-600
4.0-16.0		590-630	600-650	550-580	600-630	100-400	550-600
4.0-16.0		590-630	600-650	550-580	600-630	100-400	550-600
4.0-16.0		590-630	600-650	550-580	600-630	100-400	550-600
4.0-16.0		590-630	600-650	550-580	600-630	100-400	550-600
4.0-16.0		590-630	600-650	550-580	600-630	100-400	550-600
4.0-16.0		590-630	600-650	550-580	600-630	100-400	550-600
4.0-16.0		590-630	600-650	550-580	600-630	100-400	550-600
4.0-16.0		590-630	600-650	550-580	600-630	100-400	550-600
4.0-16.0		590-630	600-650	550-580	600-630	100-400	550-600
4.0-16.0		590-630	600-650	550-580	600-630	100-400	550-600
4.0-16.0		590-630	600-650	550-580	600-630	100-400	550-600
4.0-16.0		590-630	600-650	550-580	600-630	100-400	550-600

Grade	Filler	Sp Grav	Shrink, mils/in	Melt flow, g/10 min	Melt temp, °F	Back pres, psi	Drying temp, °F
PPS		**Fortron**			**Hoechst**		
0205B4		1.40	8.0-12.0				250-275
0205P4		1.40	8.0-12.0				250-275
0214B1		1.40	8.0-12.0				250-275
0214P1		1.40	8.0-12.0				250-275
1140A1	GLF 40	1.64	1.0- 6.0				250-275
1140A4	GLF 40	1.64	1.0- 6.0				250-275
1140E4	GLF 40	1.60	2.0- 7.0				250-275
1140E7	GLF 40	1.60	2.0- 7.0				250-275
1140E8	GLF 40	1.60	2.0- 7.0				250-275
1140L4	GLF 40	1.60	2.0- 7.0				250-275
1140L7	GLF 40	1.60	2.0- 7.0				250-275
6165A4	GMN 65	2.00	1.0- 5.0				250-275
PPS		**Lusep**			**Lucky**		
GP-2400	GLF 40	1.65	2.5				248-320
GP-4600	GMN 60	1.96	1.5				248-320
PPS		**Plaslube**			**DSM**		
J-1300/30/TF/15	GLF 30	1.65	1.0				325
J-1300/CF/20	CF 20	1.60	0.5				325
J-1300/CF/30	CF 30	1.47	0.6				325
PPS-1305/TF/20	PTF 20	1.47					250-325
PPS		**RTP Polymers**			**RTP**		
1300	GLF	1.36	10.0			25- 100	300
1300A	GLF	1.30	10.0			25- 100	300
1301	GLF	1.41	6.0			25- 100	300
1301 P-1	GLF	1.41	4.0			25- 100	300
1302	GLF	1.41	4.0			25- 100	300
1302 TFE 10	GLF	1.46	4.0			25- 100	300
1303	GLF	1.49	3.0			25- 100	300
1303 P-1	GLF	1.48	2.0			25- 100	300
1303 TFE 20	GLF	1.64	2.0			25- 100	300
1305	GLF	1.59	2.0			25- 100	300
1305 P-1	GLF	1.55	1.0			25- 100	300
1307	GLF	1.69	1.0			25- 100	300
1307 P-1	GLF	1.65	1.0			25- 100	300
1307 TFE 10	GLF	1.71	1.0			25- 100	300
1378	GLF	1.70	2.0			25- 100	300
1379	GMN	1.73	4.0			25- 100	300
1379S	GMN	1.73	4.0			25- 100	300
1381 P-1	CF	1.38	2.0			50- 100	300
1383	CF	1.42	1.5			50- 100	300
1383 P-1	CF	1.42	1.0			50- 100	300
1385	CF	1.46	1.0			50- 100	300
1385 P-1	CF	1.44	1.0			50- 100	300
1385 TFE 15	CF	1.54	1.0			50- 100	300
1387	CF	1.49	0.5			50- 100	300
1387 HM	CF	1.50	0.5			50- 100	300
1387 P-1	CF	1.47	1.0			50- 100	300
1387 TFE 10	CF	1.50	0.8			50- 100	300
1389	CF	1.51	0.5			50- 100	300
1389 HM	CF	1.52	0.5			50- 100	300
1390	CF	1.53	0.5			50- 100	300

Drying time, hr	Inj time, sec	Front temp, °F	Mid temp, °F	Rear temp, °F	Nozzle temp, °F	Mold temp, °F	Proc temp, °F
3.0- 4.0		600-640	590-640	585-635	600-640	275-325	590-640
3.0- 4.0		600-640	590-640	585-635	600-640	275-325	560-620
3.0- 4.0		600-640	590-640	585-635	600-640	275-325	590-640
3.0- 4.0		600-640	590-640	585-635	600-640	275-325	590-640
3.0- 4.0		600-640	590-640	585-635	600-640	275-325	590-640
3.0- 4.0		600-640	590-640	585-635	600-640	275-325	590-640
3.0- 4.0		600-640	590-640	585-635	600-640	275-325	590-640
3.0- 4.0		600-640	590-640	585-635	600-640	275-325	590-640
3.0- 4.0		600-640	590-640	585-635	600-640	275-325	590-640
3.0- 4.0		600-640	590-640	585-635	600-640	275-325	590-640
3.0- 4.0		600-640	590-640	585-635	600-640	275-325	590-640
3.0- 4.0		600-640	590-640	585-635	600-640	275-325	590-640
2.0- 4.0		608	590	554	626	248-302	617-626
2.0- 4.0		644	608	572	644	248-302	635-644
4.0		590-630	600-650	550-580	600-630	100-400	550-600
4.0		590-630	600-650	550-580	600-630	100-400	550-600
4.0		590-630	600-650	550-580	600-630	100-400	550-600
4.0-16.0		590-630	600-650	550-580	600-630	100-400	550-600
6.0		610-675	575-650	550-625		150-350	
6.0		610-675	575-650	550-625		150-350	
6.0		610-675	575-650	550-625		150-350	
6.0		610-675	575-650	550-625		150-350	
6.0		610-675	575-650	550-625		150-350	
6.0		610-675	575-650	550-625		150-350	
6.0		610-675	575-650	550-625		150-350	
6.0		610-675	575-650	550-625		150-350	
6.0		610-675	575-650	550-625		150-350	
6.0		610-675	575-650	550-625		150-350	
6.0		610-675	575-650	550-625		150-350	
6.0		610-675	575-650	550-625		150-350	
6.0		610-675	575-650	550-625		150-350	
6.0		610-675	575-650	550-625		150-350	
6.0		610-675	575-650	550-625		150-350	
6.0		610-675	575-650	550-625		150-350	
6.0		610-675	575-650	550-625		150-350	
6.0		610-675	575-650	550-625		150-350	
6.0		610-675	575-650	550-625		150-350	
6.0		610-675	575-650	550-625		150-350	
6.0		610-675	575-650	550-625		150-350	
6.0		610-675	575-650	550-625		150-350	
6.0		610-675	575-650	550-625		150-350	
6.0		610-675	575-650	550-625		150-350	
6.0		610-675	575-650	550-625		150-350	
6.0		610-675	575-650	550-625		150-350	
6.0		610-675	575-650	550-625		150-350	
6.0		610-675	575-650	550-625		150-350	
6.0		610-675	575-650	550-625		150-350	
6.0		610-675	575-650	550-625		150-350	

Grade	Filler	Sp Grav	Shrink, mils/in	Melt flow, g/10 min	Melt temp, °F	Back pres, psi	Drying temp, °F
1390 HM	CF	1.55	0.4			50- 100	300
1391	CF	1.55	0.5			50- 100	300
1391 HM	CF	1.56	0.3			50- 100	300
1399X24142	UNS	1.57	1.0			50- 100	300
1399X53632B	GLF	1.89	2.0			25- 100	300
EMI-1361	GLF	1.62	3.0			25- 100	300
ESD-A-1305	GLF	1.62	2.0			25- 100	300
ESD-A-1380	CF	1.53	1.0			50- 100	300
ESD-C-1305	GLF	1.63	2.0			25- 100	300
ESD-C-1380	CF	1.55	1.0			50- 100	300
PPS		**Ryton**			**Phillips**		
A-200	GLF	1.60				0- 50	300-350
R-10 5002C	GMN	1.99				0- 50	300-350
R-10 5004A	GMN	1.99				0- 50	300-350
R-10 7006A	GMN	1.99				0- 50	300-350
R-3 02	GLF	1.60					300-350
R-4	GLF	1.67				0- 50	300-350
R-4 02	GLF	1.65				0- 50	300-350
R-4 02XT	GLF	1.65				0- 50	300-350
R-4 04	GLF	1.65				0- 50	300-350
R-4 06	GLF	1.65				0- 50	300-350
R-4XT	GLF	1.65				0- 50	300-350
R-7	GMN	1.89				0- 50	300-350
R-7 02	GMN	1.89				0- 50	300-350
PPS		**Suntra**			**Sunkyong**		
1030	GLF 30	1.49	3.0		536	0- 142	266-356
1040	GLF 40	1.55	3.0		536	0- 142	266-356
1050	GLF 50	1.67	2.0		536	0- 142	266-356
1242	GLF	1.51	4.0		536	0- 142	266-356
1247	GLF	1.55	3.0		536	0- 142	266-356
2246	MN 40	1.60	5.0		536	0- 142	266-356
3030	CF 30	1.37	3.0		536	0- 142	266-356
4030E	GMN	1.78	2.0		536	0- 142	266-356
4030M	GMN	1.92	2.0		536	0- 142	266-356
4257	GMN	1.55	3.0		536	0- 142	266-356
S100				400.00	538	0- 142	266-356
S300				135.00	536	0- 142	266-356
S500				25.00	536	0- 142	266-356
PPS		**Supec**			**GE**		
G301T	GL 30	1.60	4.0- 6.0			500	285
G323	GMN 65	1.98	1.0- 2.0			500	285
G401	GL 40	1.66	3.0- 5.0			500	285
G401T	GL 40	1.66	2.0- 4.0			500	285
G402	GLF 40	1.60	1.0- 2.0			500	285
G520	GMN	1.81				500	285
G620	MN	1.81				500	285
W331	GLF 30	1.65				500	285
PPS		**Tedur**			**Miles**		
KU1-9500-110		1.35	11.0-12.9	150.00 X		50	300
KU1-9500-30		1.35	12.4-13.7	400.00 X		50	300
KU1-9500-50		1.35	11.2-13.7	385.00 X		50	300
KU1-9510-1	GLF 40	1.66	2.0- 5.0	41.00 X		50	300

Drying time, hr	Inj time, sec	Front temp, °F	Mid temp, °F	Rear temp, °F	Nozzle temp, °F	Mold temp, °F	Proc temp, °F
6.0		610-675	575-650	550-625		150-350	
6.0		610-675	575-650	550-625		150-350	
6.0		610-675	575-650	550-625		150-350	
6.0		610-675	575-650	550-625		150-350	
6.0		610-675	575-650	550-625		150-350	
6.0		610-675	575-650	550-625		150-350	
6.0		610-675	575-650	550-625		150-350	
6.0		610-675	575-650	550-625		150-350	
6.0		610-675	575-650	550-625		150-350	
6.0		610-675	575-650	550-625		150-350	
2.0- 4.0	7.0- 15.	600-650	600-650	600-625	580-650	275-300	620-650
2.0- 4.0	5.0- 10.	600-650	600-650	600-625	600-650	275-300	620-650
2.0- 4.0	5.0- 10.	600-650	600-650	600-625	600-650	275-300	620-650
2.0- 4.0	5.0- 10.	600-650	600-650	600-625	600-650	275-300	620-650
						150-300	575-650
2.0- 4.0	5.0- 10.	600-650	600-650	600-625	600-650	275-300	620-650
2.0- 4.0	5.0- 10.	600-650	600-650	600-625	600-650	275-300	620-650
2.0- 4.0	5.0- 10.	600-650	600-650	600-625	600-650	275-300	620-650
2.0- 4.0	5.0- 10.	600-650	600-650	600-625	600-650	275-300	620-650
2.0- 4.0	5.0- 10.	600-650	600-650	600-625	600-650	275-300	620-650
2.0- 4.0	5.0- 10.	600-650	600-650	600-625	600-650	275-300	620-650
2.0- 4.0	5.0- 10.	600-650	600-650	600-625	580-650	275-300	620-650
2.0- 4.0	5.0- 10.	600-650	600-650	600-625	580-650	275-300	620-650
2.0- 4.0	10.0	590-644	590-644	572-608	590-644	257-302	
2.0- 4.0	10.0	590-644	590-644	572-608	590-644	257-302	
2.0- 4.0	10.0	590-644	590-644	572-608	590-644	257-302	
2.0- 4.0	10.0	590-644	590-644	572-608	590-644	257-302	
2.0- 4.0	10.0	590-644	590-644	572-608	590-644	257-302	
2.0- 4.0	10.0	590-644	590-644	572-608	590-644	257-302	
2.0- 4.0	10.0	590-644	590-644	572-608	590-644	257-302	
2.0- 4.0	10.0	590-644	590-644	572-608	590-644	257-302	
2.0- 4.0	10.0	590-644	590-644	572-608	590-644	257-302	
2.0- 4.0	10.0	590-644	590-644	572-608	590-644	257-302	
2.0- 4.0	10.0	590-644	590-644	572-608	590-644	257-302	
2.0- 4.0	10.0	590-644	590-644	572-608	590-644	257-302	
2.0- 4.0	10.0	590-644	590-644	572-608	590-644	257-302	
3.0- 4.0		600-630	580-600	550-580	580-610	280-320	580-630
3.0- 4.0		600-630	580-600	550-580	580-610	280-320	580-630
3.0- 4.0		600-630	580-600	550-580	580-610	280-320	580-630
3.0- 4.0		600-630	580-600	550-580	580-610	280-320	580-630
3.0- 4.0		600-630	580-600	550-580	580-610	280-320	580-630
3.0- 4.0		600-630	580-600	550-580	580-610	280-320	580-630
3.0- 4.0		600-630	580-600	550-580	580-610	280-320	580-630
3.0- 4.0		600-630	580-600	550-580	580-610	280-320	580-630
3.0- 4.0		600-640	610-650	610-650	590-630	275-310	610-650
3.0- 4.0		600-640	610-650	610-650	590-630	275-310	610-650
3.0- 4.0		600-640	610-650	610-650	590-630	275-310	610-650
3.0- 4.0		600-640	610-650	610-650	590-630	275-310	610-650

Grade	Filler	Sp Grav	Shrink, mils/in	Melt flow, g/10 min	Melt temp, °F	Back pres, psi	Drying temp, °F
KU1-9511	GLF 45	1.73	1.0- 4.4			50	300
KU1-9520-1	GMN 50	1.78	2.0- 4.0	97.00 X		50	300
KU1-9523	GMN 60	1.91	3.0- 4.0	38.00 X		50	300
KU1-9530	GMN 60	1.86	7.0	287.00 X		50	300
KU1-9552	UNS	1.80	3.0- 8.0			50	300
KU1-9553	UNS	1.80	3.0- 8.0			50	300
KU1-9554	UNS	1.80	3.0- 7.0			50	300
KU1-9560	GMN 60	1.89	7.0	63.00 X		50	300
KU1-9561	GMN 60	1.82	3.0- 5.0	85.00 X		50	300
PPS		**Thermocomp**			**LNP**		
OF-1006 HI	GLF 30	1.56	2.5- 3.5		527	0- 100	250
PPSS		**Celstran S**			**Hoechst**		
PPSS6-01-4	STS 6	1.37	5.0- 7.0				270
PPSU		**Radel**			**Amoco Perform.**		
PCR		1.54	3.5	14.00			275
R-5000		1.29	7.0	10.00			300
R-5100		1.29	7.0	10.00			300
R-7000A		1.37	5.0- 8.0	10.00		50	300
PS		**Amoco Polystyrene**			**Amoco Chem.**		
G1		1.05		8.50 G			
G18		1.05		18.50 G			
G2		1.05		12.50 G			
G3		1.05		21.00 G			
H3E		1.03		2.50 G			
H4E		1.03		3.30 G			
H4R		1.03		3.80 G			
H7M		1.03		7.50 G			
R3		1.05		3.50 G			
R5		1.05		5.50 G			
R7		1.05		7.50 G			
R9		1.05		9.50 G			
PS		**DSM Polystyrene**			**DSM**		
G-35/35	GLF 35	1.34	1.0				160
J-30/20	GLF 20	1.20	2.0				160
J-30/30	GLF 30	1.29	2.0				160
PS		**Electrafil**			**DSM**		
J-30/CF/20	CF 30	1.14	0.5				160
PS		**Fiberfil**			**DSM**		
J-30/20	GLF 20	1.19	2.0				160
J-30/30	GLF 30	1.28	2.0				160
PS		**Fiberstran**			**DSM**		
G-30/20	GLF 20	1.20	1.0			10- 50	180
G-30/30	GLF 30	1.29	1.0			10- 50	180
G-35/35	GLF 35	1.34	1.0			10- 50	180
PS		**Fina Polystyrene**			**Fina**		
525				8.00 G		50- 300	
535				4.00 G		50- 300	

Drying time, hr	Inj time, sec	Front temp, °F	Mid temp, °F	Rear temp, °F	Nozzle temp, °F	Mold temp, °F	Proc temp, °F
3.0- 4.0		600-640	610-650	610-650	590-630	275-310	610-650
3.0- 4.0		600-640	610-650	610-650	590-630	275-310	610-650
3.0- 4.0		600-640	610-650	610-650	590-630	275-310	610-650
3.0- 4.0		600-640	610-650	610-650	590-630	275-310	610-650
3.0- 4.0		600-640	610-650	610-650	590-630	275-310	610-650
3.0- 4.0		600-640	610-650	610-650	590-630	275-310	610-650
3.0- 4.0		600-640	610-650	610-650	590-630	275-310	610-650
3.0- 4.0		600-640	610-650	610-650	590-630	275-310	610-650
3.0- 4.0		600-640	610-650	610-650	590-630	275-310	610-650
4.0		580-650	580-650	580-650		300-325	600-610
4.0		600	590	585	600	300	
3.0- 4.0		660	660	610		280-320	680-700
2.5		660	660	610		280-325	680-735
2.5		660	660	610		280-325	680-735
4.0		660-680	640-660	620-640	660-690	225-325	670-700
						150-190	350-450
						150-190	350-450
						150-190	350-450
						150-190	350-450
						150-190	350-450
						150-190	350-450
						150-190	350-450
						150-190	350-450
						150-190	350-450
						150-190	350-450
						150-190	350-450
						150-190	350-450
2.0		400-420	430-460	420-450	380-420	140-180	450-520
2.0		400-420	430-460	420-450	380-420	140-180	450-520
2.0		400-420	430-460	420-450	380-420	140-180	450-520
2.0- 4.0		420-450	430-460	400-420	410-440	140-180	430-480
2.0		400-420	430-460	420-450	380-420	140-180	450-520
2.0		400-420	430-460	420-450	380-420	140-180	450-520
2.0		420-450	430-460	400-420	420-450	120-160	440-470
2.0		420-450	430-460	400-420	420-450	120-160	440-470
2.0		420-450	430-460	400-420	420-450	120-160	440-470
		440	420	350	430	80-150	
		460	440	380	450	80-150	

Grade	Filler	Sp Grav	Shrink, mils/in	Melt flow, g/10 min	Melt temp, °F	Back pres, psi	Drying temp, °F
585				1.70 G		50- 300	
625				13.00 G		50- 300	
680				2.00 G		50- 300	
740				4.00 G		50- 300	
825				8.00 G		50- 300	
825E				2.70 G		50- 300	
835				4.00 G		50- 300	
945				8.50 G		50- 300	
945E				3.50 G		50- 300	
PS		**NSC**			**Thermofil**		
CS1		1.17	4.0- 6.0	3.50 G		100	158-176
PS		**RTP Polymers**			**RTP**		
400		1.06	6.0			50	180
401	GLF 10	1.13	3.0			50	180
401 FR	GLF	1.21	2.0			50	180
403	GLF 20	1.20	2.0			50	180
403 SP	GLF 20	0.96	1.0			50	180
405	GLF 30	1.28	1.0			50	180
406	GLF 35	1.33	1.0			50	180
407	GLF 40	1.38	1.0			50	180
ESD-A-400 HI		1.10	6.0				
ESD-AS-400		1.06	5.0				
ESD-AS-400 HI		1.03	6.0				
ESD-C-400 HI		1.10	6.0				
PS		**Shuman Polystyrene**				**Shuman**	
810		1.05		4.00			
811		1.05		4.00			
830		1.05		4.00			
831		1.05		4.00			
860		1.07		1.10			180-220
PS		**Stat-Kon**			**LNP**		
PDX-C-90361		1.08	3.0- 4.5			10- 100	180
PS		**Styron**			**Dow**		
6075		1.16	3.0- 6.0	5.00 G			160-180
613		1.04	3.0- 7.0	1.40 G		10- 500	
615APR		1.04	3.0- 7.0	14.00 G		10- 500	
623		1.04	3.0- 7.0	4.00 G		10- 500	
6515		1.16	4.5	7.50 G			160-180
666D		1.04	3.0- 7.0	8.00 G		10- 500	
678C		1.04	3.0- 7.0	10.00 G			
685		1.04	3.0- 7.0	2.20 G		10- 500	
685D		1.04	3.0- 7.0	1.50 G		10- 500	
LR-175		1.04		1.60 G		10- 500	
PS		**Thermofil Polystyrene**				**Thermofil**	
CS1		1.17	5.0	4.00 G			158
SE-HH-1		1.07	5.0	9.00 G		100- 225	158
PS		**Valtra**			**Chevron**		
D7023.04		1.03	4.0- 8.0	6.50			
HG200		1.03	4.0- 8.0	6.00			

Drying time, hr	Inj time, sec	Front temp, °F	Mid temp, °F	Rear temp, °F	Nozzle temp, °F	Mold temp, °F	Proc temp, °F
		480	460	390	470	80-150	
		420	380	325	410	80-150	
		470	450	380	460	80-150	
		460	440	380	450	80-150	
		440	420	350	430	80-150	
	15.0	460	440	380	450	80-150	
		460	440	380	450	80-150	
		440	420	350	430	80-150	
		460	440	380	450	80-150	
3.0- 4.0		392-464	392-464	347-392	392-464	104-140	
2.0		450-550	430-530	410-500		100-160	
2.0		450-550	430-530	410-500		100-160	
2.0		410-510	400-490	380-470		120-180	
2.0		450-550	430-530	410-500		100-160	
2.0		450-550	430-530	410-500		100-160	
2.0		450-550	430-530	410-500		100-160	
2.0		450-550	430-530	410-500		100-160	
2.0		450-550	430-530	410-500		100-160	
		400-475	400-475	400-475		100-160	
		400-450	400-450	400-450		100-160	
		400-450	400-450	400-450		100-160	
		400-475	400-475	400-475		100-160	
							400-440
							400-440
							400-440
							400-440
2.0- 4.0							425-475
3.0		425-525	425-525	425-525		100-150	475
2.0						100-150	400-475
		375-525		350-450		20-160	
		375-525		350-450		20-160	
		375-525		350-450		20-160	
2.0						100-150	400-475
		375-525		350-450		20-160	
						100	425
		375-525		350-450		20-160	
		375-525		350-450		20-160	
		375-525		350-450		20-160	
3.0		392-464	392-464	347-392	392-464	104-140	
3.0		383-392	374-392	347-374	383-392	122-158	
							430
							430

Grade	Filler	Sp Grav	Shrink, mils/in	Melt flow, g/10 min	Melt temp, °F	Back pres, psi	Drying temp, °F
PSU		**CTI Polysulfone**			**CTI**		
PF		1.24	7.0- 8.0				275
PF-000/12.5TIO2	TIO 12	1.36	7.0				275
PF-000/15T	PTF 15	1.33	7.0- 8.0				275
PF-000/18TIO2	TIO 18	1.42	7.0				275
PF-10CF/15T	CF 10	1.36	2.5- 3.0				275
PF-10GF/000	GLF 10	1.30	4.0- 5.0				275
PF-10GF/15T	GLF 10	1.36	3.0				275
PF-20GF	GLF 20	1.38	3.0				275
PF-20GF/15T	GLF 20	1.36	3.0				275
PF-30CF	CF 30	1.36	1.0				275
PF-30GF	GLF 30	1.46	2.0- 3.0				275
PF-30GF/15T	GLF 30	1.59	1.0- 2.0				275
PF-40GF	GLF 40	1.55	1.0- 2.0				275
PF-40GF/000	GLF 40	1.53	1.0- 2.0				275
PF-7CF-7GF/15T	CF 7	1.40	3.0- 3.5				275
PSU		**DSM Polysulfone**			**DSM**		
J-1500/10	GLF 10	1.31	3.0				200-235
J-1500/20	GLF 20	1.38	2.0				200-235
J-1500/20/TF/15	GLF 20	1.49	2.0				200-235
J-1500/30	GLF 30	1.46	1.0				200-235
J-1505/10	GLF 10	1.31	4.0				200-235
J-1505/20	GLF 20	1.38	2.0				200-235
J-1505/30	GLF 30	1.47	2.0				200-235
J-1505/40	GLF 40	1.56	1.0				200-235
PSU		**Electrafil**			**DSM**		
J-1500/CF/20	CF 20	1.31	0.5				200-235
J-1500/CF/30	CF 30	1.37	0.5				200-235
PSU		**Fiberfil**			**DSM**		
J-1500/10	GLF 10	1.31	3.0- 4.0				200-235
J-1500/20	GLF 20	1.38	2.0				200-235
J-1500/30	GLF 30	1.46	1.0- 2.0				200-235
J-1500/40	GLF 40	1.56	1.0- 2.0				200-235
J-1850/15	GLF 15	1.42	3.0- 3.5				180-250
J-1850/30	GLF 30	1.51	1.5				180-250
PSU		**Fiberstran**			**DSM**		
G-1500/10	GLF 10	1.31	3.0- 4.0			10- 50	200-250
G-1500/20	GLF 20	1.38	2.0			10- 50	200-250
PSU		**Mindel**			**Amoco Perform.**		
M-800	MN	1.61	5.0- 8.0	8.50		50	275-325
M-825	MN	1.48	5.0	7.00		50	275-325
S-1000		1.23	6.6	13.00		50	275
PSU		**RTP Polymers**			**RTP**		
900	GLF	1.24	7.0			25- 75	275
900 TFE 15	PTF	1.33	7.0- 8.0			50- 75	275
900Z		1.24	7.0			25- 75	275
901	GLF	1.32	3.0			25- 75	275
901 UV	GLF	1.32	3.0			25- 75	275
903	GLF	1.38	2.0			25- 75	275

Drying time, hr	Inj time, sec	Front temp, °F	Mid temp, °F	Rear temp, °F	Nozzle temp, °F	Mold temp, °F	Proc temp, °F
4.0		680	650	630	650	200	
4.0		680-750	650-680	630-650	650-570	200-300	
4.0		680	650	630	650	200	
4.0		680-750	650-680	630-650	650-570	200-300	
4.0		680-750	650-680	630-650	650-570	200-300	
4.0		680-750	650-680	630-650	650-570	200-300	
4.0		680-750	650-680	630-650	650-570	200-300	
4.0		680-750	650-680	630-650	650-570	200-300	
4.0		680-750	650-680	630-650	650-570	200-300	
4.0		680-750	650-680	630-650	650-570	200-300	
4.0		680-750	650-680	630-650	650-570	200-300	
4.0		680-750	650-680	630-650	650-570	200-300	
4.0		680-750	650-680	630-650	650-570	200-300	
4.0		680-750	650-680	630-650	650-570	200-300	
4.0		680-750	650-680	630-650	650-570	200-300	
2.0-16.0		620-700	630-700	600-620	620-680	210-300	650-750
2.0-16.0		620-700	630-700	600-620	620-680	210-300	650-750
2.0-16.0		620-700	630-700	600-620	620-680	210-300	650-750
2.0-16.0		620-700	630-700	600-620	620-680	210-300	650-750
2.0-16.0		620-700	630-700	600-620	620-680	210-300	650-750
2.0-16.0		620-700	630-700	600-620	620-680	210-300	650-750
2.0-16.0		620-700	630-700	600-620	620-680	210-300	650-750
2.0-16.0		620-700	630-700	600-620	620-680	210-300	650-750
2.0-16.0		620-700	630-700	600-620	620-680	210-300	650-750
2.0-16.0		620-700	630-700	600-620	620-680	210-300	650-750
2.0-16.0		620-700	630-700	600-620	620-680	210-300	650-750
2.0-16.0		620-700	630-700	600-620	620-680	210-300	650-750
2.0-16.0		620-700	630-700	600-620	620-680	210-300	650-750
2.0-16.0		620-700	630-700	600-620	620-680	210-300	650-750
2.0-16.0		400-450	440-480	420-450	380-420	100-200	450-480
2.0-16.0		400-450	440-480	420-450	380-420	100-200	450-480
4.0-16.0		620-670	630-680	600-620	620-670	220-300	650-700
4.0-16.0		620-670	630-680	600-620	620-670	220-300	650-700
2.0- 3.5						250-325	675-775
2.0- 3.5						250-325	675-775
4.0						190-280	600-650
4.0		625-725	600-700	575-675		200-325	
4.0		600-700	575-675	550-650		200-325	600-660
4.0		625-725	600-700	575-675		200-325	
4.0		625-725	600-700	575-675		200-325	
4.0		625-725	600-700	575-675		200-325	
4.0		625-725	600-700	575-675		200-325	

Grade	Filler	Sp Grav	Shrink, mils/in	Melt flow, g/10 min	Melt temp, °F	Back pres, psi	Drying temp, °F
905	GLF	1.46	1.0			25- 75	275
905Z	GLF 30	1.45	1.0			25- 75	275
907	GLF	1.56	1.0			25- 75	275
ESD-A-901	GLF	1.34	4.0				
ESD-A-980	CF	1.27	1.0				
ESD-C-901	GLF	1.34	4.0				
ESD-C-980	CF	1.29	1.0				
PSU		**Udel**			**Amoco Perform.**		
GF-110	GL 10	1.33	4.0	4.00		50- 100	300-325
GF-120	GL 20	1.40	3.0	7.00		50- 100	300-325
GF-130	GL 30	1.49	2.0	7.00		50- 100	300-325
GF-205		1.28	5.0	5.00		50- 100	275-300
GF-210	GLF 10	1.36	4.0	5.00		50- 100	275-300
P-1700		1.24	7.0	6.50		30- 100	275
P-1700FR		1.24	7.0	6.50			275
P-1710		1.25	7.0	6.50			275
P-1720		1.25	7.0	6.50		30- 100	275
PSU		**Ultrason**			**BASF**		
S1010		1.24	5.8- 7.1				270
S2010		1.24	5.7- 7.1				270
S2010 G2	GLF 10	1.31	4.8- 4.9				270
S2010 G4	GLF 20	1.40	3.7- 5.3				270
S2010 G6	GLF 30	1.49	3.4- 6.0				270
S3010		1.24	6.6- 8.9				270
PSU+ABS		**Mindel**			**Amoco Perform.**		
A-670		1.13	6.6	12.00		50- 350	250
PSU+Unspecified		**Mindel**			**Amoco Perform.**		
B-310	GLF	1.41	4.0- 5.0	7.50		50	275-325
B-322	GLF	1.47	3.0- 5.0	6.50		20- 50	275-325
B-340	GLF	1.66	1.5- 6.0	10.00		50	275-325
B-360	GLF	1.52	2.5- 6.0	7.00		50	275-325
B-390	MN	1.30	6.5- 9.0	8.50		50	275-325
PUR		**Baydur**			**Miles**		
726		0.88					
PUR		**Celstran**			**Hoechst**		
PUC40-01-4	CF 40	1.38					180
PUG30-01-4	GLF 30	1.43					180
PUG40-01-4	GLF 40	1.52					180
PUG50-01-4	GLF 50	1.63					180
PUG60-01-4	GLF 60	1.76					180
PUR		**CTI Polyurethane**			**CTI**		
UR-20CF/000	CF 20	1.32	3.0- 3.5				150
UR-20GF/000	GLF 20	1.39	2.0- 2.5				200-220
UR-30CF/000	CF 30	1.36	2.0- 2.5				150
UR-30GF/000	GLF 30	1.45	3.0				150
UR-40CF/000	CF 40	1.40	1.0- 1.5				150
UR-40GF/000	GLF 40	1.56	1.0- 1.5				200-220

Drying time, hr	Inj time, sec	Front temp, °F	Mid temp, °F	Rear temp, °F	Nozzle temp, °F	Mold temp, °F	Proc temp, °F
4.0		625-725	600-700	575-675		200-325	
4.0		625-725	600-700	575-675		200-325	
4.0		625-725	600-700	575-675		200-325	
		550-650	550-650	550-650		225-325	
		550-650	550-650	550-650		225-325	
		600-675	600-675	600-675		200-300	
		550-650	550-650	550-650		225-325	
3.0- 4.0						250-325	650-750
3.0- 4.0						250-325	650-750
3.0- 4.0						250-325	650-750
4.0						250-325	650-750
4.0						250-325	650-750
3.5						200-320	650-750
3.5						200	650-750
3.5						200	650-750
3.5						200-320	650-750
4.0						250-300	610-660
4.0						250-320	625-680
4.0		705-715	680-715	645-715	715-715	265-355	660-715
4.0		705-715	680-715	645-715	715-715	265-355	660-715
4.0		705-715	680-715	645-715	715-715	265-355	660-715
4.0						250-320	625-680
3.0- 4.0						160-250	500-600
3.0- 4.0						150-210	520-600
3.0- 4.0						150-210	520-600
3.0- 4.0						150-210	520-600
3.0- 4.0						150-210	520-600
3.0- 4.0						180-220	575-650
4.0		480-500	470-490	460-480	480-500	150-170	480-500
4.0		450-470	440-460	430-450	460-480	150-170	450-470
4.0		460-480	450-470	440-460	470-490	150-170	460-480
4.0		470-490	460-480	450-470	480-500	150-170	470-490
4.0		480-500	470-490	460-480	490-510	150-170	480-500
2.0- 4.0		420-460	390-420	350-390	420-450	80-150	
2.0- 4.0		420-460	390-420	350-390	420-450	80-150	
2.0- 4.0		420-460	390-420	350-390	420-450	80-150	
2.0- 4.0		420-460	390-420	350-390	420-450	80-150	
2.0- 4.0		420-460	390-420	350-390	420-450	80-150	
2.0- 4.0		420-460	390-420	350-390	420-450	80-150	

Grade	Filler	Sp Grav	Shrink, mils/in	Melt flow, g/10 min	Melt temp, °F	Back pres, psi	Drying temp, °F
PUR		**DSM Polyurethane**			**DSM**		
J-100/10	GLF 10	1.30	5.0				160
J-100/20	GLF 20	1.36	4.0				160
J-100/30	GLF 30	1.45	3.0				160
J-105/30	GLF 30	1.47	1.0				160
PUR		**Estaloc**			**BFGoodrich**		
59100	UNS	1.40	1.5			25- 100	230
59106	UNS	1.23	2.0			25- 100	230
59116	UNS	1.23	2.0			25- 100	230
59200	UNS	1.46	1.5			25- 100	230
59206	UNS	1.23	2.0			25- 100	230
59300		1.45	1.5			25- 100	230
59403	UNS	1.33	1.3			25- 100	230
61060	UNS	1.52				25- 100	230
61080	UNS	1.60				25- 100	230
61083	UNS	1.62				25- 100	230
PUR		**Fiberfil**			**DSM**		
J-100/10	GLF 10	1.30	5.0- 6.0				160
J-100/20	GLF 20	1.36	4.0- 5.0				160
J-100/30	GLF 30	1.45	3.0- 5.0				160
PUR		**Isoplast**			**Dow**		
101		1.20	5.0	8.00 S		0- 200	180-210
101LGF40BLK	GLF	1.51	1.0			0- 200	180-210
101LGF40	GLF	1.51	1.0			0- 200	180-210
101LGF60BLK	GLF	1.71	1.0			0- 200	180-210
101LGF60	GLF	1.71	1.0			0- 200	180-210
151		1.20	5.0			0- 200	200-230
201		1.20	4.0- 6.0	2.00			210-230
202		1.20	5.0	6.00 S		0- 200	260-280
202LGF40	GLF	1.50	1.0			0- 200	260-280
202LGF60	GLF	1.70	1.0			0- 200	260-280
300		1.20	4.0- 6.0			0- 200	180-210
301		1.20	4.0- 6.0			0- 200	200-230
302		1.20	4.0- 6.0			0- 200	260-280
PUR		**RTP Polymers**			**RTP**		
1200-80D		1.22	8.0				
1201-80D	GLF 10	1.29	5.0				
1203-80D	GLF 20	1.37	4.0				
1205-80D	GLF 30	1.45	3.0				
1207-80D	GLF 40	1.54	2.0				
2301A	GLF 10	1.27	2.5- 5.0				225
2301C	GLF 10	1.27	3.5- 7.0				225
2303A	GLF 20	1.34	2.0- 4.0				225
2303C	GLF 20	1.34	2.5- 5.0				225
2305A	GLF 30	1.43	1.2- 2.4				225
2305C	GLF 30	1.43	1.8- 3.6				225
2307A	GLF 40	1.52	1.0- 2.0				225
2307C	GLF 40	1.52	1.5- 3.0				225
2381A	CF 10	1.23	1.0- 2.0				225
2381C	CF 10	1.23	1.0- 2.0				225
2383A	CF 20	1.27	0.5- 1.0				225

Drying time, hr	Inj time, sec	Front temp, °F	Mid temp, °F	Rear temp, °F	Nozzle temp, °F	Mold temp, °F	Proc temp, °F
2.0		360-390	370-400	350-370	350-400	110-180	350-410
2.0		360-390	370-400	350-370	350-400	110-180	350-410
2.0		360-390	370-400	350-370	350-400	110-180	350-410
2.0		360-390	370-400	350-370	350-400	110-180	350-410
2.0- 3.0		430-450	420-440	410-440	430-450	100-140	430-450
2.0- 3.0		430-450	420-440	410-440	430-450	100-140	430-450
2.0- 3.0		430-450	420-440	410-440	430-450	100-140	430-450
2.0- 3.0		430-450	420-440	410-440	430-450	100-140	430-450
2.0- 3.0		430-450	420-440	410-440	430-450	100-140	430-450
2.0- 3.0		430-450	420-440	410-440	430-450	100-140	430-450
2.0- 3.0		430-450	420-440	410-440	430-450	100-140	430-450
2.0- 3.0		470-490	450-470	430-460	470-490	100-140	470-490
2.0- 3.0		470-490	450-470	430-460	470-490	100-140	470-490
2.0- 3.0		470-490	450-470	430-460	470-490	100-140	470-490
2.0		360-390	370-400	350-370	350-400	110-180	350-410
2.0		360-390	370-400	350-370	350-400	110-180	350-410
2.0		360-390	370-400	350-370	350-400	110-180	350-410
4.0-12.0	2.0- 10.	440	440	450	450	150-180	430-470
4.0-12.0	2.0- 10.	470	460	470	480	150-190	460-490
4.0-12.0	2.0- 10.	470	460	470	480	150-190	460-490
4.0-12.0	2.0- 10.	450	460	470	460	150-190	460-490
4.0-12.0	2.0- 10.	450	460	470	460	150-190	460-490
4.0-12.0	2.0- 10.	440	450	460	460	150-200	450-480
4.0							420-460
4.0-12.0	2.0- 10.	460	470	480	500	200-250	460-500
4.0-12.0	2.0- 10.	480	480	490	500	200-250	470-500
4.0-12.0	2.0- 10.	470	480	490	500	200-250	470-500
4.0-12.0	2.0- 10.	430	430	400	450	150-190	430-470
4.0-12.0	2.0- 10.	450	450	410	460	150-200	450-480
4.0-12.0	2.0- 10.	490	480	450	500	200-250	460-500
		380-410	370-400	360-390		60-150	
		380-410	370-400	360-390		60-150	
		380-410	370-400	360-390		60-150	
		380-410	370-400	360-390		60-150	
		380-410	370-400	360-390		60-150	
6.0		445-525	435-515	425-505	445-525	75-150	
6.0		445-525	435-515	425-505	445-525	75-150	
6.0		445-525	435-515	425-505	445-525	75-150	
6.0		445-525	435-515	425-505	445-525	75-150	
6.0		445-525	435-515	425-505	445-525	75-150	
6.0		445-525	435-515	425-505	445-525	75-150	
6.0		445-525	435-515	425-505	445-525	75-150	
6.0		445-525	435-515	425-505	445-525	75-150	
6.0		445-525	435-515	425-505	445-525	75-150	
6.0		445-525	435-515	425-505	445-525	75-150	
6.0		445-525	435-515	425-505	445-525	75-150	

Grade	Filler	Sp Grav	Shrink, mils/in	Melt flow, g/10 min	Melt temp, °F	Back pres, psi	Drying temp, °F
2383C	CF 20	1.27	1.0- 2.0				225
2385A	CF 30	1.33	0.3- 0.6				225
2385C	CF 30	1.33	0.5- 1.0				225
2387A	CF 40	1.52	1.0- 2.0				225
2387C	CF 40	1.37	0.2- 0.4				225
PUR		**Texin**			**Miles**		
5187		1.20	8.0				207-217
5265		1.17	8.0				220-230
5286		1.12	8.0				207-217
5370		1.21	8.0				220-230
PUR-capro		**Morthane**			**Morton**		
PC26-400		1.20	14.0		266-338		122
PC76-400		1.21	14.0		410-446		176
PC86-400		1.20	12.0		365-428		176
PC91-400		1.20	13.0		374-419		176
PC95-400		1.20	10.0		356-437		176
PC96-400		1.22	13.0		410-446		176
PUR-capro		**Pellethane**			**Dow**		
2102-55D		1.21	5.0- 9.0	14.00		0- 100	190-220
2102-65D		1.22	7.0- 8.0	49.00		0- 200	210-230
2102-75A		1.17	2.0- 6.0	25.00		0- 200	180-200
2102-80A		1.18	2.0- 6.0	4.00		0- 100	180-200
2102-85A		1.18	5.0- 6.0	35.00		0- 200	190-220
2102-90A		1.20	5.0- 7.0	11.00		0- 200	190-220
2102-90AE		1.20	5.0- 7.0	29.00		0- 200	190-220
2102-90AR		1.20	5.0- 8.0	15.00		0- 200	190-220
PUR-ether/MDI		**Tecothane**			**Thermedics**		
1055D		1.16		2.30			180
1065D		1.19		2.00			180
1075A		1.10		8.00			180
1080A		1.12		5.80			180
1095A		1.15		2.30			180
2055D-B20		1.36		6.00			180
2065D-B20		1.38		3.00			180
2075A-B20		1.30		9.90			180
2080A-B20		1.32		9.00			180
2095A-B20		1.35		8.00			180
PUR-Polyester		**Elastollan**			**BASF**		
C-60AW		1.18				75- 150	
C-60D		1.24				75- 150	
C-64D		1.24				75- 150	
C-70AW		1.18				75- 150	
C-74D		1.24				75- 150	
C-80A		1.19				75- 150	
C-85A		1.19				75- 150	175-195
C-90A		1.21				75- 150	175-195
C-95A		1.22				75- 150	195-220
C-98A		1.23				75- 150	195-220
C60A-10WN				11.00		120	170-175
C70A-10WN				3.00		210	170-175
S-60D		1.25				75- 150	175-195

Drying time, hr	Inj time, sec	Front temp, °F	Mid temp, °F	Rear temp, °F	Nozzle temp, °F	Mold temp, °F	Proc temp, °F
6.0		445-525	435-515	425-505	445-525	75-150	
6.0		445-525	435-515	425-505	445-525	75-150	
6.0		445-525	435-515	425-505	445-525	75-150	
6.0		445-525	435-515	425-505	445-525	75-150	
6.0		445-525	435-515	425-505	445-525	75-150	
2.0							375-395
2.0							420-440
2.0							375-395
2.0							440-460
3.0							302-365
3.0							419-446
3.0							410-437
3.0							401-437
3.0							392-428
3.0							419-446
2.0- 4.0	3.0- 10.	390-410	370-390	360-380	400-410	60-140	400-430
2.0- 4.0						80-140	410-430
2.0- 4.0						80-140	390-420
2.0- 4.0	3.0- 10.	370-390	360-380	350-370	390-410	60-140	390-420
2.0- 4.0						80-140	390-410
2.0- 4.0						80-140	400-430
2.0- 4.0						80-140	415-420
2.0- 4.0						80-140	400-430
3.0		375		365	385	80-130	435
3.0		380		365	390	50-130	445
3.0		340		330	345	50- 90	410
3.0		370		350	380	50-110	415
3.0		375		360	385	70-120	420
3.0		365		355	375	80-130	420
3.0		370		355	375	80-130	435
3.0		330		320	335	50- 90	400
3.0		360		340	370	50-110	410
3.0		365		350	375	70-120	415
		320	300	290	325	60- 90	340-360
		405	400	385	410	80-115	400-430
		410	405	385	410	80-115	400-430
		325	310	295	330	60- 90	340-360
		410	405	385	410	80-115	400-430
		370	360	350	375	70-100	360-380
2.0- 4.0		380	370	360	385	70-100	370-390
2.0- 4.0		395	385	375	400	70-110	380-410
2.0- 4.0		400	390	380	410	70-110	390-420
2.0- 4.0		410	405	385	410	80-115	400-430
2.0- 4.0		390	385	382	400	105	382
2.0- 4.0		395	390	390	405	104	393
2.0- 4.0		405	400	385	410	80-115	400-430

Grade	Filler	Sp Grav	Shrink, mils/in	Melt flow, g/10 min	Melt temp, °F	Back pres, psi	Drying temp, °F
S-64D		1.25				75- 150	175-195
S-74D		1.26				75- 150	175-195
S-80A		1.21				75- 150	175-195
S-85A		1.22				75- 150	175-195
S-90A		1.23				75- 150	175-195
S-95A		1.23				75- 150	195-220
S-98A		1.24				75- 150	195-220
PUR-Polyester		**Estane**			**BFGoodrich**		
58130		1.21				25- 50	220
58133		1.22				25- 50	220
58137		1.23				25- 50	220
58206		1.20				25- 50	220
PUR-Polyester		**Interpol**			**Cook Composites**		
47-5116/47-5205							
47-5117/47-5205							
47-5118/47-5205							
47-5119/47-5205							
47-5124/47-5205							
47-5125/47-5205							
47-5132/47-5205							
47-5138/47-5205							
47-5138/47-5 20	GL 20						
47-5138/47-5 40	GL 40						
PUR-Polyester		**Morthane**			**Morton**		
PS176-400		1.25	20.0		392-437		176
PS186-400		1.22	10.0		374-428		176
PS195-400		1.23	8.0		338-392		176
PS196-400		1.25	18.0		374-437		176
PS440-400		1.21	10.0	25.00	284-383		176
PS441-400		1.22	7.0		284-356		176
PS455-400		1.19	2.0	8.00	266-347		113-122
PS79-300		1.19	5.0	20.00	266-356		140-149
PUR-Polyester		**Texin**			**Miles**		
4215		1.21	8.0			0- 200	180-220
PUR-Polyether		**Elastollan**			**BASF**		
1154D		1.16				75- 150	175-195
1160D		1.17				75- 150	175-195
1164D		1.18				75- 150	175-195
1174D		1.19				75- 150	175-195
1175A-10W		1.14					170-175
1180A		1.11				75- 150	175-195
1185A		1.12				75- 150	175-195
1190A		1.13				75- 150	175-195
1195A		1.14				75- 150	195-220
MP-100		1.12					
PUR-Polyether		**Estane**			**BFGoodrich**		
58881		1.13				25- 50	220
PUR-Polyether		**Morthane**			**Morton**		
PE111-400		1.18	5.0		356-410		176

Drying time, hr	Inj time, sec	Front temp, °F	Mid temp, °F	Rear temp, °F	Nozzle temp, °F	Mold temp, °F	Proc temp, °F
2.0- 4.0		410	405	385	410	80-115	400-430
2.0- 4.0		410	405	385	410	80-115	400-430
2.0- 4.0		370	360	350	375	70-100	360-380
2.0- 4.0		380	370	360	385	70-100	370-390
2.0- 4.0		395	385	375	400	70-110	380-410
2.0- 4.0		400	390	380	410	70-110	390-420
2.0- 4.0		400	390	385	410	70-110	390-420
2.0		420	400	390	420	50- 85	420
2.0		420	400	390	420	50- 85	420
2.0		420	400	390	420	50- 85	425
2.0		390	370	350	400	50- 85	400
						130-150	
						130-150	
						130-150	
						130-150	
						130-150	
						130-150	
						130-150	
						130-150	
						130-150	
						130-150	
3.0							401-419
3.0							383-428
3.0							374-428
3.0							419-437
3.0							329-392
3.0							329-392
3.0							302-338
3.0							338-392
1.0- 3.0		440-460	440-460	430-450	450-475	80-110	465
2.0- 4.0		405	395	385	410	75-115	400-420
2.0- 4.0		405	400	385	410	80-115	400-430
2.0- 4.0		410	405	385	410	80-115	400-430
2.0- 4.0		410	405	385	410	80-115	400-430
2.0- 4.0		335	330	325	340	70-100	340-360
2.0- 4.0		370	360	350	375	70-100	360-380
2.0- 4.0		380	370	360	385	70-100	370-390
2.0- 4.0		395	385	375	400	70-110	380-410
2.0- 4.0		400	390	380	410	70-110	390-420
		380	370	360	385		370-390
2.0		380	370	350	380	50- 85	380
3.0							392-446

Grade	Filler	Sp Grav	Shrink, mils/in	Melt flow, g/10 min	Melt temp, °F	Back pres, psi	Drying temp, °F
PE36-400		1.08	10.0		266-356		140
PE47-400		1.13	2.0		374-428		140
PE50-400		1.06	10.0	7.00	212-338		122
PE74-400		1.03	4.0	18.00	167-284		104
PUR-Polyether		**Pellethane**			**Dow**		
2103-55D		1.15	5.0- 9.0	15.00		0- 100	190-220
2103-65D		1.17	5.0-10.0	35.00		0- 200	210-230
2103-70A		1.06	3.0- 8.0	11.00		0- 200	180-200
2103-80AE		1.13	2.0- 8.0	40.00		0- 100	180-200
2103-80AEF		1.13	2.0- 7.0	13.00		0- 100	180-200
2103-80AEN		1.13	5.0- 7.0	20.00		0- 100	180-200
2103-80PF		1.10	2.0- 7.0	39.00		0- 100	180-200
2103-85AE		1.14	5.0- 7.0	24.00		0- 200	190-220
2103-90A		1.14	7.0- 8.0	23.00		0- 200	190-220
2103-90AE		1.14	4.0- 7.0	7.00		0- 200	160-220
2103-90AEF		1.14	4.0- 8.0	3.00		0- 200	190-220
2103-90AEN		1.14	6.0-10.0	9.00		0- 200	190-220
2103-90AS		1.14	7.0- 8.0	30.00		0- 200	190-220
PUR-Polyether		**Tecoflex**			**Thermedics**		
EG100A		1.09					150
EG60D		1.09					150
EG65D		1.10					150
EG68D		1.10					150
EG72D		1.11					150
EG80A		1.04					150
EG85A		1.05					150
EG93A		1.08					150
PUREL-TP		**Pellethane**			**Dow**		
2351-85AE		1.13		25.00		0- 200	190-220
2354-45D		1.19	5.0- 7.0	20.00		0- 200	190-220
2354-55D		1.21	5.0- 7.0	22.00		0- 100	190-220
2354-65D		1.22	7.0- 8.0	24.00		0- 100	210-230
2355-75A		1.19	1.0- 7.0	28.00		0- 200	180-200
2355-80AE		1.18	2.0- 6.0	7.00		0- 100	180-200
2355-85ABR		1.18	1.0- 6.0	52.00		0- 200	190-220
2363-55D		1.15	4.0- 8.0	10.00		0- 100	190-220
2363-55DE		1.15	6.0- 8.0	30.00		0- 200	190-220
2363-65D		1.17	7.0- 9.0	40.00		0- 200	210-230
2363-75D		1.21	3.0- 8.0	28.00		0- 200	210-230
2363-80A		1.13		23.00		0- 100	180-200
2363-80AE		1.12	1.0- 6.0	10.00		0- 100	180-200
2363-80AR0120		1.30	1.0- 8.0			0- 100	180-200
2363-90A		1.14	5.0- 6.0	30.00		0- 200	190-220
2363-90AE		1.14	5.0- 8.0	32.00		0- 200	190-220
PUREL-TP		**PermaStat**			**RTP**		
1200		1.15	6.0- 9.0				
PUREL-TP		**RTP Polymers**			**RTP**		
1200.5-90A	GLF 5	1.23	2.0				
1201-90A	GLF 10	1.27	1.0				
1202-90A	GLF 15	1.30	1.0				
1203-90A	GLF 20	1.34	1.0				

Drying time, hr	Inj time, sec	Front temp, °F	Mid temp, °F	Rear temp, °F	Nozzle temp, °F	Mold temp, °F	Proc temp, °F
3.0							302-392
3.0							401-446
3.0							257-356
3.0							230-302
2.0- 4.0	3.0- 10.	390-410	370-390	360-380	400-410	60-140	380-410
2.0- 4.0						80-140	410-440
2.0- 4.0						80-140	380-410
2.0- 4.0	3.0- 10.	370-390	360-380	350-370	390-410	60-140	360-410
2.0- 4.0	3.0- 10.	370-390	360-380	350-370	390-410	60-140	360-390
2.0- 4.0	3.0- 10.	370-390	360-380	350-370	390-410	60-140	370-400
2.0- 4.0	3.0- 10.	370-390	360-380	350-370	390-410	60-140	370-400
2.0- 4.0						80-140	370-400
2.0- 4.0						80-140	400-430
2.0- 4.0						80-140	380-410
2.0- 4.0						80-140	380-410
2.0- 4.0						80-140	380-410
2.0- 4.0						80-140	400-430
4.0		340-360		315-335	350-370	65- 85	
4.0		365-385		350-370	370-390	65- 85	
4.0		380-400		365-385	390-410	65- 85	
4.0		380-400		365-385	390-410	65- 85	
4.0		400-420		365-385	400-420	65- 85	
4.0		315-335		300-320	325-345	65- 85	
4.0		315-335		315-335	325-345	65- 85	
4.0		315-335		315-335	325-345	65- 85	
2.0- 4.0						80-140	370-400
2.0- 4.0						80-140	410-440
2.0- 4.0	3.0- 10.	390-410	370-390	360-380	400-410	60-140	410-440
2.0- 4.0						80-140	410-440
2.0- 4.0						80-140	370-400
2.0- 4.0	3.0- 10.	370-390	360-380	350-370	390-410	60-140	380-400
2.0- 4.0						80-140	340-360
2.0- 4.0	3.0- 10.	390-410	370-390	360-380	400-410	60-140	410-440
2.0- 4.0	3.0- 10.	390-410	370-390	360-380	400-410	60-140	390-420
2.0- 4.0						80-140	410-440
2.0- 4.0						80-140	410-440
2.0- 4.0	3.0- 10.	370-390	360-380	350-370	390-410	60-140	380-410
2.0- 4.0	3.0- 10.	370-390	360-380	350-370	390-410	60-140	370-400
2.0- 4.0	3.0- 10.	370-390	360-380	350-370	390-410	60-140	380-410
2.0- 4.0						80-140	400-430
2.0- 4.0						80-140	380-410
		350-400	350-400	350-400		60-150	
		380-410	370-400	360-390		60-150	
		380-410	370-400	360-390		60-150	
		380-410	370-400	360-390		60-150	
		380-410	370-400	360-390		60-150	

Grade	Filler	Sp Grav	Shrink, mils/in	Melt flow, g/10 min	Melt temp, °F	Back pres, psi	Drying temp, °F
PUREL-TP		**Texin**			**Miles**		
DP7-1077		1.12					200-220
DP7-1078		1.13					200-220
PVC		**ACP**			**Alpha Chemical**		
800HX		1.22		8.50 F			
PVC		**Alpha PVC**			**Alpha Chemical**		
2212-100		1.30		87.00 F			
2212-110		1.32		72.00 F			
2212-67		1.18		10.00 E		50	
2212/7-114		1.29		29.00 F			
2212/7-118		1.33		23.00 F			
2212C2B-118		1.33		34.00 F			
2212RHT-118		1.33		35.00 F			
2212RRM-118		1.34		15.00 F			
2212RSM-100		1.31		77.00 F			
2212RSM-110		1.32		62.00 F			
2227N/X-75IM		1.26		2.50 F		50	
3006-70		1.20		13.00 F		50	
3006-95		1.28		1.10 F		50	
3006/1-60		1.18		18.00 F		50	
3006/2A1-82		1.23		1.50 F		50	
3006/X-65		1.34		5.00 F		50	
3006R-65		1.18		14.00 F			
3006R-75		1.21		8.00 F		50	
3006R-85		1.25		5.00 F		50	
3018-90		1.53		164.00 F			
3018/20-90		1.53				50	
3019-40/45		1.13		215.00 F		50	
3624XFS-40		1.29		187.00 E		50	
PVC		**Dural**			**Alpha Chemical**		
211A2		1.33					
PVC		**Fiberloc**			**Geon**		
803GR10	GL 10	1.37				50- 100	160-180
803GR20	GL 20	1.43				50- 100	160-180
803GR30	GL 30	1.53				50- 100	160-180
811GR10	GL 10	1.43				50- 100	160-180
811GR20	GL 20	1.50				50- 100	160-180
811GR30	GL 30	1.57				50- 100	160-180
812GR10	GL 10	1.46				50- 100	160-180
812GR20	GL 20	1.54				50- 100	160-180
812GR30	GL 30	1.62				50- 100	160-180
823GR10	GL 10	1.43				50- 100	160-180
823GR20	GL 20	1.50				50- 100	160-180
823GR30	GL 30	1.56				50- 100	160-180
881GR10	GL 10	1.31				50- 100	160-180
881GR20	GL 20	1.40				50- 100	160-180
881GR30	GL 30	1.49				50- 100	160-180
883GR10	GL 10	1.29				50- 100	160-180
883GR20	GL 20	1.38				50- 100	160-180
883GR30	GL 30	1.47				50- 100	160-180

Drying time, hr	Inj time, sec	Front temp, °F	Mid temp, °F	Rear temp, °F	Nozzle temp, °F	Mold temp, °F	Proc temp, °F
2.0		370-400	370-390	360-380	380-410	40- 90	385
2.0		370-400	370-400	360-390	390-410	40- 90	400
							360
							320-340
							350
		340	320	300	340		320-340
							340-360
							340-360
							355
							355
							360
							345
							345
		340	320	300	340		320-340
		340	320	300	340		320-340
		340	320	300	340		320-340
		340	320	300	340		320-340
		340	320	300	340		320-340
		340	320	300	340		320-340
							335
		340	320	300	340		320-340
		340	320	300	340		320-340
							350
		340	320	300	340		320-340
		340	320	300	340		320-340
		340	320	300	340		320-340
							355
2.0		350-360	350-360	350-360	380	70-150	385-400
2.0		350-360	350-360	350-360	380	70-150	385-400
2.0		350-360	350-360	350-360	380	70-150	385-400
2.0		350-360	350-360	350-360	380	70-150	385-400
2.0		350-360	350-360	350-360	380	70-150	385-400
2.0		350-360	350-360	350-360	380	70-150	385-400
2.0		350-360	350-360	350-360	380	70-150	385-400
2.0		350-360	350-360	350-360	380	70-150	385-400
2.0		350-360	350-360	350-360	380	70-150	385-400
2.0		350-360	350-360	350-360	380	70-150	385-400
2.0		350-360	350-360	350-360	380	70-150	385-400
2.0		350-360	350-360	350-360	380	70-150	385-400
2.0		350-360	350-360	350-360	380	70-150	385-400
2.0		350-360	350-360	350-360	380	70-150	385-400
2.0		350-360	350-360	350-360	380	70-150	385-400
2.0		350-360	350-360	350-360	380	70-150	385-400
2.0		350-360	350-360	350-360	380	70-150	385-400
2.0		350-360	350-360	350-360	380	70-150	385-400

Grade	Filler	Sp Grav	Shrink, mils/in	Melt flow, g/10 min	Melt temp, °F	Back pres, psi	Drying temp, °F
PVC		**Flexalloy**			**Teknor Apex**		
3500-35-NT		1.10	10.0-25.0			50- 150	
3500-45-NT		1.12	10.0-25.0			50- 150	
3500-55-NT		1.15	10.0-25.0			50- 150	
3500-65-NT		1.17	10.0-25.0			50- 150	
3500-75-NT		1.20	10.0-25.0			50- 150	
9100-35		1.07	10.0-25.0			50- 150	
9100-45		1.10	10.0-25.0			50- 150	
9100-55		1.12	10.0-25.0			50- 150	
9100-65		1.15	10.0-25.0			50- 150	
9100-75		1.18	10.0-25.0			50- 150	
9200-35-BL		1.07	10.0-25.0			50- 150	
9200-45-BL		1.10	10.0-25.0			50- 150	
9200-55-BL		1.13	10.0-25.0			50- 150	
9200-65-BL		1.15	10.0-25.0			50- 150	
9200-75-BL		1.18	10.0-25.0			50- 150	
9300-60		1.22	10.0-25.0			50- 150	
9300-70		1.24	10.0-25.0			50- 150	
9400-35		1.16	10.0-25.0			50- 150	
9400-50		1.19	10.0-25.0			50- 150	
9400-70		1.24	10.0-25.0			50- 150	
PVC		**Geon**			**Geon**		
2042		1.33					
82024		1.20					
82611		1.34					
83008		1.34					
83332		1.36					
83421		1.28					
83457		1.22					
83463		1.25					
83528		1.34					
83718		1.39					
83741		1.36					
83791		1.30					
83794		1.42					
85856	TIN	1.47	2.0- 5.0			50- 100	170-180
85890	TIN	1.45	2.0- 5.0			50- 100	170-180
85891	TIN	1.47	2.0- 5.0			50- 100	170-180
86054		1.35					
86153		1.18					
86154		1.20					
86155		1.24					
86188		1.28					
86270		1.13			390	50- 100	170-180
86272		1.16			390	50- 100	170-180
86274		1.19			390	50- 100	170-180
86276		1.23			395	50- 100	170-180
86278		1.25			395	50- 100	170-180
86279		1.29			405	50- 100	170-180
86280		1.14			390	50- 100	170-180
86282		1.16			395	50- 100	170-180
86284		1.18			395	50- 100	170-180
86286		1.21			395	50- 100	170-180
86288		1.30			405	50- 100	170-180

Drying time, hr	Inj time, sec	Front temp, °F	Mid temp, °F	Rear temp, °F	Nozzle temp, °F	Mold temp, °F	Proc temp, °F
		320-350	320-350	320-350		75-125	
		320-350	320-350	320-350		75-125	
		340-370	340-370	340-370		75-125	
		350-380	350-380	350-380		75-125	
		360-390	360-390	360-390		75-125	
		320-350	320-350	320-350		75-125	
		320-350	320-350	320-350		75-125	
		340-370	340-370	340-370		75-125	
		350-380	350-380	350-380		75-125	
		360-390	360-390	360-390		75-125	
		320-350	320-350	320-350		75-125	
		320-350	320-350	320-350		75-125	
		340-370	340-370	340-370		75-125	
		350-380	350-380	350-380		75-125	
		360-390	360-390	360-390		75-125	
		350-380	350-380	350-380		75-125	
		360-390	360-390	360-390		75-125	
		320-350	320-350	320-350		75-125	
		340-370	340-370	340-370		75-125	
		360-390	360-390	360-390		75-125	
							350-360
							320-335
							335-350
							360-375
							340-355
							360-380
							335-350
							350-360
							375-382
							350-355
							335-350
							370-380
							345-360
2.0		320	320	320	330	70-100	390-410
2.0		320	320	320	330	70-100	390-410
2.0		320	320	320	330	70-100	390-410
							370-380
							340-350
							345-355
							350-360
							345-380
2.0		320	320	320	330	70-100	390-410
2.0		320	320	320	330	70-100	390-410
2.0		320	320	320	330	70-100	390-410
2.0		320	320	320	330	70-100	390-410
2.0		320	320	320	330	70-100	390-410
2.0		320	320	320	330	70-100	390-410
2.0		320	320	320	330	70-100	390-410
2.0		320	320	320	330	70-100	390-410
2.0		320	320	320	330	70-100	390-410
2.0		320	320	320	330	70-100	390-410
2.0		320	320	320	330	70-100	390-410

Grade	Filler	Sp Grav	Shrink, mils/in	Melt flow, g/10 min	Melt temp, °F	Back pres, psi	Drying temp, °F
86289		1.30			405	50- 100	170-180
86290		1.20			390	50- 100	170-180
86292		1.24			390	50- 100	170-180
86294		1.27			390	50- 100	170-180
86296		1.33			395	50- 100	170-180
86298		1.36			400	50- 100	170-180
86299		1.40			405	50- 100	170-180
86340		1.27			390	50- 100	170-180
86342		1.30			390	50- 100	170-180
86344		1.34			390	50- 100	170-180
86346		1.34			395	50- 100	170-180
86348		1.42			395	50- 100	170-180
86349		1.52			405	50- 100	170-180
86372		1.24			390	50- 100	170-180
86374		1.26			395	50- 100	170-180
86376		1.27			395	50- 100	170-180
86382		1.24			395	50- 100	170-180
86384		1.25			395	50- 100	170-180
86386		1.27			402	50- 100	170-180
86392		1.31			390	50- 100	170-180
86394		1.34			390	50- 100	170-180
86396		1.36			395	50- 100	170-180
87241	TIN	1.33	2.0- 5.0			50- 100	170-180
87319	TIN	1.31	2.0- 5.0			50- 100	170-180
87321	TIN	1.42	2.0- 5.0			50- 100	170-180
87322	TIN	1.33	2.0- 5.0			50- 100	170-180
87371	TIN	1.38	2.0- 5.0			50- 100	170-180
87395		1.32				50- 100	
87402	TIN	1.31	2.0- 5.0			50- 100	170-180
87407	TIN	1.32	2.0- 5.0			50- 100	170-180
87409		1.32				50- 100	
87420	TIN	1.36	2.0- 5.0			50- 100	170-180
87447	TIN	1.36	2.0- 5.0			50- 100	170-180
87457	TIN	1.35	2.0- 5.0			50- 100	170-180
87498	TIN	1.41	1.0- 4.0			50- 100	170-180
87508	TIN	1.29	2.0- 5.0			50- 100	170-180
87537		1.31				50- 100	
8804FR		1.38					
8812		1.24					
8813		1.27					
8814		1.32					
8815		1.34					
8857		1.42					
8883		1.25					

PVC Novatemp Novatec

Grade	Filler	Sp Grav	Shrink, mils/in	Melt flow, g/10 min	Melt temp, °F	Back pres, psi	Drying temp, °F
8100		1.17	5.0			0- 50	170

PVC Polyvin A. Schulman

Grade	Filler	Sp Grav	Shrink, mils/in	Melt flow, g/10 min	Melt temp, °F	Back pres, psi	Drying temp, °F
4507		1.23					
5819		1.28					
6820		1.30					
7974		1.35					
9020		1.30					
9605		1.32					

Drying time, hr	Inj time, sec	Front temp, °F	Mid temp, °F	Rear temp, °F	Nozzle temp, °F	Mold temp, °F	Proc temp, °F
2.0		320	320	320	330	70-100	390-410
2.0		320	320	320	330	70-100	390-410
2.0		320	320	320	330	70-100	390-410
2.0		320	320	320	330	70-100	390-410
2.0		320	320	320	330	70-100	390-410
2.0		320	320	320	330	70-100	390-410
2.0		320	320	320	330	70-100	390-410
2.0		320	320	320	330	70-100	390-410
2.0		320	320	320	330	70-100	390-410
2.0		320	320	320	330	70-100	390-410
2.0		320	320	320	330	70-100	390-410
2.0		320	320	320	330	70-100	390-410
2.0		320	320	320	330	70-100	390-410
2.0		320	320	320	330	70-100	390-410
2.0		320	320	320	330	70-100	390-410
2.0		320	320	320	330	70-100	390-410
2.0		320	320	320	330	70-100	390-410
2.0		320	320	320	330	70-100	390-410
2.0		320	320	320	330	70-100	390-410
2.0		320	320	320	330	70-100	390-410
2.0		320	320	320	330	70-100	390-410
2.0		320	320	320	330	70-100	390-410
2.0		320	320	320	330	70-100	390-410
2.0		320	320	320	330	70-100	390-410
2.0		320	320	320	330	70-100	390-410
2.0		320	320	320	330	70-100	390-410
2.0		320	320	320	330	70-100	390-410
							395-405
2.0		320	320	320	330	70-100	390-410
2.0		320	320	320	330	70-100	390-410
							395-405
2.0		320	320	320	330	70-100	390-410
2.0		320	320	320	330	70-100	390-410
2.0		320	320	320	330	70-100	390-410
2.0		320	320	320	330	70-100	390-410
2.0		320	320	320	330	70-100	390-410
							395-405
							360-370
							335-350
							345-360
							345-360
							360-375
							350-360
							360-370
2.0- 4.0		345-360	330-345	310-325	350-370	140-175	395-415
							330-390
							330-390
							330-390
							330-390
							330-390
							330-390

Grade	Filler	Sp Grav	Shrink, mils/in	Melt flow, g/10 min	Melt temp, °F	Back pres, psi	Drying temp, °F
PVC		**Roscom**			**Roscom**		
201-45/50		1.16					
201-60		1.17					
201-65 CLEAR		1.21					
201-68		1.19					
201-70 CLEAR		1.21					
201-75 CLEAR		1.27					
201-80 CLEAR		1.23					
201LF-50		1.14					
201UV-90		1.27					
303-50		1.14					
303-60 CLEAR		1.18					
303-75 CLEAR		1.23					
PVC		**Satinflex**			**Alpha Chemical**		
1200-60		1.16					
1200-70		1.19					
1200-80		1.21					
1200-85		1.23					
PVC		**Superkleen**			**Alpha Chemical**		
2213-100		1.32				50	
2223-80		1.25				50	
2223C-70		1.20				50	
3003-65		1.20		11.00 E		50	
3003-85		1.25				50	
3003-90		1.30				50	
PVC		**Unichem**			**Colorite**		
1114-05					370-380	100- 300	
8511 G-05		1.25					
8511-02		1.25					
8512A-02		1.25					
8512B-02		1.25					
8512N-02		1.25	20.0-25.0				
PVC		**Vistel**			**Vista Chemical**		
9101		1.40				0- 100	170
9102		1.40				0- 100	170
9111		1.45				0- 100	170
9112		1.44				0- 100	170
9115		1.46				0- 100	170
9121		1.36				0- 100	170
9122		1.39				0- 100	170
PVC Alloy		**Geon HTX**			**Geon**		
6110		1.27	3.0- 6.0				
6120		1.38	4.0- 6.0				
6130		1.39	4.0- 6.0				
6140		1.41	4.0- 6.0				
6210		1.28	4.0- 6.0				
6220		1.22	4.0- 6.0				
PVC Elastomer		**Sunprene**			**Sunprene**		
FA-63014		1.24	8.0-25.0				176-180

Drying time, hr	Inj time, sec	Front temp, °F	Mid temp, °F	Rear temp, °F	Nozzle temp, °F	Mold temp, °F	Proc temp, °F
							320-340
							320-340
							320-340
							320-340
							320-340
							320-340
							320-340
							275-295
							320-340
							275-295
							320-340
							320-340
							360-380
							360-380
							360-380
							360-380
		340	320	300	340		320-340
		340	320	300	340		320-340
		340	320	300	340		320-340
		340	320	300	340		320-340
		340	320	300	340		320-340
		340	320	300	340		320-340
		350	340	290	350		370-380
							300-340
							300-340
							300-340
							300-340
							300-340
2.0- 4.0		350	330	290	360	100-140	380-400
2.0- 4.0		350	330	290	360	100-140	380-400
2.0- 4.0		350	330	290	360	100-140	380-400
2.0- 4.0		350	330	290	360	100-140	380-400
2.0- 4.0		350	330	290	360	100-140	380-400
2.0- 4.0		350	330	290	360	100-140	380-400
2.0- 4.0		350	330	290	360	100-140	380-400
							405
							405
							410
							410
							405
							405
2.0- 3.0	10.0- 15.	365	338	311	347	104-113	

Grade	Filler	Sp Grav	Shrink, mils/in	Melt flow, g/10 min	Melt temp, °F	Back pres, psi	Drying temp, °F
FA-64014		1.19	8.0-25.0				176-180
FA-65014		1.20	8.0-25.0				176-180
FA-66084		1.24	8.0-25.0				176-180
FA-66104		1.26	8.0-25.0				176-180
FA-66114		1.38	8.0-25.0				176-180

PVC+NBR Alloy — Vynite — Alpha Chemical

Grade	Filler	Sp Grav	Shrink, mils/in	Melt flow, g/10 min	Melt temp, °F	Back pres, psi	Drying temp, °F
XPI-70		1.17				50	

PVC+NR Alloy — Vynite — Alpha Chemical

Grade	Filler	Sp Grav	Shrink, mils/in	Melt flow, g/10 min	Melt temp, °F	Back pres, psi	Drying temp, °F
XPI-65		1.19				50	

PVC+PUR Alloy — Vythene — Alpha Chemical

Grade	Filler	Sp Grav	Shrink, mils/in	Melt flow, g/10 min	Melt temp, °F	Back pres, psi	Drying temp, °F
11-68		1.20		34.00 E		100	140
13X-58		1.19				100	140
27D-90A		1.53		47.00 E		100	140

PVDF — Foraflon — Atochem

Grade	Filler	Sp Grav	Shrink, mils/in	Melt flow, g/10 min	Melt temp, °F	Back pres, psi	Drying temp, °F
1000 HD		1.78		10.00	338		
2500 HD		1.78		25.00	338		
4000 HD		1.78		40.00	338		
6000 HD		1.78		60.00	338		
9000 HD		1.78		90.00	338		

PVDF — Hylar — Ausimont

Grade	Filler	Sp Grav	Shrink, mils/in	Melt flow, g/10 min	Melt temp, °F	Back pres, psi	Drying temp, °F
460		1.75			311-320		
461		1.75			311-320		
710		1.77		20.00	329-338		
711		1.77		20.00	329-338		
720		1.77		10.00	332-338		
721		1.77		10.00	332-338		
740		1.77		23.00	329-338		
741		1.77		23.00	332-338		
760		1.77		10.00	329-338		
761		1.77		10.00	329-338		

PVDF — Kynar — Atochem

Grade	Filler	Sp Grav	Shrink, mils/in	Melt flow, g/10 min	Melt temp, °F	Back pres, psi	Drying temp, °F
320	CP	1.84		2.00	329-338		
460		1.75			311-320		
720		1.77		10.00 S	329-338		
740		1.77		23.00	329-338		
760		1.77		10.00	329-338		

PVDF — Kynar Flex — Atochem

Grade	Filler	Sp Grav	Shrink, mils/in	Melt flow, g/10 min	Melt temp, °F	Back pres, psi	Drying temp, °F
2800		1.76		3.00	285-293		

SAN — Cevian-N — Hoechst

Grade	Filler	Sp Grav	Shrink, mils/in	Melt flow, g/10 min	Melt temp, °F	Back pres, psi	Drying temp, °F
NF012A		1.11	3.0- 5.0	18.00		50- 100	176-185

SAN — CTI SAN — CTI

Grade	Filler	Sp Grav	Shrink, mils/in	Melt flow, g/10 min	Melt temp, °F	Back pres, psi	Drying temp, °F
SN-30GF/000 FR	GLF 30	1.63	0.6				180
SN-40GF/000	GLF 30	1.40	1.0				180

SAN — DSM SAN — DSM

Grade	Filler	Sp Grav	Shrink, mils/in	Melt flow, g/10 min	Melt temp, °F	Back pres, psi	Drying temp, °F
J-40/20	GLF 20	1.22	1.0				160
J-40/35	GLF 35	1.35	1.0				160

Drying time, hr	Inj time, sec	Front temp, °F	Mid temp, °F	Rear temp, °F	Nozzle temp, °F	Mold temp, °F	Proc temp, °F
2.0- 3.0	10.0- 15.	365	338	311	347	104-113	
2.0- 3.0	10.0- 15.	365	338	311	347	104-113	
2.0- 3.0	10.0- 15.	365	338	311	347	104-113	
2.0- 3.0	10.0- 15.	365	338	311	347	104-113	
2.0- 3.0	10.0- 15.	365	338	311	347	104-113	
		340	320	300	340		320-340
		340	320	300	340		320-340
2.0- 3.0		340	320	300	340	80- 90	350
2.0- 3.0		340	320	300	340	80- 90	350
2.0- 3.0		340	320	300	340	80- 90	350
							356-572
							356-572
							356-572
							356-572
							356-572
		410-490		390-450	400-490	120-200	
		410-480		395-450	400-490	120-200	
		375-440		360-410	360-440	120-200	
		375-440		360-410	360-440	120-200	
		375-450		360-410	360-450	120-200	
		375-450		360-410	360-450	120-200	
		400-470		380-430	380-470	120-200	
		400-470		380-430	380-470	120-200	
		410-480		395-450	400-490	120-200	
		410-480		395-450	400-490	120-200	
		430-500		380-450	450-510	120-200	
		410-490		390-450	400-490	120-200	
		375-450		360-410	360-450	120-200	
		400-470		380-430	380-470	120-200	
		410-480		395-450	400-490	120-200	
		400-480		375-410	380-490	120-200	
2.0- 4.0		410	392	356		104-140	356-446
						150	440-480
						150	450-520
2.0		400-430	450-500	430-460	380-420	160-180	450-500
2.0		400-430	450-500	430-460	380-420	160-180	450-500

Grade	Filler	Sp Grav	Shrink, mils/in	Melt flow, g/10 min	Melt temp, °F	Back pres, psi	Drying temp, °F
SAN		**Fiberfil**			**DSM**		
J-40/20	GLF 20	1.22	1.0- 2.0				160
J-40/35	GLF 35	1.35	1.0- 2.0				160
SAN		**Fiberfil VO**			**DSM**		
G-40/20/VO	GLF 20	1.40	1.5			25- 50	160
G-40/30/VO (Black)	GLF 30	1.47	1.0			25- 50	160
G-40/30/VO	GLF 30	1.47	1.0			25- 50	160
SAN		**Fiberstran**			**DSM**		
G-40/20	GLF 20	1.22	1.0- 2.0			10- 50	180
G-40/20/VO	GLF 20	1.40	1.0				160
G-40/30/VO	GLF 30	1.47	1.0				160
G-40/35	GLF 35	1.35	1.0- 2.0			10- 50	180
SAN		**Lupan**			**Lucky**		
GP-2205	GLF 20	1.20	1.0- 3.0			0- 568	176-194
GP-2305	GLF 30	1.30	1.0- 2.0			0- 568	176-194
HF-2208	GLF 20	1.20	1.0- 3.0			0- 568	176-164
HF-2358	GLF 35	1.30	1.0- 2.0			0- 568	176-194
HR-2207	GLF 20	1.20	1.0- 3.0			0- 568	176-194
SAN		**Luran**			**BASF**		
358N		1.08	3.0- 7.0				176
358N Trans. 77741		1.08	3.0- 7.0				176
368R		1.08	3.0- 7.0				176
368R Trans. 77741		1.08	3.0- 7.0				176
378P		1.08	3.0- 7.0				176
388S		1.08	3.0- 7.0				176
SAN		**Lustran SAN**			**Monsanto**		
31		1.07	3.0- 4.0	8.00 I			180
SAN		**Lustran Sparkle**			**Monsanto**		
Sparkle		1.07		12.00 I			180
SAN		**Network Polymers SAN**				**Network Poly.**	
260		1.08		2.00 G			176
268		1.07		3.00 G			176
278		1.08		2.00 G			176
SAN		**RTP SAN**			**RTP**		
500		1.08	6.0			50	180
501	GLF 10	1.15	4.0			50	180
501 FR	GLF	1.39	2.0			50	180
503	GLF 20	1.22	2.0			50	180
503 FR	GLF	1.46	1.0			50	180
505	GLF 30	1.31	1.0			50	180
505 FR	GLF	1.53	1.0			50	180
506	GLF 35	1.35	1.0			50	180
507	GLF 40	1.40	1.0			50	180
SAN		**Tyril**			**Dow**		
1000B		1.08	5.0	8.00 I		10- 500	160-180
1011		1.08	5.0	7.00 I		10- 500	160-180

Drying time, hr	Inj time, sec	Front temp, °F	Mid temp, °F	Rear temp, °F	Nozzle temp, °F	Mold temp, °F	Proc temp, °F
2.0		400-430	450-500	430-460	380-420	160-180	450-500
2.0		400-430	450-500	430-460	380-420	160-180	450-500
2.0		440-470	450-480	430-460	400-450	120-180	440-470
2.0		440-470	450-480	430-460	400-450	120-180	440-470
2.0		440-470	450-480	430-460	400-450	120-180	440-470
2.0		440-470	450-480	420-440	440-470	120-200	450-480
2.0		400-430	450-500	430-460	380-420	160-190	450-500
2.0		400-430	450-500	430-460	380-420	160-190	450-500
2.0		440-470	450-480	420-440	440-470	120-200	450-480
2.0- 3.0		410-446	392-428	374-410	410-446	140-194	428-464
2.0- 3.0		410-446	392-428	374-410	410-446	140-194	428-464
2.0- 3.0		410-446	392-428	374-410	410-446	140-194	428-464
2.0- 3.0		410-446	392-428	374-410	410-446	140-194	428-464
2.0- 3.0		446-482	428-464	410-446	446-482	140-194	464-500
2.0- 4.0						104-176	392-482
2.0- 4.0						104-176	392-482
2.0- 4.0						104-176	428-518
2.0- 4.0						104-176	428-518
2.0- 4.0						104-176	392-482
2.0- 4.0						104-176	428-518
2.0							450
2.0		380-450	380-450	380-450		120-175	400-425
2.0- 4.0		428-500	428-500	428-500		104-158	
2.0- 4.0		428-500	428-500	428-500		104-158	
2.0- 4.0		428-500	428-500	428-500		104-158	
2.0		460-550	450-530	430-520		100-175	
2.0		460-550	450-530	430-520		100-175	
2.0		460-500	440-530	420-510		100-175	
2.0		460-550	450-530	430-520		100-175	
2.0		460-500	440-530	420-510		100-175	
2.0		460-550	450-530	430-520		100-175	
2.0		460-500	440-530	420-510		100-175	
2.0		460-550	450-530	430-520		100-175	
2.0		460-550	450-530	430-520		100-175	
2.0		375-550		300-400		30-190	
2.0		375-550		300-400		30-190	

Grade	Filler	Sp Grav	Shrink, mils/in	Melt flow, g/10 min	Melt temp, °F	Back pres, psi	Drying temp, °F
125		1.08		25.00 I		10- 500	160-180
880		1.08	5.0	3.50 I		10- 500	160-180
880B		1.08	5.0	3.50 I		10- 500	160-180
990		1.07	5.0	8.70 I		10- 500	160-180
SB		**K-Resin**			**Phillips**		
KR01		1.01		8.00 G			140
KR03		1.01		8.00 G			140
SBS Copolymer		**C-Flex**			**Concept Polymer**		
R70-058 CLHR70		0.94	3.0-22.0	0.30 E			150-200
R70-067 CLHR40		0.92	3.0-22.0	9.90 E			150-200
R70-068 CLHR50		0.93	3.0-22.0	1.80 E			150-200
SBS Copolymer		**Dynaflex**			**GLS**		
D-3202		1.00	3.0- 5.0	14.00 E		80	
D-3204		1.01	1.0- 4.0	6.00 E		80	
D-3226		1.00	4.0- 6.0	20.00 E		80	
SBS Copolymer		**Finaclear**			**Fina**		
520		1.01		7.60 G			
530		1.02		11.00 G			
SBS Copolymer		**Kraton**			**Shell**		
D-2103		0.95		15.00 G		50- 100	
D-2104		0.92		22.00 G		50- 100	
D-2109		0.94		15.00 G		50- 100	
D-2122X		0.93				50- 100	
D-3226		1.00	4.0- 6.0	20.00 E		50- 100	
D-5298		0.95		23.00 E		50- 100	
D-5999X		1.13				50- 100	
D-7340		1.05				50- 100	
DX-1118		0.94				50- 100	
SEBS Copolymer		**C-Flex**			**Concept Polymer**		
R70-001 EM50A		0.90	3.0-22.0	0.25 E			150-200
R70-002 EM65A		0.90	3.0-22.0	1.90 E			150-200
R70-003 EM70A		0.90	3.0-22.0	1.80 E			150-200
R70-005 EM30A		0.90	3.0-22.0	1.10 E			150-200
R70-006 EM60A		0.90	3.0-22.0				150-200
R70-020 HR50A		0.90	3.0-22.0				150-200
R70-026 EM90A		0.90	3.0-22.0				150-200
R70-028 EM35A		0.90	3.0-22.0				150-200
R70-040 EM60CL		0.90	3.0-22.0	0.25 E			150-200
R70-041 HR55A		0.90	3.0-22.0				150-200
R70-042 LS55A		0.90	3.0-22.0				150-200
R70-046 EM35CL		0.90	3.0-22.0	1.30 E			150-200
R70-050 EM50CL		0.90	3.0-22.0	0.60 E			150-200
R70-051 EM70CL		0.90	3.0-22.0	2.70 E			150-200
R70-081 LSHR45A		0.90	3.0-22.0				150-200
R70-085 TLS50A		0.90	3.0-22.0				150-200
SEBS Copolymer		**Dynaflex**			**GLS**		
G-2701		0.90		16.00 E		200- 300	
G-2703		0.90	15.0-20.0			100- 700	
G-2706		0.89	25.0-30.0			80- 150	

Drying time, hr	Inj time, sec	Front temp, °F	Mid temp, °F	Rear temp, °F	Nozzle temp, °F	Mold temp, °F	Proc temp, °F
2.0		375-550		300-400		30-190	
2.0		375-550		300-400		30-190	
2.0		375-550		300-400		30-190	
2.0		375-550		300-400		30-190	
1.0						80-120	380-450
1.0						80-120	380-450
		300-425	300-425	300-425			
		300-425	300-425	300-425			
		300-425	300-425	300-425			
		360		350	365	75	
		385		370	395	75	
		345		335	350	75	
	10.0					100-150	450
		400	380	360	400	125-150	
		320-390	310-380	300-370	330-400	50-105	330-400
		320-390	310-380	300-370	330-400	50-105	330-400
		320-390	310-380	300-370	330-400	50-105	330-400
	5.0	320-390	310-380	300-370	330-400	50-105	
		320-390	310-380	300-370	330-400	50-105	330-400
		320-390	310-380	300-370	330-400	50-105	330-400
		320-390	310-380	300-370	330-400	50-105	330-400
		320-390	310-380	300-370	330-400	50-105	330-400
		320-390	310-380	300-370	330-400	50-105	330-400
		300-425	300-425	300-425			
		300-425	300-425	300-425			
		300-425	300-425	300-425			
		300-425	300-425	300-425			
		300-425	300-425	300-425			
		300-425	300-425	300-425			
		300-425	300-425	300-425			
		300-425	300-425	300-425			
		300-425	300-425	300-425			
		300-425	300-425	300-425			
		300-425	300-425	300-425			
		300-425	300-425	300-425			
		300-425	300-425	300-425			
		300-425	300-425	300-425			
		300-425	300-425	300-425			
		300-425	300-425	300-425			
		370-400		360-380	370-410	100-140	
		380-400		370-390	400-420	75	
		370-390		360-380	380-400	100-140	

Grade	Filler	Sp Grav	Shrink, mils/in	Melt flow, g/10 min	Melt temp, °F	Back pres, psi	Drying temp, °F
G-2712		0.88	15.0-20.0	28.00 G		100- 200	
G-6713		0.90		16.50 G		50	
G-7731		1.04	16.0-29.0			50- 100	
G-7745		1.15	12.0-29.0			50- 100	
G-7760		1.02	11.0			50- 100	
G-7770		1.06	9.0			50- 100	
G-7780		1.01	10.0			50- 100	
G-7790		1.13	9.0			50- 100	
SEBS Copolymer		**Kraton**			**Shell**		
G-2701		0.90		16.00 E		50- 100	
G-2703		0.90				50- 100	
G-2705		0.90	16.0-22.0	28.00 G		50- 100	
G-2706		0.89				50- 100	
G-2712		0.88				50- 100	
G-2712X		0.90		28.00		50- 100	
G-2732X		0.92				50- 100	
G-7150		1.95	8.0-12.0			50- 100	
G-7155		1.83	8.0-12.0			50- 100	
G-7160		1.90	8.0-12.0			50- 100	
G-7410		0.92				50- 100	
G-7430		0.94				50- 100	
G-7450		1.01				50- 100	
G-7523X		0.92				50- 100	
G-7528X		0.91				50- 100	
G-7680		1.03				50- 100	
G-7702X		1.07				50- 100	
G-7705		1.18	15.0-20.0			50- 100	
G-7715		0.91				50- 100	
G-7720		1.19	17.0-21.0			50- 100	
G-7722X		1.08				50- 100	
G-7820		1.14	8.0-15.0			50- 100	
G-7821X		1.24	15.0			50- 100	
G-7827		1.15	9.0-13.0			50- 100	
G-7890		0.92	12.0-15.0			50- 100	
G-7890X		0.91				50- 100	
G-7940		0.97				50- 100	
SMA		**Cadon**			**Monsanto**		
112		1.07	6.0- 9.0	0.70 I		0- 200	200
127		1.07	5.0- 7.0	1.80 I		0- 200	200
140		1.07	4.0- 6.0	3.00 I		0- 200	200
160		1.07	4.0- 6.0	2.40 I		0- 200	200
161		1.07	4.0- 6.0	2.60 I		0- 200	200
FRX		1.24	4.0- 6.0	3.10 I		0- 200	180
G2110	GLF 10	1.15	2.0- 4.0	0.80 I		0- 200	200
G2120	GLF 20	1.23	2.0- 4.0	0.70 I		0- 200	200
G2310	GLF 10	1.15	4.0			50- 100	180-200
G2320	GLF 20	1.23	2.5			50- 100	180-200
SMA		**Celstran**			**Hoechst**		
SMAG30-01-4	GLF 30	1.33					190
SMAG40-01-4	GLF 40	1.42					190
SMAG50-01-4	GLF 50	1.54					190

Drying time, hr	Inj time, sec	Front temp, °F	Mid temp, °F	Rear temp, °F	Nozzle temp, °F	Mold temp, °F	Proc temp, °F
		380-400		370-390	400-420	75	
		360-380	330-350	310-330	370-390	70-110	
		360-390		350-380	370-400	70-100	
		360-390		350-380	370-400	70-100	
		360-390		350-380	370-400	70-100	
		360-390		350-380	370-400	70-100	
		360-390		350-380	370-400	70-100	
		360-390		350-380	370-400	70-100	
		420-465	410-455	400-445	430-475	95-150	430-500
		420-465	410-455	400-445	430-475	95-150	430-500
		420-465	410-455	400-445	430-475	95-150	430-500
		420-465	410-455	400-445	430-475	95-150	430-500
		420-465	410-455	400-445	430-475	95-150	430-500
		420-465	410-455	400-445	430-475	95-150	430-500
	5.0	420-465	410-455	400-445	430-475	95-150	
		420-465	410-455	400-445	430-475	95-150	430-500
		420-465	410-455	400-445	430-475	95-150	430-500
		420-465	410-455	400-445	430-475	95-150	430-500
		420-465	410-455	400-445	430-475	95-150	430-500
		420-465	410-455	400-445	430-475	95-150	430-500
		420-465	410-455	400-445	430-475	95-150	430-500
	5.0	420-465	410-455	400-445	430-475	95-150	
		420-465	410-455	400-445	430-475	95-150	430-500
		420-465	410-455	400-445	430-475	95-150	430-500
		420-465	410-455	400-445	430-475	95-150	430-500
		420-465	410-455	400-445	430-475	95-150	430-500
		420-465	410-455	400-445	430-475	95-150	430-500
		420-465	410-455	400-445	430-475	95-150	430-500
		420-465	410-455	400-445	430-475	95-150	430-500
		420-465	410-455	400-445	430-475	95-150	430-500
		420-465	410-455	400-445	430-475	95-150	430-500
		420-465	410-455	400-445	430-475	95-150	430-500
		420-465	410-455	400-445	430-475	95-150	430-500
		420-465	410-455	400-445	430-475	95-150	430-500
		420-465	410-455	400-445	430-475	95-150	430-500
2.0		420-440	420-440	420-440	400-440	120-180	490-510
2.0		420-440	420-440	420-440	400-440	120-180	490-510
2.0		420-440	420-440	420-440	400-440	120-180	490-510
2.0		420-440	420-440	420-440	400-440	120-180	490-510
2.0		420-440	420-440	420-440	400-440	120-180	490-510
2.0		390-430	390-430	390-430	390-430	120-180	470-490
2.0		420-440	420-440	420-440	400-440	120-180	490-510
2.0		420-440	420-440	420-440	400-440	120-180	490-510
2.0	5.0- 8.	490-520	460-480	440-460	470-490	150-175	510-530
2.0	5.0- 8.	490-520	460-480	440-460	470-490	150-175	510-530
2.0		490	480	470	500	150	
2.0		490	480	470	500	150	
2.0		490	490	480	510	150	

Grade	Filler	Sp Grav	Shrink, mils/in	Melt flow, g/10 min	Melt temp, °F	Back pres, psi	Drying temp, °F
SMA		**Dylark**			**Arco**		
132		1.08	5.0	1.50 L		100	
232		1.08	5.0	1.70 L		100	
238		1.08	5.0	1.80 L		100	
238F20	GLF 20	1.23	3.0			50	
238P20	GLF 20	1.22	3.0	0.90 L		50	
250		1.06	6.0	0.90 L		100	
250P10	GLF 10	1.13	4.0	0.60 L		50	
250P20	GLF 20	1.20	3.0	0.35 L		50	
332		1.10	5.0	1.90 L		100	
350		1.06	6.0			100	
378		1.08	6.0	1.00 L		100	
378F20	GLF 20	1.22	3.0			50	
378P10	GLF 10	1.13		0.70 L		50	
378P20	GLF 20	1.21	3.0	0.40 L		50	
700		1.06	6.0	0.50 L		100	
SMA		**Stapron S**			**DSM**		
SG320	GLF 10	1.14	3.0- 4.0	6.00			176
SG324F	GLF 12	1.15	3.0- 4.0	9.00			176
SG340	GLF 20	1.25	2.0- 3.0	6.00			176
SG340F	GLF 20	1.24	2.0- 3.0	7.00			176
SG360F	GLF 30	1.28	2.0- 3.0	6.00			176
SG440	GLF 20	1.23	2.0- 3.0	5.00			176
SG460	GLF 30	1.32	2.0- 3.0	4.00			176
SG480	GLF 40	1.42	2.0- 3.0	2.00			176
SG540	GLF 20	1.27	2.0- 3.0	4.00			176
SM200		1.05	5.0- 8.0	9.00			176
SM300		1.08	5.0- 8.0	10.00			176
SM400		1.10	5.0- 8.0	13.00			176
SM500		1.12		17.00			176
SMA+PBT Alloy		**Dylark**			**Arco**		
DPN-501		1.19	5.2- 5.6	2.50 L		50- 100	230
DPN-540	GLF 30	1.40		2.00 L		50- 100	230
DPN-560		1.39		2.50 L		50- 100	230
SMMA Copolymer		**NAS**			**Novacor**		
21		1.08	2.0- 6.0	2.10 G			180
30		1.09	2.0- 6.0	2.20 G			180
35		1.09	2.0- 6.0	2.20 G			180
36		1.09	2.0- 6.0	2.20 G			180
50		1.12	2.0- 6.0	1.30 I			180
55		1.12	2.0- 6.0	1.30 I			180
SMMA Copolymer		**Network SMMA**			**Network Polymers**		
MS 100		1.08	2.0- 6.0	0.60 G			180
MS 300		1.09	2.0	2.40 G			180
MS 300 SF		1.09	2.0	2.40 G			180
MS 600		1.13	2.0- 6.0	1.80 G			180
Styrene Elast.		**RTP Polymers**			**RTP**		
ESD-A-2700		1.03	12.0				
ESD-C-2700		1.06	12.0				

Drying time, hr	Inj time, sec	Front temp, °F	Mid temp, °F	Rear temp, °F	Nozzle temp, °F	Mold temp, °F	Proc temp, °F
		425	400	375		100-130	425
		425	400	375		100-130	425
		475	450	425		120-150	475
		525	510	490		120-150	525
		525	510	490		120-150	525
		475	450	425		120-150	475
		525	510	490		120-150	525
		525	510	490		120-150	525
		425	400	375		100-130	425
		475	450	425		120-150	475
		475	450	425		120-150	475
		525	510	490		120-150	525
		525	510	490		120-150	525
		525	510	490		120-150	525
		475	450	425		120-150	475
2.0		460-480	480	460	460-480	140-176	470-520
2.0		460-480	480	460	460-480	140-176	470-520
2.0		460-480	480	460	460-480	140-176	470-520
2.0		460-480	480	460	460-480	140-176	470-520
2.0		460-480	480	460	460-480	140-176	470-520
2.0		460-480	480	460	460-480	140-176	470-520
2.0		460-480	480	460	460-480	140-176	470-520
2.0		460-480	480	460	460-480	140-176	470-520
2.0		460-480	480	460	460-480	140-176	470-520
2.0		460-480	480	460	460-480	140-176	470-520
2.0		460-480	480	460	460-480	140-176	470-520
2.0		460-480	480	460	460-480	140-176	470-520
2.0		460-480	480	460	460-480	140-176	470-520
2.0		480-495	470-490	455-470	465-500	150-180	475-500
2.0		480-495	470-490	455-470	465-500	150-180	475-500
2.0		480-495	470-490	455-470	465-500	150-180	475-500
2.0		420	400	370		130	420
2.0		420	400	370		130	420
2.0		420	400	370		130	420
2.0		420	400	370		130	420
2.0- 4.0		460	440	400		120	460
2.0- 4.0		460	440	400		120	460
2.0						120-170	430-460
2.0						120-140	400-430
2.0						120-140	400-430
2.0						120	430-460
		325-450	325-450	325-450		60-150	
		325-450	325-450	325-450		60-150	

Grade	Filler	Sp Grav	Shrink, mils/in	Melt flow, g/10 min	Melt temp, °F	Back pres, psi	Drying temp, °F
SVA		**Suprel SVA**			**Vista Chemical**		
9301		1.22	4.0- 6.0			50- 100	160-170
9302		1.19	4.0- 6.0			50- 100	160-170
9401		1.21	4.0- 6.0			50- 100	160-170
9410		1.20	4.0- 6.0			50- 100	160-170
9412		1.20	4.0- 6.0			50- 100	160-170
9420		1.24	4.0- 6.0			50- 100	160-170
9502		1.16	4.0- 6.0			50- 100	160-170
9503		1.15	4.0- 6.0			50- 100	160-170
9601		1.29	4.0- 6.0			50- 100	160-170
TPE		**Alphatec**			**Alpha Chemical**		
301-50		0.96		10.00 F			
TPE		**Arnitel**			**DSM**		
EL550		1.20			396		
EL630		1.23			415		
EL740		1.27			430		
EM400		1.12			383	0- 50	170
EM460		1.16			365	0- 50	170
PL380		1.16			419		
PL580		1.23			424		
PL720		1.28			433		
UL550		1.25			392	0- 50	170
UL740		1.27			423		
UM550		1.25					195
TPE		**Elexar**			**Teknor Apex**		
3707		0.90	5.0-20.0	1.50 G		100	150
3720		1.00	5.0-20.0	1.60 G		100	150
3954-45		1.18	5.0-20.0	1.30 G		100	150
3954-50		1.18	5.0-20.0	1.30 G		100	150
3954-62		1.18	5.0-20.0	1.30 G		100	150
3954-70		1.18	5.0-20.0	1.40 G		100	150
3954-80		1.18	5.0-20.0	1.40 G		100	150
3954-90		1.18	5.0-20.0	1.40 G		100	150
8313		1.25	5.0-20.0	3.50 G		100	150
8421		0.98	5.0-20.0	13.50 G		100	150
8431		0.93	5.0-20.0	2.00 G		100	150
8451		1.00	5.0-20.0	16.66 G		100	150
8614		1.15	5.0-20.0	3.00 G		100	150
8623		1.15	5.0-20.0	2.10 G		100	150
8712		1.09	5.0-20.0	1.90 G		100	150
8730		1.24	5.0-20.0	9.70 G		100	150
8730X		1.27					
8921		0.97	13.0				220
TPE		**Flexprene**			**Teknor Apex**		
6100-35		1.11		15.60 L		80	150
6100-50		1.11		10.00 L		80	150
6100-65		1.11		6.50 L		80	150
6100-75		1.11		6.50 L		80	150
6100-90		1.11		2.50 L		80	150
6200-45		0.95		14.00 L		80	150
6200-55		0.95		7.30 L		80	150

Drying time, hr	Inj time, sec	Front temp, °F	Mid temp, °F	Rear temp, °F	Nozzle temp, °F	Mold temp, °F	Proc temp, °F
1.0		360-375	360-375	360-375	360-375	80-110	385-410
1.0		360-375	360-375	360-375	360-375	80-110	385-410
1.0		360-375	360-375	360-375	360-375	80-110	385-410
1.0		360-375	360-375	360-375	360-375	80-110	385-410
1.0		360-375	360-375	360-375	360-375	80-110	385-410
1.0		360-375	360-375	360-375	360-375	80-110	385-410
1.0		360-375	360-375	360-375	360-375	80-110	385-410
1.0		360-375	360-375	360-375	360-375	80-110	385-410
1.0		360-375	360-375	360-375	360-375	80-110	385-410
							350-400
		392-482	392-482	392-482		68-122	428-500
		392-482	392-482	392-482		68-122	428-500
		392-482	392-482	392-482		68-122	428-500
2.0- 4.0		410	400	390	420	70-120	430
2.0- 4.0		420	410	410	430	70-120	435
		392-482	392-482	392-482		68-122	428-482
		392-482	392-482	392-482		68-122	428-482
		392-482	392-482	392-482		68-122	428-482
2.0- 4.0		465	455	445	470	70-120	480
		392-482	392-482	392-482		68-122	428-482
1.0		445	435	420	455		
2.0- 4.0		430	415	390	430	85-150	420-430
2.0- 4.0		435	425	415	440	85-150	430-440
2.0- 4.0		430	415	390	430	85-150	420-430
2.0- 4.0		430	415	390	430	85-150	420-430
2.0- 4.0		430	415	390	430	85-150	420-430
2.0- 4.0		430	415	390	430	85-150	420-430
2.0- 4.0		430	415	390	430	85-150	420-430
2.0- 4.0		430	415	390	430	85-150	420-430
2.0- 4.0		440	430	420	445	85-150	435-445
2.0- 4.0		430	415	390	430	85-150	420-430
2.0- 4.0		430	415	390	430	85-150	420-430
2.0- 4.0		430	415	390	430	85-150	420-430
2.0- 4.0		435	425	415	440	85-150	430-440
2.0- 4.0		435	425	415	440	85-150	430-440
2.0- 4.0		430	415	390	430	85-150	420-430
2.0- 4.0		430	415	390	430	85-150	420-430
							425-450
							490-520
2.0- 4.0		385	380	370	380	50-105	
2.0- 4.0		385	380	370	380	50-105	
2.0- 4.0		385	380	370	380	50-105	
2.0- 4.0		395	390	380	400	50-105	
2.0- 4.0		395	390	380	400	50-105	
2.0- 4.0		385	380	370	390	50-105	
2.0- 4.0		385	380	370	390	50-105	

Grade	Filler	Sp Grav	Shrink, mils/in	Melt flow, g/10 min	Melt temp, °F	Back pres, psi	Drying temp, °F
6200-65		0.95		9.00 L		80	150
6200-75		0.95		12.00 L		80	150
6200-85		0.95		5.00 L		80	150
6200-90		0.95		4.70 L		80	150
6300-30-NT		0.91		48.00 L		80	150
6400-50-NT		0.94		3.20 L		80	150
6400-55-NT		0.94		3.00 L		80	150
6400-60-NT		0.94		2.10 L		80	150
6400-65-NT		0.94		1.70 L		80	150
TPE		**Geolast**			**Adv. Elastomer**		
701-70		1.00				0- 100	175
701-80		0.99				0- 100	175
701-80 W-183		0.99				0- 100	175
701-87		0.98				0- 100	175
703-40		0.97				0- 100	175
703-50		0.96				0- 100	175
TPE		**HiFax**			**Himont**		
XL42D01		0.94	8.0-12.0				160
XL50D01		0.94	8.0-12.0				160
XL60D01		0.94	8.0-12.0				160
XL65A01		0.91	8.0-12.0				160
XL75A01		0.91	8.0-12.0				160
XL80A01		0.92	8.0-12.0				160
XL85A01		0.93	8.0-12.0				160
TPE		**Kraton**			**Shell**		
D-2120X		0.95		6.40 E		25	
D-2121X		0.95		7.40 E		30	
D-3202		1.00	3.0- 5.0	14.00 E		50- 100	
D-3204		1.02		35.00 G		50- 100	
D-5119		1.09		27.00 E		50- 100	
D-5122		1.08		25.00 E		50- 100	
D-5250		1.10		23.00 E		50- 100	
G-2730X		0.93		7.80 G		50	
G-2731X		0.90		7.80 G		50	
TPE		**Lomod**			**GE**		
AE2020A		1.20	17.0-19.0		424	75- 175	230
AE2040A		1.20	14.0-16.0		453	75- 175	230
AE2040B		1.20	14.0-16.0		453	75- 175	230
AE2060A		1.20	16.0-18.0		430	75- 175	230
AE2060B		1.20	16.0-18.0		430	75- 175	230
HG 3015A		2.15	16.0-17.0			50- 100	210-230
NBE063		1.14	16.0-17.0			50- 100	230
TPE		**Ontex**			**D&S Plastics**		
120 APX		0.95	13.0				
60 APX		0.96	13.0				
TPE		**PermaStat**			**RTP**		
1500-35D		1.15	25.0-30.0				
1500-70D		1.24	20.0-25.0				
2700		0.94	10.0-20.0				

Drying time, hr	Inj time, sec	Front temp, °F	Mid temp, °F	Rear temp, °F	Nozzle temp, °F	Mold temp, °F	Proc temp, °F
2.0- 4.0		395	390	380	400	50-105	
2.0- 4.0		395	390	380	400	50-105	
2.0- 4.0		395	390	380	400	50-105	
2.0- 4.0		395	390	380	400	50-105	
2.0- 4.0		395	390	380	400	50-105	
2.0- 4.0		395	390	380	400	50-105	
2.0- 4.0		395	390	380	400	50-105	
2.0- 4.0		395	390	380	400	50-105	
2.0- 4.0		395	390	380	400	50-105	
3.0- 4.0	0.5- 4.	360-380	355-375	350-370	370-390	140-210	380-400
3.0- 4.0	0.5- 4.	360-380	355-375	350-370	370-390	140-210	380-400
3.0- 4.0	0.5- 4.	360-380	355-375	350-370	370-390	140-210	380-400
3.0- 4.0	0.5- 4.	370-390	365-385	360-380	375-395	140-210	385-405
3.0- 4.0	0.5- 4.	370-390	365-385	360-380	380-400	140-210	390-410
3.0- 4.0	0.5- 4.	370-390	365-385	360-380	380-400	140-210	390-410
2.0- 3.0						70-100	360-400
2.0- 3.0						70-100	360-400
2.0- 3.0						70-100	360-400
2.0- 3.0						70-100	360-400
2.0- 3.0						70-100	360-400
2.0- 3.0						70-100	360-400
2.0- 3.0						70-100	360-400
	6.0	380	350	340	390	150	400
	3.0	380	370	360	410	100	410
		320-390	310-380	300-370	330-400	50-105	330-400
		320-390	310-380	300-370	330-400	50-105	330-400
		320-390	310-380	300-370	330-400	50-105	330-400
		320-390	310-380	300-370	330-400	50-105	330-400
		320-390	310-380	300-370	330-400	50-105	330-400
	4.0	410	390	380	430	120	440
	3.0	370	360	350	380	100	390
2.0- 4.0		480-500	460-480	440-460	480-510	60-140	490-520
2.0- 4.0		480-500	460-480	440-460	480-510	60-140	490-520
2.0- 4.0		480-500	460-480	440-460	480-510	60-140	490-520
2.0- 4.0		480-500	460-480	440-460	480-510	60-140	490-520
2.0- 4.0		480-500	460-480	440-460	480-510	60-140	490-520
2.0- 4.0		455-465	445-455	425-445	455-465	90-120	425-465
2.0- 4.0		440-460	440-460	420-440	460-480	100-120	440-480
							370-500
							370-500
		370-450	370-450	370-450		70-100	
		370-450	370-450	370-450		70-100	
		325-450	325-450	325-450		60-150	

Grade	Filler	Sp Grav	Shrink, mils/in	Melt flow, g/10 min	Melt temp, °F	Back pres, psi	Drying temp, °F
TPE		**Riteflex**			**Hoechst**		
372		1.23		14.00	470-490		250
372ZS		1.24		14.00			250
540		1.15		16.00		0- 50	225
540ZS		1.15		16.00		0- 50	225
547		1.17		11.00		0- 50	225
547		1.41				0- 50	225
547ZS		1.17		11.00		0- 50	225
555		1.20		10.00		0- 50	225
555HS		1.20		3.00		0- 50	225
555ZS		1.20		10.00		0- 50	225
572ZS		1.24		28.00		0- 50	250
635		1.14	9.0	12.50	327	0- 50	225
640		1.15	10.0	10.00	356	0- 50	225
647		1.17	13.0	10.00	376	0- 50	225
655		1.19	15.0	9.00	392	0- 50	225
663		1.24	19.0	10.00	406	0- 50	225
672		1.26	20.0	12.50	417	0- 50	225
BP 8929		1.18	15.0-20.0	7.00		0- 50	250
BP 9056		1.14	13.0-18.0	8.00		0- 50	225
BP 9057		1.14	13.0-18.0	8.00		0- 50	225
BP 9086		1.16	15.0-20.0	8.00		0- 50	235
TPE		**RTP Polymers**			**RTP**		
ESD-A-1500-50D		1.27	12.0				
ESD-A-2800		1.00	15.0				
ESD-C-1500-50		1.29	12.0				
ESD-C-2800		1.02	15.0				
TPE		**Santoprene**			**Adv. Elastomer**		
101-55		0.97					175
101-64		0.97					175
101-73		0.98					175
101-80		0.97					175
101-87		0.97					175
103-40		0.97					175
103-50		0.97					175
111-45		0.10					175
181-55		0.97				0- 150	150-170
181-64		0.97					150-170
201-55		0.97					175
201-64		0.97					175
201-73		0.98					175
201-80		0.97					175
201-87		0.97					175
203-40		0.97					175
203-50		0.97					175
211-45		0.10					175
251-80		1.24				0- 150	175
251-85		1.14				0- 150	175
251-92		1.30				0- 150	175
253-50		1.10				0- 150	175
271-55		0.97					175
271-64		0.97					175
271-73		0.97					175

Drying time, hr	Inj time, sec	Front temp, °F	Mid temp, °F	Rear temp, °F	Nozzle temp, °F	Mold temp, °F	Proc temp, °F
4.0							
4.0							
4.0	10.0- 20.	360-400	350-390	340-380	350-390	75-125	350-390
4.0	10.0- 20.	360-400	350-390	340-380	350-390	75-125	350-390
4.0	10.0- 20.	370-410	360-400	360-400	370-410	75-125	350-390
4.0	10.0- 20.	370-410	360-400	360-400	370-410	75-125	350-390
4.0	10.0- 20.	370-410	360-400	360-400	370-410	75-125	370-410
4.0	8.0- 15.	390-430	380-420	370-410	390-430	75-125	370-410
4.0	8.0- 15.	390-430	380-420	370-410	390-430	75-125	370-410
4.0	8.0- 15.	390-430	380-420	370-410	390-430	75-125	370-410
4.0	8.0- 15.	460-500	450-490	450-480	460-500	75-125	460-500
4.0		340-360	340-360	310-340	340-370	75-125	340-370
4.0		360-400	360-390	325-360	360-400	75-125	350-400
4.0		390-420	390-410	370-390	390-420	75-125	390-420
4.0		420-460	420-450	390-420	420-460	75-125	430-460
4.0		420-450	420-450	390-420	420-450	75-125	430-460
4.0		420-460	420-450	390-420	420-460	75-125	430-460
4.0		460-500	450-490	450-480	460-500	75-125	460-500
4.0		450-490	445-485	445-475	450-490	75-125	450-490
4.0		450-490	445-485	445-475	450-490	75-125	450-490
4.0		455-495	445-485	445-475	455-495	75-125	455-495
		370-420	370-420	370-420		70-100	
		350-450	350-450	350-450		60-150	
		370-420	370-420	370-420		70-100	
		350-450	350-450	350-450		60-150	
3.0- 4.0	0.5- 1.	340-380	340-380	340-380	340-380	50-175	360-400
3.0- 4.0	0.5- 1.	340-380	340-380	340-380	340-380	50-175	365-405
3.0- 4.0	0.5- 1.	350-370	350-370	350-370	350-370	50-175	360-400
3.0- 4.0	0.5- 1.	350-370	350-370	350-370	350-370	50-175	360-400
3.0- 4.0	0.5- 1.	360-380	355-375	350-370	370-390	50-175	380-420
3.0- 4.0	0.5- 1.	350-390	350-390	350-390	360-400	50-175	390-430
3.0- 4.0	0.5- 1.	350-390	350-390	350-390	360-400	50-175	390-430
3.0- 4.0	0.5- 3.	390	370	350	400	50-175	380-420
3.0- 4.0	0.5- 1.	340-380	340-380	340-380	340-380	50-175	360-400
3.0- 4.0	0.5- 1.	340-380	340-380	340-380	340-380	50-175	365-405
3.0- 4.0	1.0- 2.	340-380	340-380	340-380	340-380	50-175	360-400
3.0- 4.0	1.0- 2.	340-380	340-380	340-380	340-380	50-175	360-400
3.0- 4.0	0.5- 1.	350-370	350-370	350-370	350-370	50-175	360-400
3.0- 4.0	0.5- 1.	350-370	350-370	350-370	350-370	50-175	360-400
3.0- 4.0	0.5- 1.	360-380	355-375	350-370	370-390	50-175	380-420
3.0- 4.0	0.5- 1.	350-390	350-390	350-390	360-400	50-175	390-430
3.0- 4.0	0.5- 1.	350-390	350-390	350-390	360-400	50-175	390-430
3.0- 4.0	1.0- 3.	390	370	350	400	50-175	380-420
3.0- 4.0	2.0- 5.	350	345	340	370	100	375-385
3.0- 4.0	2.0- 5.	350	345	340	370	100	375-385
3.0- 4.0	2.0- 5.	350	345	340	370	100	375-385
3.0- 4.0	2.0- 5.	370	370	365	390	100	400-410
3.0- 4.0	0.5- 1.	350-390	350-390	350-390	360-400	50-175	390-430
3.0- 4.0	0.5- 1.	350-390	350-390	350-390	360-400	50-175	390-430
3.0- 4.0	0.5- 1.	350-390	350-390	350-390	360-400	50-175	390-430

Grade	Filler	Sp Grav	Shrink, mils/in	Melt flow, g/10 min	Melt temp, °F	Back pres, psi	Drying temp, °F
271-87		0.96					175
281-55		0.97					150-170
281-64		0.97					150-170
281-87		0.96					150-170
283-40		0.95					150-170
TPE		**Sarlink**			**DSM**		
1140		1.18		200.00 E		10- 50	
1150		1.18		200.00 E		10- 50	
1160		1.18		133.00 E		10- 50	
1170		1.20		25.00 E		10- 50	
1180		1.20				10- 50	
1260		1.15				10- 50	
1360		1.24				10- 50	
1370		1.22				10- 50	
1455		1.19				10- 50	
1560		1.20				10- 50	
1570		1.19				10- 50	
1580		1.19				10- 50	
2160		0.95		102.00 E		10- 150	
2170		0.99		138.00 E		10- 150	
2180		0.99		192.00 E		10- 150	
2440		0.95				10- 150	
2450		0.95				10- 150	
2460		0.98				10- 150	
3145D		0.94				10- 150	
3150		0.96				10- 150	
3160		0.95				10- 150	
3165		0.95				10- 150	
3170		0.95				10- 150	
3180		0.95				10- 150	
3190		0.95				10- 150	
3240		0.94				50- 150	
3245D		0.94				50- 150	
3250		0.96				50- 150	
3260		0.95				50- 150	
3270		0.95				50- 150	
3280		0.95				50- 150	
3290		0.95				50- 150	
3380		1.29				10- 150	160
3660B		0.96				10- 150	160
TPE		**Tekron**			**Teknor Apex**		
3954-45		1.18		1.30 L		100	150
3954-50		1.18		1.30 L		100	150
3954-62		1.18		1.30 L		100	150
3954-70		1.18		1.40 L		100	150
3954-80		1.18		1.40 L		100	150
3954-90		1.18		1.40 L		100	150
4000-25		1.18		4.80 L		100	150
4000-35		1.18		0.80 L		100	150
4000-45		1.18		0.80 L		100	150
4000-60		1.18		0.30 L		100	150
4000-80		1.18		0.80 L		100	150
4000-90		1.18		0.60 L		100	150
4200-30		0.91		2.10 L		100	150

Drying time, hr	Inj time, sec	Front temp, °F	Mid temp, °F	Rear temp, °F	Nozzle temp, °F	Mold temp, °F	Proc temp, °F
3.0- 4.0	0.5- 1.	350-390	350-390	350-390	360-400	50-175	390-430
3.0- 4.0	0.5- 1.	340-380	340-380	340-380	340-380	50-175	360-400
3.0- 4.0	0.5- 1.	340-380	340-380	340-380	340-380	50-175	365-405
3.0- 4.0	0.5- 1.	360-380	355-375	350-370	370-390	50-175	380-420
3.0- 4.0	0.5- 1.	350-390	350-390	350-390	360-400	50-175	390-430
	3.0- 5.	330-360	320-340	310-330	340-370	70-120	330-370
	3.0- 5.	330-360	320-340	310-330	340-370	70-120	330-370
	3.0- 5.	330-360	320-340	310-330	340-370	70-120	330-370
	3.0- 5.	330-360	320-340	310-330	340-370	70-120	330-370
		370-390	350-370	320-350	370-390	70-120	370-390
		330-360	320-340	310-330	340-370	70-120	330-370
		330-360	320-340	310-330	340-370	70-120	330-370
		330-360	320-340	310-330	340-370	70-120	330-370
		330-360	320-340	310-330	340-370	70-120	330-370
		350-390	350-380	340-360	370-390	70-120	370-390
		330-360	320-340	310-330	340-370	70-120	330-370
		350-390	340-370	320-350	360-390	70-120	350-390
	2.0- 5.	360-420	360-420	360-420	370-430	50-130	360-430
	2.0- 5.	360-420	360-420	360-420	370-430	50-130	360-430
	2.0- 5.	360-420	360-420	360-420	370-430	50-130	360-430
		360-410	360-410	360-410	370-420	50-130	360-420
		360-410	360-410	360-410	370-420	50-130	360-420
		360-420	360-420	360-420	370-430	50-130	360-430
		350-420	350-420	350-420	370-430	50-150	360-430
		360-410	360-410	360-410	370-420	50-150	360-420
		360-420	360-420	360-420	370-430	50-150	360-430
		350-420	350-420	350-420	370-430	50-150	360-430
		350-420	350-420	350-420	370-430	50-150	360-430
		350-420	350-420	350-420	370-430	50-150	360-430
		350-420	350-420	350-420	370-430	50-150	360-430
		360-410	360-410	360-410	370-420	50-150	360-420
		350-420	350-420	350-420	370-430	50-150	360-430
		360-410	360-410	360-410	370-420	50-150	360-420
		360-420	360-420	360-420	370-430	50-150	360-430
		350-420	350-420	350-420	370-430	50-150	360-430
		350-420	350-420	350-420	370-430	50-150	360-430
		350-420	350-420	350-420	370-430	50-150	360-430
4.0		350-410	350-400	350-400	370-420	50-150	360-420
4.0		350-410	350-410	350-410	370-420	60-150	350-420
2.0- 4.0		410	400	390	420	85-150	
2.0- 4.0		410	400	390	420	85-150	
2.0- 4.0		410	400	390	420	85-150	
2.0- 4.0		430	420	410	440	85-150	
2.0- 4.0		430	420	410	440	85-150	
2.0- 4.0		430	420	410	440	85-150	
2.0- 4.0		410	400	390	420	85-150	
2.0- 4.0		410	400	390	420	85-150	
2.0- 4.0		410	400	390	420	85-150	
2.0- 4.0		410	400	390	420	85-150	
2.0- 4.0		430	420	410	440	85-150	
2.0- 4.0		430	420	410	440	85-150	
2.0- 4.0		410	400	390	420	85-150	

Grade	Filler	Sp Grav	Shrink, mils/in	Melt flow, g/10 min	Melt temp, °F	Back pres, psi	Drying temp, °F
4200-45		0.91		2.40 L		100	150
4200-60		0.91		5.20 L		100	150
4200-75		0.91		5.90 L		100	150
4200-90		0.91		8.50 L		100	150
5000-30-NT		0.91		2.10 L		100	150
5000-45-NT		0.91		2.40 L		100	150
5000-55-NT		0.91		4.10 L		100	150
5000-60-NT		0.91		5.20 L		100	150
5000-75-NT		0.89		5.90 L		100	150
5000-80-NT		0.89		5.20 L		100	150
TPE		**Telcar**			**Teknor Apex**		
1000-100		0.90				50- 150	140-158
1000-105		0.92				50- 150	140-158
1000-90		0.87				50- 150	140-158
1000-95		0.90				50- 150	140-158
1000-96		0.89				50- 150	140-158
1000-98		0.89				50- 150	140-150
1025-65		1.00		0.20 L		50- 150	140-158
1025-75		1.00		0.20 L		50- 150	140-158
1025-85		1.00		0.30 L		50- 150	140-158
1025-90		1.00		0.30 L		50- 150	140-158
1050-65		1.00		0.60 L		50- 150	140-158
1050-75		1.00		0.50 L		50- 150	140-158
1050-85		0.99		0.80 L		50- 150	140-158
1050-90		0.99		0.60 L		50- 150	140-158
1075-105		1.23		1.50 L		50- 150	140-158
1075-106		1.30		1.50 L		50- 150	140-158
TPE		**Trefsin**			**Adv. Elastomer**		
3201-50		0.94					158-176
3201-60		0.97					158-176
3201-70		0.98					158-176
3281-50		0.94					158-176
3281-60		0.97					158-176
3281-70		0.98					158-176
TPE		**VistaFlex**			**Adv. Elastomer**		
9101-55W900		1.00				0- 100	175
9101-65W900		1.00				0- 100	175
9101-75W900		0.99				0- 100	175
9101-85W900		0.97				0- 100	175
9103-45		0.90				0- 100	175
9103-45W900		0.95				0- 100	175
9103-54W900		0.94				0- 100	175
TPE		**Vyram**			**Adv. Elastomer**		
6111-60		1.04	18.0				175
6111-70		1.03	17.0				175
6111-80		1.02	16.0				175
6111-90		1.01	15.0				175
6113-50		0.97					175
TPO		**Dexflex**			**D&S Plastics**		
1036		0.90	10.0-13.0	3.80 L			
1046		0.90	11.0-14.0	5.00 L			
1066		0.90	12.0-15.0	6.00 L			

Drying time, hr	Inj time, sec	Front temp, °F	Mid temp, °F	Rear temp, °F	Nozzle temp, °F	Mold temp, °F	Proc temp, °F
2.0- 4.0		410	400	390	420	85-150	
2.0- 4.0		410	400	390	420	85-150	
2.0- 4.0		430	420	410	440	85-150	
2.0- 4.0		430	420	410	440	85-150	
2.0- 4.0		410	400	390	420	85-150	
2.0- 4.0		410	400	390	420	85-150	
2.0- 4.0		410	400	390	420	85-150	
2.0- 4.0		410	400	390	420	85-150	
2.0- 4.0		430	420	410	440	85-150	
2.0- 4.0		430	420	410	440	85-150	
2.0- 4.0		400		390	410	50-150	
2.0- 4.0		400		390	410	50-150	
2.0- 4.0		360		350	370	50-150	
2.0- 4.0		400		390	410	50-150	
2.0- 4.0		360		350	370	50-150	
2.0- 4.0		400		390	410	50-150	
2.0- 4.0		360		350	370	50-150	
2.0- 4.0		360		350	370	50-150	
2.0- 4.0		360		350	370	50-150	
2.0- 4.0		360		350	370	50-150	
2.0- 4.0		360		350	370	50-150	
2.0- 4.0		360		350	370	50-150	
2.0- 4.0		360		350	370	50-150	
2.0- 4.0		360		350	370	50-150	
2.0- 4.0		380		370	390	50-150	
2.0- 4.0		380		370	390	50-150	
3.0	0.5- 3.	365-385	360-380	340-360	365-385	85-140	370-390
3.0	0.5- 3.	365-385	360-380	340-360	365-385	85-140	370-390
3.0	0.5- 3.	365-385	360-380	340-360	365-385	85-140	370-390
3.0	0.5- 3.	365-385	360-380	340-360	365-385	85-140	370-390
3.0	0.5- 3.	365-385	360-380	340-360	365-385	85-140	370-390
3.0	0.5- 3.	365-385	360-380	340-360	365-385	85-140	370-390
3.0- 4.0	0.5- 4.	385	380	370	395	100-120	400
3.0- 4.0	0.5- 4.	385	380	370	395	100-120	400
3.0- 4.0	0.5- 4.	385	380	370	395	100-120	400
3.0- 4.0	0.5- 4.	385	380	370	395	100-120	400
3.0- 4.0	0.5- 4.	385	380	370	395	100-120	400
3.0- 4.0	0.5- 4.	385	380	370	395	100-120	400
3.0- 4.0	0.5- 4.	385	380	370	395	100-120	400
3.0- 4.0	0.5- 1.	365	360	355	365	100-120	375
3.0- 4.0	0.5- 1.	365	360	355	365	100-120	375
3.0- 4.0	0.5- 1.	365	360	355	365	100-120	375
3.0- 4.0	0.5- 1.	365	360	355	375	100-120	375
3.0- 4.0	0.5- 1.	365	360	355	375	100-120	375
							370-500
							370-500
							370-500

Grade	Filler	Sp Grav	Shrink, mils/in	Melt flow, g/10 min	Melt temp, °F	Back pres, psi	Drying temp, °F
1203		0.93	14.0	6.00 L			
440		0.93	12.0-15.0	6.00 L			
475		0.93	10.0-13.0	6.00 L			
760		0.99	10.0-15.0	8.00 L			
777		1.10	8.0-15.0	7.00 L			
813		0.93	10.0-14.0	9.00 L			
815		0.97	10.0-13.0	6.00 L			
850		0.90	9.5-12.5	8.00 L			
D-45		0.91	10.0-16.0	6.50			
D-60		0.93	9.0-12.0	6.00 L			
D-6040		0.91	9.5-12.5	10.00 L			
D-64		0.93	8.5-11.5	6.00 L			
SA-1020		1.25	10.0	4.00 L			
TPO		**Ektar FB**			**Eastman**		
TG116	GLF 30	1.14	2.0- 6.0			50- 200	150-225
TG120	GLF 15	1.02		5.50 L		50- 200	150-225
TG122	GLF 15	1.02		5.50 L		50- 200	150-225
TG124	GLF 15	1.02		5.50 L		50- 200	150-225
TG130	GLF 30	1.14		4.00 L		50- 200	150-225
TG132	GLF 30	1.14		4.00 L		50- 200	150-225
TG134	GLF 30	1.14		4.00 L		50- 200	150-225
TG217	GLF 30	1.14	2.0- 6.0			50- 200	150-225
TG217-9203T	GLF 30	1.14	2.0- 6.0			50- 200	150-225
TG220	GLF 15	1.01		4.50 L		50- 200	150-225
TG222	GLF 15	1.01		4.50 L		50- 200	150-225
TG224	GLF 15	1.01		4.50 L		50- 200	150-225
TG230	GLF 30	1.13		3.90		50- 200	150-225
TG232	GLF 30	1.13		3.90		50- 200	150-225
TG234	GLF 30	1.13		3.90		50- 200	150-225
TPO		**Polytrope**			**A. Schulman**		
TPP 341		0.93	9.0-11.0	9.00 L		50	
TPP 503		0.92	10.0-12.0	8.00		50	
TPP 504		0.92	10.0-12.0	9.00		50	
TPP 508		0.91	11.0-13.0	6.00		50	
TPP 510		0.91	11.0-13.0	7.00		50	
TPP 512		0.91	11.0-13.0	7.00		50	
TPP 514	MN	0.97	8.0-10.0	5.00		50	
TPP 517	MN	1.06	8.0-10.0	7.00		50	
TPP 524	MN	1.07	8.0-10.0	5.00		50	
TPP 608		0.99	11.0-13.0	9.00		50	
TPP 615		1.08	6.0- 8.0	8.00	400-440	50	
TPP 620		1.09	8.0-10.0	8.00		50	
UHMWPE		**Nylatron**			**DSM**		
Ultrawear 5902WT		0.94	25.0			0- 100	
Urea Compound		**PMC Urea**			**Plastics Mfg.**		
Urea Molding Comp.	CE	1.48	7.0-12.0				
Urea Formal.		**Beetle**			**Cytec**		
1342	AC	1.50	8.0-10.0				
75	AC	1.50	8.0-10.0				
77	AC	1.50	8.0-10.0				
8023	AC	1.50	8.0-10.0				
8053	AC	1.50	8.0-10.0				

Drying time, hr	Inj time, sec	Front temp, °F	Mid temp, °F	Rear temp, °F	Nozzle temp, °F	Mold temp, °F	Proc temp, °F
							370-500
							370-500
							370-500
							370-500
							370-500
							370-500
							370-500
							370-500
							370-500
							370-500
							370-500
							370-500
							370-500
4.0- 6.0		400-470	400-470	400-470	360-470	100-150	420-500
4.0- 6.0		400-470	400-470	400-470	360-470	100-150	420-500
4.0- 6.0		400-470	400-470	400-470	360-470	100-150	420-500
4.0- 6.0		400-470	400-470	400-470	360-470	100-150	420-500
4.0- 6.0		400-470	400-470	400-470	360-470	100-150	420-500
4.0- 6.0		400-470	400-470	400-470	360-470	100-150	420-500
4.0- 6.0		400-470	400-470	400-470	360-470	100-150	420-500
4.0- 6.0		400-470	400-470	400-470	360-470	100-150	420-500
4.0- 6.0		400-470	400-470	400-470	360-470	100-150	420-500
4.0- 6.0		400-470	400-470	400-470	360-470	100-150	420-500
4.0- 6.0		400-470	400-470	400-470	360-470	100-150	420-500
4.0- 6.0		400-470	400-470	400-470	360-470	100-150	420-500
4.0- 6.0		400-470	400-470	400-470	360-470	100-150	420-500
4.0- 6.0		400-470	400-470	400-470	360-470	100-150	420-500
4.0- 6.0		400-470	400-470	400-470	360-470	100-150	420-500
		420-440	410-430	400-420	410-430	80-120	400-440
		420-440	410-430	400 420	410-430	80-120	400-440
		420-440	410-430	400-420	410-430	80-120	400-440
		420-440	410-430	400-420	410-430	80-120	400-440
		420-440	410-430	400-420	410-430	80-120	400-440
		420-440	410-430	400-420	410-430	80-120	400-440
		420-440	410-430	400-420	410-430	80-120	400-440
		420-440	410-430	400-420	410-430	80-120	400-440
		420-440	410-430	400-420	410-430	80-120	400-440
		420-440	410-430	400-420	410-430	80-120	400-440
		420-440	410-430	400-420	410-430	80-120	400-440
		420-440	410-430	400-420	410-430	80-120	400-440
		520-560	530-570	540-570	560	75-150	530-560
						300-330	
	5.0- 10.	175-210		80-160	180-220	290-320	175-240
	5.0- 10.	175-210		80-160	180-220	290-320	175-240
	5.0- 10.	175-210		80-160	180-220	290-320	175-240
	5.0- 10.	175-210		80-160	180-220	290-320	175-240
	5.0- 10.	175-210		80-160	180-220	290-320	175-240

Supplier Directory

Name	Full Name, State, Telephone
3M	3M Performance Polymers & Additives (MN) 612-736-9700
A. Schulman	A. Schulman Inc. (OH) 216-666-3751
Ablestik	Ablestick Laboratories (CA) 310-764-4600
Adv. Elastomer	Advanced Elastomer Systems (MO) 800-352-7866
Albis	Albis Corp. (TX) 713-342-3311
Albis Canada	Albis Canada Inc. (ON) 800-263-7582
Allied Signal	Allied Signal, Inc. (NJ) 800-446-1800
Alpha Chemical	Alpha Chemical & Plastics Corp. (NC) 704-554-8675
Amoco Chemical	Amoco Chemical Co. (IL) 800-621-4590
Amoco Perform.	Amoco Performance Products, Inc. (GA) 800-621-4557
Arco	Arco Chemical Co. (PA) 800-321-7000
Ashland	Ashland Chemical Co. (OH) 614-889-3333
Ashley	Ashley Polymers, Inc. (NY) 718-851-8111
AtoHaas	AtoHaas North America Inc. (PA) 800-523-7500
Atochem	Atochem North America, Inc. (NJ) 800-225-7788

NAME	FULL NAME, STATE, TELEPHONE
Ausimont	Ausimont, Inc. (NJ) 800-323-2874
BASF	BASF Corp. (NJ) 800-BC-RESIN
BFGoodrich	BFGoodrich Co. (OH) 800-331-1144
Bay Resins	Bay Resins, Inc. (MD) 410-928-3083
Bulk Molding	Bulk Molding Compounds (BMC), Inc. (IL) 708-377-1065
CTI	Compounding Technology, Inc. (CA) 800-325-1564
Chem Polymer	Chem Polymer Corp. (FL) 813-337-0400
Chevron	Chevron Chemical Co. (TX) 800-231-3826
Colorite	Colorite Plastics Co. (NJ) 201-941-2900
ComAlloy	ComAlloy International (TN) 615-333-3453
Concept Polymer	Concept Polymer Technologies (FL) 800-541-6880
Cook Composites	Cook Composites & Polymers (MO) 816-391-6000
Cosmic Plastics	Cosmic Plastics, Inc. (CA) 800-423-5613
Custom Resins	Custom Resins (KY) 800-626-7050
Cyro	Cyro Industries (CT) 800-631-5384
Cytec	Cytec Industries, Inc. (CT) (formerly American Cyanamid) 800-243-6874
D & S Plastics	D & S Plastics International (NC) 704-553-0046

Name	Full Name, State, Telephone
DSM	DSM Engineering Plastics (IN) (formerly Akzo) 800-333-4237
Diamond	Diamond Polymers, Inc. (CT) 203-787-0607
Dow	Dow Chemical U.S.A. (MI) 800-441-4369
Du Pont	Du Pont de Nemours & Co. Inc. (DE) 302-999-4592
EMS	EMS-American Grilon Inc. (SC) 803-481-6171
Eastman	Eastman Chemical Co. (TN) 800-327-8626
EniChem	EniChem America Inc. (NY) 212-382-6300
Exxon	Exxon Chemical Co. (TX) 713-870-6000
Ferro	Ferro Corp. (IN) 812-423-5218
Fina	Fina Oil & Chemical Corp. (TX) 800-344-3462
GE	General Electric Co. (MA) 800-437-5278
Geon	Geon Co. (OH) 800-438-4366
GLS	GLS Plastics (IL) 800-GLS-TPRS
Himont	Himont, Inc. (DE) 800-458-1416
Hitachi	Hitachi Chemical Co. Ltd. (Japan) 03-346-3111
Hoechst	Hoechst Celanese Corp. (NJ) 800-526-4960
Kolon	Kolon America (NY) 212-736-0120

Name	Full Name, State, Telephone
LNP	LNP Engineering Plastics Inc. (PA) 800-854-8774
Lucky	Lucky (NJ) 201-816-2300
Miles	Miles, Inc. (PA) 412-777-2000
Mitsubishi Gas	Mitsubishi Gas Chemical America, Inc. (NY) 212-752-4620
Mitsubishi Rayo	Mitsubishi Rayon America Inc. (NY) 212-759-5605
Mitsui Toatsu	Mitsui Toatsu Chemicals (NY) 212-867-6330
Mobil	Mobil Chemical Co. (NJ) 908-321-3500
MonTor	MonTor Performance Plastics (MI) 313-377-6290
Monsanto	Monsanto Co. (MO) 800-984-8400
Morton	Morton International (NH) 800-828-9878
Multibase	Multibase, Inc. (OH) 800-343-5626
Nan Ya Plastics	Nan Ya Plastics Corp. (Taiwan) 886-2-727-8254
Network Polymer	Network Polymers, Inc. (OH) 216-773-2700
Nippon Zeon	Nippon Zeon Co., Ltd. (Japan) 03 3769 8638
Norold Compos.	Norold Composites Inc. (ON) 800-563-2089
Novacor	Novacor Chemicals Inc. (MA) 800-225-8063
Novamont	Novamont North America Inc. (NY) 212-997-7035

NAME	FULL NAME, STATE, TELEPHONE
Novatec	Novatec Plastics & Chemicals Co. Inc. (NJ) 800-782-6682
Nytex Compos.	Nytex Composites Co., Ltd. (USA) (CA) 818-287-7282
OxyChem	Occidental Chemical Corp. (TX) 800-800-4327
PMC	Plastic Materials Co., Inc. (DE) 302-422-3021
Perstorp	Perstorp Compounds, Inc. (MA) 413-584-2472
Phillips	Phillips Chemical Co. (TX) 800-537-3746
Plastics Mfg.	Plastics Manufacturing Co. (TX) 214-330-8671
Plenco	Plastics Engineering Co. (WI) 414-458-2121
Polymerland	Polymerland, Inc. (WV) 800-752-7842
Polymers Intl.	Polymers International Inc. (SC) 803-579-2729
Quantum	Quantum Chemical Corp. (OH) 513-530-6500
RTP	RTP Co. (MN) 800-433-4787
Resinoid	Resinoid Engineering Corp. (IL) 708-637-1050
Rexene	Rexene (TX) 214-450-9000
Rogers	Rogers Corp. (CT) 203-646-5500
Roscom	Roscom Inc. (NJ) 609-393-4200
Rotuba	Rotuba Plastics (NJ) 908-486-1000

NAME	FULL NAME, STATE, TELEPHONE
Sam Yang	Sam Yang Co., Ltd. (Korea) 02 740-7114
Shell	Shell Chemical Co. (TX) 713-241-2396
Shinkong	Shinkong Synthetic Fiber Corp. (Taiwan) 02-507-1251
Shuman Plastics	Shuman Plastics, Inc. (NY) 716-685-2121
Solvay	Solvay Polymers, Inc. (TX) 800-231-6313
Sunprene	Sunprene Co. (OH) 419-483-2931
Sunkyong	Sunkyong Industries (Korea) 02-273-3131
Teijin Chemical	Teijin Chemicals Ltd. (Japan) 03 506-4772
Teknor Apex	Teknor Apex Co. (RI) 800-523-9242
Texapol	Texapol Corp. (PA) 800-523-9242
Thai Petrochem	Thai Petrochemical Industry Co., Ltd. (Thailand) 238-4000
Thermedics	Thermedics (MA) 617-938-3786
Thermofil	Thermofil, Inc. (MI) 800-444-4408
Tong Yang	Tong Yang Nylon Co., Ltd. (NY) 212-736-7100
Toray	Toray Marketing & Sales (America), Inc. (NY) 212-697-8150
Ube	Ube Industries, Ltd. (Japan) 03-505-9300

Name	Full Name, State, Telephone
UVtec	UVtec, Inc. (TX) 817-640-5600
Vista Chemical	Vista Chemical Co. (TX) 713-588-3000
Washington Penn	Washington Penn Plastic Co. Inc. (PA) 412-228-1260
Wellman	Wellman, Inc. (SC) 800-821-6022

Notes

TRADENAME DIRECTORY

TRADENAME	RESIN	SUPPLIER
ACP	PVC	Alpha Chemical
AVP	ABS, Nylon 6/6, PBT PC, PC+PBT Alloy	Polymerland
Ablebond	Epoxy	Ablestik
Acctuf	PP	Amoco Chemical
Acetron	Acetal	DSM
Acrylite	Acrylic	Cyro
Acrylite Plus	Acrylic	Cyro
Affinity	Plastomer	Dow
Akulon	Nylon 6, 6/6	DSM
Akuloy RM	Nylon 6, 6/6 Alloys	DSM
Alathon	HDPE HDPE Copolymer	OxyChem
Alphatec	TPE	Alpha Chemical
Amodel	PPA	Amoco Perform.
Apec	PC	Miles
Aqualoy	Nylon 6/6, 6/12, PP	ComAlloy
Ardel	Polyarylate	Amoco Perform.
Arnite	PBT, PET	DSM
Arnitel	TPE	DSM
Ashlene	Nylon 6, 6/6, 6/12	Ashley
Aurum	Polyimide (TP)	Mitsui Toatsu
Bayblend	ABS+PC Alloy	Miles
Baydur	PUR	Miles
Beetle	Urea Formal.	Cytec
Bexloy	PET	Du Pont
C-Flex	SBS Copolymer SEBS Copolymer	Concept Polymer
Cadon	SMA	Monsanto
Calibre	PC	Dow
Capron	Nylon 6, 6/6, Nylon 6/6 Copolymer	Allied Signal
Cefor	PP Copolymer	Shell

TRADENAME	RESIN	SUPPLIER
Celanex	PBT	Hoechst
Celcon	Acetal Copolymer	Hoechst
Celstran	ABS+PC Alloy Acetal, HDPE Nylon 6, 6/6 PBT, PC, PET PP, PPS, PUR, SMA	Hoechst
Celstran S	ABS, Nylon 6/6 PBT, PC, PPS, PPSS	Hoechst
Centrex	ASA, ASA/AES	Monsanto
Cevian	ABS, ABS+PBT Alloy	Hoechst
Cevian-N	SAN	Hoechst
Cevian-V	ABS	Hoechst
Comshield	PP	ComAlloy
Comtuf	Nylon 6, 6/6, 6/12 PBT, PBT Alloy, PP PP Copolymer Polyester Alloy	ComAlloy
Crystalor	PMP	Phillips
Cycolac	ABS, ABS+PBT Alloy	GE
Cycolin	ABS+PBT Alloy	GE
Cycoloy	ABS+PC Alloy	GE
Cyglas	Polyester, TS	Cytec
Cymel	Mel Formaldehyde	Cytec
Cyrolite	Acrylic	Cyro
Delrin	Acetal	Du Pont
Delrin II	Acetal	Du Pont
Delrin P	Acetal	Du Pont
Dexflex	TPO	D&S Plastics
Dimension	Nylon 6, 6/6 Alloy	Allied Signal
Dowlex	LLDPE	Dow
Duraflex	Polybutylene	Shell
Dural	PVC	Alpha Chemical
Durel	Polyarylate	Hoechst
Durethan	Nylon 6	Miles
Dylark	SMA, SMA+PBT Alloy	Arco

TRADENAME	RESIN	SUPPLIER
Dynaflex	SBS, SEBS Copolymers	GLS
Ecdel	Polyester, ElCo	Eastman
Ektar	PCTG, PET, PETG	Eastman
Ektar FB	PCT, PCTG	Eastman
	PET, PP, TPO	
Ektar MB	PC+Polyester, PCTG	Eastman
Elastollan	PUR-Polyester	BASF
	PUR-Polyether	
Electrafil	ABS, Acetal, HDPE,	DSM
	Nylon 6, 6/6, 6/12	
	PBT, PC, PEEK, PEI	
	PET, PP, PP Copolymer	
	PS, PSU	
Elexar	TPE	Teknor Apex
Elvax	EVA	Du Pont
Emac	EMA	Chevron
Enathene	EnBA	Quantum
Engage	Polyolefin Ela	Dow
Escalloy	PP	ComAlloy
Escorene	PP, PP Copolymer	Exxon
Estaloc	PUR	BFGoodrich
Estane	PUR-Polyester	BFGoodrich
	PUR-Polyether	
Exact	Plastomer	Exxon
FR-PC	PC	Lucky
FTPE	Fluorelast	3M
Ferrene	PE, Polyolefin	Ferro
Ferrex	PP	Ferro
Ferrocon	Polyolefin	Ferro
Ferropak	PP	Ferro
Fiberfil	ABS, Acetal, HDPE	DSM
	Nylon 6, 6/6, 6/10	
	Nylon 6/12, PC, PEI	
	PES, PP, PPO, PPS	
	PS, PSU, PUR, SAN	
Fiberfil TN	Nylon 6, 6/6, 6/12	DSM

TRADENAME	RESIN	SUPPLIER
Fiberfil VO	ABS, Nylon 6, 6/6 Nylon 6/10, 6/12 PBT, PC, PP, PPO, SAN	DSM
Fiberloc	PVC	Geon
Fiberstran	ABS, HPDE, Nylon 6 Nylon 6/6, 6/10, 6/12 PC, PES, PP, PPO PS, PSU, SAN	DSM
Finaclear	SBS Copolymer	Fina
Flexalloy	PVC	Teknor Apex
Flexprene	TPE	Teknor Apex
Foraflon	PVDF	Atochem
Formion	Ionomer	A. Schulman
Fortron	PPS	Hoechst
Gapex AY	Nylon	Ferro
Geloy	ASA, ASA+PC Alloy ASA+PVC Alloy	GE
Geolast	TPE	Adv. Elastomer
Geon HTX	PVC Alloy	Geon
Grilamid	Nylon 12 Nylon 12 Elast.	EMS
Grilon	Nylon 6, 6/6 Nylon 6 Elast.	EMS
Grilpet	PET	EMS
Grivory	Nylon	EMS
Halar	PECTFE	Ausimont
Halon	ETFE	Ausimont
HiFax	PP, TPE	Himont
Hiloy	Nylon 6, 6/6, 6/6 Alloy, Nylon 6/12, PBT, PBT Alloy PET, PP	ComAlloy
Hitachi ASA	ASA	Hitachi
Hostacom	PP	Hoechst
Hostalloy	Polyolefin	Hoechst
Hylar	PVDF	Ausimont

TRADENAME	RESIN	SUPPLIER
Hytrel	Polyester, ElCo	Du Pont
Impet	PET	Hoechst
Interpol	PUR-Polyester	Cook Composites
Isoplast	PUR	Dow
Iupiace	PPE	Mitsubishi Gas
Iupilon	ABS+PC Alloy, PC	Mitsubishi Gas
Iupital	Acetal	Mitsubishi Gas
Ixef	Polyaryl Amide	Solvay
K-Resin	SB	Phillips
Kapton	Polyimide (TS)	Du Pont
Koblend	ABS+PC Alloy	EniChem
Kodapak	PET	Eastman
Kodar	ASA/AES, PETG	Eastman
Kopa	Nylon 6, 6/6	Kolon
Kraton	SBS	Shell
	SEBS Copolymers, TPE	
Kynar	PVDF	Atochem
Kynar Flex	PVDF	Atochem
Lexan	PC, PPC	GE
Lexan SP	PC	GE
Lomod	Polyester, ElCo; TPE	GE
Lubricomp	Acetal, Nylon 6/6, PC	LNP
Lubrilon	Nylon 6, 6/6, 6/12	ComAlloy
Lubrilon	PBT, PBT Alloy	ComAlloy
Lucel	Acetal Copoly	Lucky
Lucet	Acetal Copoly	Lucky
Lumax	PBT Alloy	Lucky
Lupan	SAN	Lucky
Lupol	Polyolefin	Lucky
Lupon	Nylon 6/6	Lucky
Lupos	ABS	Lucky
Lupox	PBT	Lucky
Lupoy	ABS+PBT Alloy	Lucky
Luran	SAN	BASF
Luran S	ASA	BASF
Lusep	PPS	Lucky

TRADENAME	RESIN	SUPPLIER
Lustran	ABS, SAN	Monsanto
Lustran Elite	ABS	Monsanto
Lustran Sparkle	SAN	Monsanto
Lustran Ultra	ABS	Monsanto
Magnum	ABS	Dow
Makroblend	PC+PET Alloy	Miles
Makrolon	PC	Miles
Marlex	HDPE, PP	Phillips 66
Mater-Bi	Bio Syn Poly	Novamont
Microthene	PE	Quantum
Mindel	PSU, PSU+ABS	Amoco Perform.
Minlon	Nylon 6/6	Du Pont
MonTor Nylon	Nylon 6, 6/6	MonTor
Morthane	PUR-Polyester PUR-Polyether PUR-Capro	Morton
Multi-Pro	PP	Multibase
NAS	SMMA Copolymer	Novacor
NSC	Nylon, PS	Thermofil
Norsophen	Phenolic	Norold
Nortuff	HDPE, PP	Polymerland
Noryl	PPO, PPO Alloy	GE
Noryl GTX	PPO	GE
Novapol	LLDPE	Novacor
Novatemp	PVC	Novatec
Nybex	Nylon 6/12	Ferro
Nylafil	Nylon 6, 6/6, 6/12	DSM
Nylatron	Nylon 4/6, 6/6 6/6 Alloy, UHMWPE	DSM
Nyloy	Nylon 6/6, 6/6 Alloy PC, PP	Nytex Compos.
Nypel	Nylon 6	Allied Signal
Nytron	Nylon 6/6	Nytex Compos.
Ontex	TPE	D&S Plastics
Optema	EMA	Exxon
Panlite	PC	Teijin Chemical

TRADENAME	RESIN	SUPPLIER
Pebax	PEBA	Atochem
Pellethane	PUR-Polyether	Dow
	PUR-Capro, PUREL-TP	
PermaStat	ABS, ABS+PC Alloy	RTP
	Acetal, PBT, PP, PPA	
	PUREL-TP, TPE	
Petlon	PET	Albis
Petra	PET	Allied Signal
Petrothene	EVA, HDPE, LDPE	Quantum
	LLDPE, PP	
	PP Copolymer	
Plaslube	Acetal	DSM
	Nylon 6, 6/6, 6/12	
	PC, PPS	
Plexiglas	Acrylic	AtoHaas
Pocan	PBT	Albis
Polyfort	PP, PP Copolymer	A. Schulman
Polytrope	TPO	A. Schulman
Polyvin	PVC	A. Schulman
Porene	ABS	Thai Petrochem
Prevail	ABS+TPU Alloy	Dow
Prevex	PPE	GE
Pro-fax	PP, PP Copolymer	Himont
Pulse	ABS+PC Alloy	Dow
Radel	PPSU, Polyarylsulfone	Amoco Perform.
Radiflam	Nylon 6, 6/6	Polymers Intl.
Radilon	Nylon 6, 6/6	Polymers Intl.
Reny	Nylon MXD6	Mitsubishi Gas
Resinoid	Phenolic	Resinoid
Retain	PE	Dow
Rilsan	Nylon 11, 12	Atochem
Riteflex	TPE	Hoechst
Ronfalin	ABS	DSM
Rynite	PBT, PBT	Du Pont
Ryton	PPS	Phillips
Sabre	PC+Polyester	Dow

TRADENAME	RESIN	SUPPLIER
Safe-FR	Polyolefin	UVtec
Santoprene	TPE	Adv. Elastomer
Sarlink	TPE	DSM
Satinflex	PVC	Alpha Chemical
Schulamid 6	Nylon 6	A. Schulman
Schulamid 6.6	Nylon 6/6	A. Schulman
Selar PA	Nylon	Du Pont
Selar PT	PET	Du Pont
Shinite	PBT	Shinkong
ShinkoLite-P	PMMA	Mitsubishi Rayo
Stanyl	Nylon 4/6	DSM
Stapron C	ABS+PC Alloy	DSM
Stapron S	SMA	DSM
Stat-Kon	ABS, Nylon 6, PS Polyester/Ela Polyester, TP	LNP
Styron	PS	Dow
Suntra	PPS	Sunkyong
Supec	PPS	GE
Superkleen	PVC	Alpha Chemical
Suprel SVA	SVA	Vista Chemical
Surlyn	Ionomer	Du Pont
Tecoflex	PUR-Polyether	Thermedics
Tecothane	PUR-ether/MDI	Thermedics
Tedur	PPS	Miles
Teflon	FEP, PFA	Du Pont
Tefzel	ETFE	Du Pont
Tekron	TPE	Teknor Apex
Telcar	TPE	Teknor Apex
Tempalloy	PP	ComAlloy
Tenite	PP	Eastman
Terluran	ABS	BASF
Terlux	Acrylic+ABS Aly	BASF
Texalon	Nylon 6, 6/6 Nylon 6/10, 6/12	Texapol

Tradename	**Resin**	**Supplier**
Texin	PC+PUR Alloy, PUR	Miles
	PUR-Polyester	
	PUREL-TP	
	Polyester, TP, Polyether	
Thermocomp	Nylon 6, 6/6, PC	LNP
	PES, PPA, PPS	
Topex	PBT	Tong Yang
Torlon	Polyamide-Imide	Amoco Perform.
Toyolac	ABS, AS Copolymer	Toray
Trefsin	TPE	Adv. Elastomer
Triax	ABS+PA Alloy	Monsanto
	ABS+PC Alloy	
	ABS+PVC Alloy	
Trirex	PC	Sam Yang
Tufrex	ABS	Monsanto
Tyril	SAN	Dow
Udel	PSU	Amoco Perform.
Ultem	PEI, PEI+PC Alloy	GE
Ultradur	PBT	BASF
Ultraform	POM	BASF
Ultramid	Nylon 6, 6/6,	BASF
	6/66 Copolymer	
	Nylon 6/6T, 6/10	
Ultramid T	Nylon 6/6T	BASF
Ultrapek	PAEK	BASF
Ultrason	PES, PSU	BASF
Ultrathene	EVA	Quantum
Unichem	PVC	Colorite
Valox	PBT, PBT+PET Alloy	GE
	PCT, PET	
Valtra	PS	Chevron
Vectra	LCP	Hoechst
Verton	Nylon 6, 6/6, 6/10	LNP
VistaFlex	TPE	Adv. Elastomer
Vistel	PVC	Vista Chemical

Tradename	Resin	Supplier
Voloy	Nylon 6, 6/6, 6/12	ComAlloy
	PBT, PBT Alloy	
	PET, PP	
Vybex	Polyester, TP	Ferro
Vydyne	Nylon 6/6,	Monsanto
	6/66 Copolymer, 6/9	
Vynite	PVC+NBR Alloy	Alpha Chemical
	PVC+NR Alloy	
Vyram	TPE	Adv. Elastomer
Vythene	PVC+PUR Alloy	Alpha Chemical
Wellamid	Nylon 6, 6/6	Wellman
	Nylon 6/66 Copolymer	
XT Polymer	Acrylic	Cyro
Xenoy	PC+PBT Alloy	GE
	PC+PET Alloy	
Xydar	LCP	Amoco Perform.
Zeonex	Polyolefin	Nippon Zeon
Zylar	Acrylic	Novacor
	MBS Terpolymer	
Zylar ST	MBS Terpolymer	Novacor
Zytel	Nylon 6/6	Du Pont
	Nylon 6/12, 12/12	

TROUBLESHOOTING GUIDE

This troubleshooting guide is divided into three parts. The first part will help you identify defects. The second part is a chart that shows you a recommended course of action to fix any of these defects. The numbers stand for the sequence in which remedies should be tried, and the arrows indicate either an increase in the setting, a decrease in the setting, or the necessity to balance the setting. The final section is simply a narrative version of the chart.

IDENTIFYING MOLDING DEFECTS

Black Specks: Tiny black particles on the surface of an opaque part and visible throughout a transparent part.

Blister: Defect on the surface of a molded part caused by gases trapped within the part during curing.

Blush: Discoloration generally appearing at gates, around inserts, or other obstruction along the flow path. Usually indicates weak points.

Brittleness: Tendency of a molded part to break, crack, shatter, etc. under conditions which it would not normally do so.

Burn Marks: Black marks or scorch marks on surface of molded part; usually on the side of the part opposite the gate or in a deep cavity.

Cracking: Fracture of the plastic material in an area around a boss, projection, or molded insert.

Crazing: Fine cracks in part surface. May extend in a network over the surface or through the part.

Delamination (Skinning): Surface of the finished part separates or appears to be composed of layer of solidified resin. Strata or fish scale type appearance where the layers may be separated.

Discoloration: Refers to any nonuniform coloration, whether a general brown color such as that caused by overheating or streaky discoloration resulting from contamination.

Excessive Warpage/Shrinkage: Excessive dimensional change in a part after processing, or the excessive decrease in dimension in a part through cooling.

Flash: Excess plastic that flows into the parting line of the mold beyond the edges of the part and freezes to form thin, sheet-like protrusions from the part.

Flow Marks: Marks visible on the finished item that indicate the direction of flow in the cavity.

Gels (Clear Spots): Surface imperfections resulting from usage of unplasticized pellets.

Jetting ("Snake Flow"): Turbulence in the resin melt flow caused by undersized gate, abrupt change in cavity volume, or too high injection pressure.

Poor Surface Finish (Gloss): Surface roughness resulting from high speed fill which causes surface wrinkling as the polymer melt flows along the wall of the mold.

Poor Weld Lines (Knit Lines): Inability of two melt fronts to knit together in a homogeneous fashion during the molding process, resulting in weak areas in the part of varying severity.

Short Shot: Injection of insufficient material to fill the mold.

Sink Marks: Depression in a molded part caused by shrinking or collapsing of the resin during cooling.

Splay Marks (Silver Streaking, Splash Marks): Marks or droplet type imperfections formed on the surface of a finished part.

Voids (Bubbles): An unfilled space within the part.

TROUBLESHOOTING CHART

Numbers indicate sequence of making corrective steps;
Arrows indicate
↑ - Increase
↓ - Decrease
↕ - Balance/Vary

		MACHINE DEFECTS					
		Excessive Flash	Oversized Part	Part Sticking	Short Shot	Sprue Sticking	Undersized Part
MACHINE VARIABLES	Back Pressure	5↓			8↑		
	Inj. Forward (Booster) Time		2↓	3↓		3↓	3↑
	Clamp Pressure	3↑	8↑				
	Cylinder Temperature	2↓	5↓	6↓			6↑
	Holding Pressure		4↓	2↓			4↑
	Injection Hold Time	4↓		7↓	9↑	2↓	5↑
	Injection Pressure	1↓	3↓	1↓	2↑	1↓	2↑
	Injection Speed	6↓	1↓	8↓	6↑	5↓	1↑
	Shot Size (Material Feed)				1↑		
	Melt Temperature				3↑		
	Mold Cooling Time			4↑		4↓	9↓
	Mold Temperature	7↓	6↑	5↓	5↑		7↓
	Nozzle Temperature				4↑	6↕	
	Overall Cycle Time		7↓				
	Screw Speed						
MOLD VAR.	Change Gate Location						
	Size of Gate				11↑		8↑
	Size of Sprue/Runner				10↑		
	Size of Vent	10↕			7↑		
OTHER ACTION	Check for Material Contamination						
	Check Fit of Mold Faces	9					
	Clean Cavity Surface						
	Clean Mold Faces	8					
	Clean Vents				12		
	Dry Materials						
	Regrind Quantity						
	Purge/Clean Screw & Barrel						

PART DEFECTS

Black Spots, Brown Streaks	Blisters	Brittleness	Burn Marks	Cracking, Crazing	Delamination	Discoloration	Flow Marks	Jetting	Poor Surface Finish (Gloss)	Poor Weld Lines	Silver Streaks, Splay, Splash	Sink Marks	Voids	Warping
5↓	2↑	2↓				4↓						7↑		
			4↓											
				2↑		2↓	1↑		3↑		3↓			
												3↑	3↑	
										3↑				3↑
		4↓	5↓	5↓			4↑		2↑	1↑		2↑	2↑	4↕
			1↓	1↕	4↕		5↕	1↓	4↑	5↑	4↓	6↑	6↓	
												1↑	1↑	
3↓	4↓	1↑	2↓		2↑			2↑		2↑		5↓	5↓	5↕
									8↑					2↑
	3↑		3↓	4↑	1↑		2↑	3↑	1↑	4↑	6↑	4↓	4↑	1↕
				3↑		3↓	3↑				5↓			
4↓						5↓					7↓			
	1↓	3↓										8↑		
	6							5	7	7		13	11	7
			6↑				6↑	4↑			8↑	10↑	9↑	6↑
							7↑					9↑	8↑	
	7↑		8↑			7↑			5↑	6↑				
2		7			3	6					2			
									6					
			7										7	
	5	5		6	5					8	1	11	10	
		6↓			6↓							12↓		
1						1								

TROUBLESHOOTING STEPS

BLACK SPOTS, BROWN STREAKS:

- Purge and/or clean the screw and barrel
- Check the material for contamination
- Decrease the melt temperature
- Decrease the overall cycle time

BLISTERS:

- Decrease screw speed
- Increase back pressure
- Increase mold temperature
- Decrease melt temperature
- Dry material
- Relocate gate
- Provide additional mold vents
- Insure regrind is not too coarse

BRITTLENESS:

- Increase melt temperature
- Decrease back pressure
- Decrease screw speed
- Decrease injection pressure
- Dry material
- Decrease amount of regrind used
- Check for material contamination

BUBBLES:

- Dry material further
- Increase number and/or size of vents
- Decrease injection temperature
- Increase shot size
- Increase injection pressure
- Decrease injection speed

BURN MARKS:

- Decrease injection speed
- Decrease melt and/or mold temperature
- Decrease booster time
- Decrease injection pressure
- Alter gate position and/or increase gate size
- Improve mold cavity venting
- Check for heater malfunction

CRACKING, CRAZING:

- Modify injection speed
- Increase cylinder temperature
- Increase nozzle temperature
- Increase mold temperature
- Decrease injection pressure
- Dry material

DELAMINATION:

- Increase mold temperature
- Increase melt temperature
- Check for material contamination
- Adjust injection speed
- Dry material

DISCOLORATION:

- Purge heating cylinder
- Decrease melt temperature
- Decrease nozzle temperature
- Decrease back pressure
- Shorten overall cycle
- Check hopper and feed zone for contamination
- Provide additional vents in mold
- Move mold to smaller shot-size press

Excessive Flash:

- Decrease injection pressure
- Decrease cylinder temperature
- Increase clamp pressure
- Decrease injection hold time
- Decrease back pressure
- Decrease injection speed
- Decrease mold temperature

Flow, Halo, Blush Marks:

- Increase melt temperature
- Increase mold temperature
- Increase nozzle temperature
- Increase injection pressure
- Decrease injection speed
- Increase size of sprue/runner/gate
- Increase cold slug area in size or number

Gels:

- Increase plasticating capacity of machine or use machine with larger plasticating capacity
- Increase cylinder temperature
- Increase overall cycle time
- Increase back pressure
- Change screw speed
- Use higher compression screw

Jetting:

- Decrease injection speed
- Increase melt and/or mold temperature
- Increase gate size and/or change gate location

Oversized Part:

- Decrease injection speed
- Decrease booster time
- Decrease injection pressure
- Decrease holding pressure
- Decrease cylinder temperature
- Increase mold temperature
- Decrease overall cycle time
- Increase clamp pressure

Part Sticking:

- Decrease injection pressure
- Decrease injection-hold
- Decrease booster time
- Increase mold-close time
- Decrease mold cavity temperature
- Decrease cylinder and nozzle temperature
- Check mold for undercuts and/or insufficient draft
- Check resin lubricant level

Short Shot:

- Increase shot size and confirm cushion
- Increase injection pressure
- Increase melt temperature
- Increase nozzle temperature
- Increase mold temperature
- Increase injection speed
- Make sure mold is vented correctly and vents are clear
- Increase back pressure
- Increase injection-hold

Sink Marks:

- Increase shot length
- Increase injection pressure
- Increase injection-hold
- Decrease mold temperature
- Decrease melt temperature
- Increase injection speed
- Increase back pressure
- Increase screw speed
- Increase size of sprue and/or runners and/or gates
- Dry material
- Decrease the amount of regrind used
- Relocate gates on or as near as possible to thick sections

Splay Marks, Silver Streaks, Splash Marks:

- Dry resin pellets before use
- Check for contamination
- Decrease melt temperature
- Decrease injection speed
- Decrease nozzle temperature
- Raise mold temperature
- Shorten overall cycle
- Open the gates

Sprue Sticking:

- Decrease injection pressure
- Decrease injection-hold
- Decrease booster time
- Decrease mold-close time
- Decrease injection speed
- Decrease nozzle temperature
- Increase core temperature
- Check mold for undercuts and/or insufficient draft
- Check resin lubricant level

Surface Finish (Low Gloss):

- Increase mold temperature
- Increase injection pressure
- Increase cylinder temperature
- Increase injection speed
- Make sure venting is adequate
- Clean mold surfaces
- Change gate location

Surface Finish (Scars, Ripples, Wrinkles):

- Increase injection pressure
- Increase injection speed
- Decrease back pressure
- Increase cylinder temperature
- Increase overall cycle time
- Increase shot size
- Decrease booster time
- Decrease nozzle temperature
- Inspect mold for surface defects

Undersized Parts:

- Increase Injection speed
- Increase injection pressure
- Increase booster time
- Increase holding pressure
- Increase hold-time
- Increase cylinder temperature
- Decrease mold temperature
- Increase size of gate

Voids:

- Increase shot length
- Increase injection pressure
- Increase injection-hold
- Increase mold temperature

VOIDS (CONT'D.):

- Decrease melt temperature
- Decrease injection speed
- Clean vents
- Increase size of sprue and/or runners and/or gates
- Dry material
- Relocate gates on or as near as possible to thick sections

WARPING, PART DISTORTION:

- Equalize/balance mold temperature of both halves
- Increase mold cooling time
- Increase injection-hold
- Try increasing and decreasing injection pressure
- Adjust melt temperature (increase to relieve molded-in stress, decrease to avoid over packing)
- Check gates for proper location and adequate size
- Check mold knockout mechanism for proper design and operation
- Make sure part contains no sharp variations in cross-sections

WELD LINES:

- Increase injection pressure
- Increase melt temperature
- Increase injection-hold
- Increase mold temperature
- Increase injection speed
- Vent cavity in the weld area
- Change gate location to alter flow pattern

NOTES

Index to Materials

Generic Material — **Page**

ABS 2
ABS+PA Alloy 14
ABS+PBT Alloy 16
ABS+PC Alloy 18
ABS+PTFE Alloy 20
ABS+PVC Alloy 20
ABS+TPU Alloy 20
Acetal 20
Acetal Copolymer 26
Acrylic 30
Acrylic + ABS Alloy 32
AS Copolymer 32
ASA 32
ASA+PC Alloy 32
ASA+PVC Alloy 34
ASA/AES 34

Bio Syn Poly 34

CA 34

DAP 34

EMA 36
EnBA 36
Epoxy 36
ETFE 38
EVA 38

FEP 38

GENERIC MATERIAL **PAGE**
Fluoroelastomer 38

HDPE.. 38
HDPE Copolymer........................ 46
HIPS... 46

Ionomer 46

LCP.. 48
LDPE 50
LLDPE 56

MBS Terpolymer........................ 60
Mel Formaldehyde 60
Mel Phenolic 60

Nylon... 60
Nylon 11 62
Nylon 12 64
Nylon 12 Elastomer 66
Nylon 12/12 66
Nylon 4/6 66
Nylon 6 66
Nylon 6 Alloy 88
Nylon 6 Copoly......................... 88
Nylon 6 Elast............................. 88
Nylon 6/10 88
Nylon 6/12 90
Nylon 6/6 94
Nylon 6/6 Alloy 122
Nylon 6/66 Cop.......................... 124
Nylon 6/6T 124

GENERIC MATERIAL **PAGE**
Nylon 6/9 124
Nylon MXD6 124

PAEK 124
PBT 126
PBT Alloy 138
PBT+PET Alloy 140
PC 140
PC+PBT Alloy 166
PC+PET Alloy 166
PC+Polyester 168
PC+PUR Alloy 168
PCT 168
PCTG 170
PE 170
PEBA 172
PECTFE 172
PEEK 172
PEI 172
PEI+PC Alloy 176
PEKEKK 178
PES 178
PET 180
PETG 184
PFA 184
Phenolic 184
Plastomer 188
PMMA 188
PMP 188
Polyamide-Imide 190
Polyaryl Amide 190
Polyarylate 190

Generic Material	**Page**
Polyarylsulfone	190
Polybutylene	192
Polyester	192
Polyester Alloy	192
Polyester, Elastomer	192
Polyester, ElCo	192
Polyester, TP	194
Polyester, TS	198
Polyether	198
Polyimide, TP	200
Polyolefin	200
Polyolefin elastomer	202
POM	202
PP	202
PP Copolymer	224
PP Homopolymer	232
PPA	236
PPC	238
PPE	238
PPO	238
PPO Alloy	244
PPS	244
PPSS	250
PPSU	250
PS	250
PSU	254
PSU+ABS	256
PSU+Unspecified	256
PUR	256
PUR-capro	260
PUR-ether/MDI	260
PUR-Polyester	260

GENERIC MATERIAL	PAGE
PUR-Polyether	262
PUREL-TP	264
PVC	266
PVC Alloy	272
PVC Elastomer	272
PVC+NBR Alloy	274
PVC+NR Alloy	274
PVC+PUR Alloy	274
PVDF	274
SAN	274
SB	278
SBS Copolymer	278
SEBS Copolymer	278
SMA	280
SMA+PBT Alloy	282
SMMA Copolymer	282
Styrene Elastomer	282
SVA	284
TPE	284
TPO	292
UHMWPE	294
Urea Compound	294
Urea Formaldehyde	294

NOTES

Ordering Information

Additional copies of *Pocket Specs for Injection Molding* are available for $34.95 each, plus $4.00 for postage and handling of the first copy, $1.00 per additional copy in the same order. Payment may be made by check or by credit card.

To order, call:

1-800-788-4668

Or, you may copy this page and fax it to **307-745-9339.**

Please send me ____ copies of *Pocket Specs for Injection Molding* at $34.95 each, plus $4.00 for postage and handling of the first copy, $1.00 per addl. copy (same order).

___ I enclose a check for ________________

___ Please charge my credit card:

___ Amex ___ MasterCard ___ Visa

Card No. _____________________Exp. Date_______

Name

Company

Address

City State Zip

Phone Fax